QUANTUM PHYSICS FOR SCIENTISTS AND TECHNOLOGISTS

QUANTUM PHYSICS FOR SCIENTISTS AND TECHNOLOGISTS

Fundamental Principles and Applications for Biologists, Chemists, Computer Scientists, and Nanotechnologists

Paul Sanghera
Senior Research Scientist
Infonential, Inc.
Former Research Scientist at CERN and Cornell

WILEY

A JOHN WILEY & SONS, INC., PUBLICATION

The following images in this book are covered by the GNU License (see http://www.gnu.org/copyleft/fdl.html *for details): Figures 1.17, 1.18, 1.19, 4.1, 4.5, 4.6, 6.0, 7.4, 8.5, 8.8, 8.10, and 11.13.*

Published by John Wiley & Sons, Inc., Hoboken, New Jersey
Published simultaneously in Canada

Library of Congress Cataloging-in-Publication Data is available.

ISBN: 978-0-470-29452-9

10 9 8 7 6 5 4 3 2 1

MIX
Paper from responsible sources
FSC® C013604

*To scores of my teachers, thousands of my students,
and hundreds of my colleagues all across the globe from whom
I have learned a great deal. Here are only a few to mention:*

Physics Teachers: *Dr. Richard Hemingway, Dr. Hans Mes, Dr. Robert Carnegie, Dr. Richard Enns, Dr. Bruce Clayman, Dr. R.P. Bajpai, Dr. Sukhdev, and Dr. Sham Singh Chandel*

Chemistry Teachers: *Dr. Victor Tam, Mr. Mousa Ghanma, Dr. Richard Daley, Dr. Kei-Lee Kelley Liu, and Mr. Gurpal Singh Maan*

Biology Teachers: *Dr. Karen Erickson, Dr. Rachel Janowicz, and Dr. Nita Sharma*

Mathematics Teachers: *Dr. S.N. Dubey, Dr. Rick Martinez, and Mr. Tilak Raj*

Colleagues: *Dr. Siegfried Bethke, Dr. Richard Stroynowski, Dr. David Miller, Dr. Nari Mistry, Dr. Prabhu Krishan Raina, Dr. Shri Yogesh, Dr. Gerald Pauler, Delon Dotson, Dr. Denis Dumas, and Dr. Alix Pouladdej*

Contents

Acknowledgments

Each time I get a book published, I re-learn a lesson that transforming an idea into a finished book is a project and it takes a project team. As they say—well, if they don't say it anymore, they should—*first things first.* Let me begin by thanking Paul Petralia, Senior Editor at Wiley, for initiating this project, and Simone Taylor to see it through to completion. With two thumbs up, I thank Michael Christian and Ms. Anastasia Wasko for their excellent coordination of the pre-production stage of this project.

Before a book can be put into your hands, the production team plays the crucial role of converting the text manuscript and illustrations into printer-ready pages. It is my pleasure to thank Kristen Parrish, Senior Production Editor at Wiley, for an excellent job in leading the production team, and Jeannette Stiefel for a diligent job in copy editing. Thanks are also due to other members of the production team, including Stephanie Sakson, of Best-set Premedia, and Dean Gonzalez, the illustrations manager at Wiley.

My special thanks go to Dr. John Serri, the technical editor of this book, for carefully reviewing all the chapters and taking the book to the next level through his valuable feedback. This acknowledgment would be incomplete without extending my gratitude to some great minds in science, including Niels Bohr, Erwin Schrödinger, Richard Feynman, Max Delbrück, Francis Crick, James Watson, Richard Smalley, and Johnjoe McFadden, without whose off-the-track contributions to science, I and many others would not be able to clearly see the quantum thread running through all sciences. I also thank in advance all the readers who will be able to recognize or appreciate this thread a little better after reading this book.

Last but not least, my appreciation (along with my heart) goes to my wife, Renee, and my son, Adam, for their continual support through this and other projects.

About the Author

Dr. Paul Sanghera, an educator, scientist, technologist, and an entrepreneur, has a diverse background in major fields on which nanoscience and nanotechnology is based including physics, chemistry, biology, computer science, and math. He holds a Ph.D. in Physics from Carleton University, Canada; a Master degree in Computer Science from Cornell University, U.S.A; and a B.Sc. from India with triple major: physics, chemistry, and math. He has authored and co-authored more than 100 research papers on the subatomic particles of matter published in well-reputed European and American research journals. At the world class laboratories, such as CERN in Europe and Nuclear Lab at Cornell, he has participated in designing and conducting experiments to test the quantum theories and models of subatomic particles. His current research interests involve topics in biotechnology, biochemistry, and molecular biology. In computer science, Dr. Sanghera has contributed to building such world class technologies as Netscape Communicator and Novell's NDS. As an engineering manager, he has been at the ground floor of several startups. A former lecturer at San Jose State University, he has taught a wide spectrum of courses at institutions all across the globe including India, Canada, and the United States. Dr. Sanghera is the author of several best selling books in the fields of science, technology, and project management. He lives in Silicon Valley, California.

About the Tech Editor

Dr. John Serri has held a variety of positions from basic physics research to application software development. His broad industrial and academic experience enabled him to develop a rigorous yet practical perspective to physics. Serri received his BS degree in Math and Physics from the State University of New York (SUNY) at Albany and earned a Ph.D. in Physics from the Massachusetts Institute of Technology where he developed novel techniques for using lasers to probe the dynamics of intermolecular collisions. The techniques he developed have led to numerous applications in basic physics including laser cooling, which led one of his mentors, Dr. William Phillips, to the Nobel Prize in Physics in 1997. After leaving MIT, Dr. Serri joined the staff at Bell Laboratories, where he conducted basic research in surface physics and applied research related to the effects of nuclear weapons on telecommunication systems. Later he joined the staff of Loral Aerospace and was one of the designers of the Globalstar System. At Globalstar, he led the development of the systems to control and manage the Globalstar network. Serri currently serves on the adjunct faculty of the Mathematics and Computer Science Department at California State University East Bay and is Vice President of Strategic Inititives at Manhattan Software Inc., a major provider of Advanced Integrated enterprise software.

Periodic Table of the Elements

Period	Group 1 IA 1A	2 IIA 2A	3 IIIB 3B	4 IVB 4B	5 VB 5B	6 VIB 6B	7 VIIB 7B	8	9 VIII 8	10	11 IB 1B	12 IIB 2B	13 IIIA 3A	14 IVA 4A	15 VA 5A	16 VIA 6A	17 VIIA 7A	18 vIIIA 8A
1	1 H 1.008																	2 He 4.003
2	3 Li 6.941	4 Be 9.012											5 B 10.81	6 C 12.01	7 N 14.01	8 O 16.00	9 F 19.00	10 Ne 20.18
3	11 Na 22.99	12 Mg 24.31											13 Al 26.98	14 Si 28.09	15 P 30.97	16 S 32.07	17 Cl 35.45	18 Ar 39.95
4	19 K 39.10	20 Ca 40.08	21 Sc 44.96	22 Ti 47.88	23 V 50.94	24 Cr 52.00	25 Mn 54.94	26 Fe 55.85	27 Co 58.47	28 Ni 58.69	29 Cu 63.55	30 Zn 65.39	31 Ga 69.72	32 Ge 72.59	33 As 74.92	34 Se 78.96	35 Br 79.90	36 Kr 83.80
5	37 Rb 85.47	38 Sr 87.62	39 Y 88.91	40 Zr 91.22	41 Nb 92.91	42 Mo 95.94	43 Tc (98)	44 Ru 101.1	45 Rh 102.9	46 Pd 106.4	47 Ag 107.9	48 Cd 112.4	49 In 114.8	50 Sn 118.7	51 Sb 121.8	52 Te 127.6	53 I 126.9	54 Xe 131.3
6	55 Cs 132.9	56 Ba 137.3	57 La* 138.9	72 Hf 178.5	73 Ta 180.9	74 W 183.9	75 Re 186.2	76 Os 190.2	77 Ir 190.2	78 Pt 195.1	79 Au 197.0	80 Hg 200.5	81 Tl 204.4	82 Pb 207.2	83 Bi 209.0	84 Po (210)	85 At (210)	86 Rn (222)
7	87 Fr (223)	88 Ra (226)	89 Ac (227)	104 Rf (257)	105 Db (260)	106 Sg (263)	107 Bh (262)	108 Hs (265)	109 Mt (266)	110 — ()	111 — ()	112 — ()		114 — ()		116 — ()		118 — ()

Lanthanide Series	58 Ce 140.1	59 Pr 140.9	60 Nd 144.2	61 Pm (147)	62 Sm 150.4	63 Eu 152.0	64 Gd 157.3	65 Tb 158.9	66 Dy 162.5	67 Ho 164.9	68 Er 167.3	69 Tm 168.9	70 Yb 173.0	71 Lu 175.0
Actinide Series	90 Th 232.0	91 Pa (231)	92 U (238)	93 Np (237)	94 Pu (242)	95 Am (243)	96 Cm (247)	97 Bk (247)	98 Cf (249)	99 Es (254)	100 Fm (253)	101 Md (256)	102 No (254)	103 Lr (257)

Fundamental Physical Constants[a]

Quantity	Symbol	Value(s)
Atomic mass unit or Dalton	amu, u, Da	$1.660538782 \times 10^{-27}$ kg
		931.494028×10^{6} eV/c^2
Avogadro number	L, N_A	$6.02214179 \times 10^{23}$ mol^{-1}
Boltzmann constant	k	$1.3806504 \times 10^{-23}$ J K^{-1}
Elementary charge	e	$1.602176487 \times 10^{-19}$ C
Gravitational constant	G	6.67428×10^{-11} m^3 kg^{-1} s^{-2}
Mass of electron	m_e	$9.10938215 \times 10^{-31}$ kg
		$8.18710438 \times 10^{-14}$ J/c^2
		0.510998910 MeV/c^2
		$5.4857990943 \times 10^{-4}$ u
Mass of proton	m_p	$1.672621637 \times 10^{-27}$ kg
		$1.503277359 \times 10^{-10}$ J/c^2
		938.272013 MeV
		1.00727646677 u
Molar gas constant	R	8.314472 J K^{-1} mol^{-1}
		0.08205746 L atm K^{-1} mol^{-1}
		K for kelvin
Permeability of vacuum	μ_0	$4\pi \times 10^{-7}$ N A^{-2}
Magnetic constant		N for newton and A for ampere.
Permittivity of vacuum	ϵ_0	$8.854187817 \times 10^{-12}$ F m^{-1}
Electric constant		
Planck constant	h	$6.62606896 \times 10^{-34}$ J s
		$4.13566733 \times 10^{-15}$ ev s
		$6.62606896 \times 10^{-27}$ erg s
Speed of light in vacuum	c	2.99792458×10^{8} m/s
Stefan-Boltzmann constant	σ	5.670400×10^{-8} W m^{-2} K^{-4}
		W for watt

[a] Courtesy of the National Institute of Standards and Constants (NIST).

Important Combinations of Physical Constants

Quantity	Symbol	Value(s)
Bohr magneton	$\mu_B = \dfrac{e\hbar}{2m_e}$	$927.400915 \times 10^{-26}\,\mathrm{J\,T^{-1}}$
Bohr radius	$a_0 = \dfrac{4\pi\varepsilon_0\hbar^2}{m_e e^2}$	$0.52917720859 \times 10^{-10}\,\mathrm{m}$
Compton wavelength	$\lambda_c = \dfrac{h}{m_e c}$	$2.4263102175 \times 10^{-12}\,\mathrm{m}$
Molar gas constant	$R = N_A k$	$8.314472\,\mathrm{J\,K^{-1}\,mol^{-1}}$ $0.08205746\,\mathrm{L\,atm\,K^{-1}\,mol^{-1}}$
Rydberg constant	$\Re_\infty = \dfrac{me^4}{4\pi c\hbar^3(4\pi\varepsilon_0)^2}$	$1.0973731568527 \times 10^{-7}\,\mathrm{m^{-1}}$
Speed of light	$c = \dfrac{1}{\sqrt{\varepsilon_0\mu_0}}$	$2.99792458 \times 10^8\,\mathrm{m/s}$
—	$\dfrac{1}{4\pi\varepsilon_0}$	$8.98755 \times 10^9\,\mathrm{m\,F^{-1}}$ F for farad, a unit of capacitance
—	$\hbar = \dfrac{h}{2\pi}$	$1.054571628 \times 10^{-34}\,\mathrm{J\,s}$ $6.58211899 \times 10^{-16}\,\mathrm{eV\,s}$
—	hc	$1.9864 \times 10^{-25}\,\mathrm{J\,m}$ $1239.8\,\mathrm{eV\,nm}$
—	$\hbar c$	$3.1615 \times 10^{-26}\,\mathrm{J\,m}$ $197.33\,\mathrm{eV\,nm}$

Preface
Science, Technology, and Quantum Physics: Mind the Gap

Quantum physics thus reveals a basic oneness of the universe.
Erwin Schrodinger

Figure P.1 Celestial map from the seventeenth century, by the Dutch cartographer Frederik de Wit (1630–1698).

In This Preface

To an artist, in Shakespeare's words, All the world's a stage. Taking this metaphor to another level, to a scientist, the whole universe is a colossal party with a cosmic dance on dance floors at all levels, ranging from an expanding universe with swirling galaxies, to planets revolving around their suns, to organisms of all shapes and sizes dancing through their life cycles, to molecules in action inside living and nonliving systems, to atoms making and breaking bonds to make molecules and crystals, to electrons dancing around the nucleus of atoms, and so on. The universe and everything in it, living and nonliving, originally started (and still starts) at the microscopic level, a level too small for human senses to resolve. In this book, we focus on the concept of micro in contrast to that of macro; micro means anything small enough not to be seen by the naked eye, including the size scales of micrometers, nanometers, and smaller items. Because the universe and everything in it is comprised of microentities, to fully understand the macro we need to understand the micro.

Humans, the macrobeings, have evolved to a relatively advanced level. For better or for worse we are late comers to this party of life on Earth. On the scale of a 12-h clock, if the party started at the hour zero (i.e., midnight) with the solidification of the earth's crust, and if it is noon now, we have just arrived at 11:59:59. However, now that we have arrived, we are the most curious and ambitious creatures at the party. We want to know everything about everyone and everything in the party: Who (or what) are you? What are you made of? How did you get here? Where are you from, no I mean where are you originally from? Where are you headed, that is, what is your future?

All the answers accumulated from our age old queries have uncovered the three most important higher-level secrets of nature, important and general enough that they should become part of the thinking of scientists and students of all sciences.

P.1 THREE SECRETS OF NATURE

At first glance, the diversity around us is obvious and ubiquitous. However, during the entire history of the development of science, the history of discoveries and inventions, nature has taught us a very powerful lesson over and over again: Look for the underlying unity behind apparent diversity of things and phenomena, and therefore behind the laws governing those things and phenomena. In my opinion, the three most salient of all the secrets of nature that science has discovered so far are the following:

There Is an Underlying Unity Behind Apparent Diversity. This concept is a key point to understanding not only the things and phenomena around us, but also their diversity. For example, all life is made up of the same basic building block of life: the cell. Most great discoveries and breakthroughs in the history of science have revealed this secret over and over again: unity behind diversity.

All Macroscopic Things Are Composed of Smaller Building Blocks. This truth exists at various levels. For example, all materials around us are made up of molecules or crystals, all molecules and crystals are made of atoms, which in turn are made of subatomic particles (proton, neutron, and electron), protons and neutrons are made of quarks, and so on. This is also true about living organisms. For example, we are made of organs, organs are made of tissues, tissues are made of cells, cells are made of and run by the molecules of life and atoms or ions, and the molecules of life are made of atoms, and so on. This structural hierarchy is also apparent in our own inventions, for example, there are rooms contained in buildings, the buildings are contained in neighborhoods, and the neighborhoods make up a City, and so on.

Nature Does Its Most Important and Fundamental Work on a Smaller Scale. This statement means that if you really want to understand a macroscopic structure, you will need to understand its smaller building blocks. The bottom line is we cannot have the macroworld without the existence of a microworld

The underlying unity behind diversity has been the key to major breakthroughs in the sciences. The history of physics can be told in terms of discovering the unification of multiple forces. For example, the terrestrial force that keeps us bound to the Earth and the celestial force that keeps the planets bound to the Sun are the same force: gravity. This realization helped Newton formulate the law of universal gravitation that applies to earth-bound problems, as well as explains the empirical laws of planetary motion discovered by Kepler. Unifying theories of electric and magnetic forces into a single theory of a force, called the electromagnetic (EM) force, was another great feat in physics, and the classical theory of EM is based on this unification. We now know from physics that our universe is shaped and governed by four fundamental forces: gravitational, EM, weak nuclear, and strong nuclear forces. For many, this may be unity enough, but the attempts to discover further unification continued, and there is great scientific evidence that even these four forces are different low-energy manifestations of a single force that was in action at the very beginning of the creation of the universe.

This history of unification of forces is intertwined with the history of discovering and understanding the smaller and smaller building blocks of matter. For example, physicists (and chemists) discovered that all materials around us are made of smaller components called molecules; the molecules are made of even smaller components called atoms; and atoms in turn are made of yet smaller constituents called neutrons, protons, and electrons. Physicists have probed neutrons and protons, and have discovered that they are made of even smaller and more fundamental particles called quarks. Therefore, the path of development that physics, and as a result other disciplines of science, have taken is the path from the macro (large) to the micro (small). As mentioned earlier, the reason for this is that we, the humans, are macrobeings and we started observing things with the most primitive tools, our five senses (e.g., our eyes), which were the only tools easily available to us. These tools are limited to the macroworld. However, to our advantage, we humans had a capability that other organisms did not have, and that is, to develop tools and techniques: say technology. The gradual development of technology, and hence better tools (e.g., microscopes), enabled us to continue our journey of understanding and exploration from the macroworld of planets and visible objects around us to the microworld of molecules, atoms, and subatomic particles. In fact, by extending this idea we find there is a feedback loop between technology and basic fundamental science. As science evolves, new technologies emerge and in turn these new technologies allow for the creation of better and more powerful scientific tools. These tools in turn allow for new observations that lead to an even better understanding of the microscopic world, and so the cycle goes. For example, the colossal colliders at the particle physics lab at CERN (near Geneva, Switzerland) would not be possible without the capability of tracking particles to the precision of micrometers and microseconds provided by this technology, which itself is based on physics.

Combining the three secrets of nature, the truths of unification and the unity behind diversity are essential to our journey from the macro to the micro. This relationship among these three secrets of nature is the reason it has taken us this long to find them: the whole history of the development of science on this planet.

Rather ironically, when we had less information and knowledge than we have today, it appeared as if one person could know everything, and there was only one discipline called *natural philosophy*, the study of nature and the physical universe. Today's physics is actually the modern version of *natural philosophy*.

P.2 FROM NATURAL PHILOSOPHY TO PHYSICS

In olden times, natural philosophy referred to the study of nature and the physical universe. It served as a precursor to developing fields of natural sciences led by physics. Rest aside physics, modern notions of *science* and *scientists* date only to the nineteenth century. Before then, the word "science" simply meant knowledge and the title *scientist* did not exist. For centuries, scientists were called natural philosophers, and these practitioners often pursued a wide variety of interests. A remnant of that heritage is a doctorate in physics, or other sciences, which is still called a Ph.D., and is an abbreviation for doctor of philosophy. The title of Isaac Newton's scientific treatise published in 1687, commonly known as the Principia, is called *The Mathematical Principles of Natural Philosophy*. So, it is not a coincidence that long-established Chairs of Natural Philosophy at older universities are currently occupied largely by physics professors. The connection between physics and other disciplines of science can also be understood from the fact that all other fields of science are historically developed from natural philosophy or its descendants. During the fourteenth and fifteenth centuries, the term *natural philosophy* was referred to what is now called physical science. By the mid-nineteenth century, with the increase of information and knowledge, it became increasingly clear that it was not generally practical for scientists to contribute to numerous areas. Thus specialization occurred and now there were physicists, chemists, biologists, geologists, and so on.

P.3 PHYSICS THE MOST FUNDAMENTAL SCIENCE

Being the most fundamental of all the sciences, physics has profoundly affected all science fields. Given its fundamental nature, students in many fields are required to take some physics courses in order to understand the fundamental principles underlying many phenomena across many science disciplines. In this sense, physics is intimately related to other sciences, and also to applied fields, such as engineering, medicine, and now nanotechnology. This is because principles discovered in physics apply to all natural sciences. For example, the principle of conservation of energy is common to all physical and chemical systems.

Chemistry is the scientific study of matter and materials and the changes that they undergo; matter is anything that exists, has mass, and occupies space; whereas a material is any kind of matter that can be used for something, such as glass. Starting with the macroworld as physics did, early chemistry largely dealt with substances, such as extracting metals from their ores, making pigments for cosmetics and painting, and fermenting beer and wine. Chemistry is the science that not only is most affected by physics, but also overlaps or collaborates with physics in many areas of study, such as atoms, molecules, and collections of atoms and molecules. A great example of the interaction between physics and chemistry is the quantum mechanical explanation of the relationships among chemical elements and the rules of their interaction (reactions) with one another captured in the periodic table.

Biology is the scientific study of life; that is, the study of living things and the phenomena related to them. Like physics and chemistry, biology also began its journey in the macroworld: finding and studying things that you can see with your naked eye, doing the counting, classifying them, and so on. By now, however, biology has progressed into the microworld in different disciplines, such as microbiology, biochemistry, and molecular biology. There are an enormous number of physical phenomena involved in developing a single cell into an organism and then maintaining the organism. In order to fully understand life, all these phenomena will be understood in terms of the principles of physics. Many key advances in the biological sciences already have been made possible by physics-based techniques. Some examples being used to study the molecules of life are the optical microscope to study cells, X-ray diffraction used to discover the structure of the deoxyribonucleic acid (DNA) molecule, and the electron microscope. Nuclear magnetic resonance (NMR), a quantum mechanical phenomenon, is revolutionizing the field of medicine in the form of magnetic resonance imaging (MRI). With developments in genetics and molecular biology, an overwhelming amount of data is being generated in biology. The only way biologists will be able to fully make sense of these data is by developing a theoretical framework based on physics principles. Quite possibly in this process, we will discover new physics laws, partly because biological systems are generally different in some important ways from the physical systems that led to the current physics laws. After all, during its early days, biology helped physics in the discovery of the principle of conservation of energy. This principle was first realized and proposed by Dr. Julius Robert Mayer, a practicing physician, in terms of the amount of heat taken in and given out by a living body.

Needless to say, physics is also at the foundation of engineering and technology. Understanding the application of physics laws is necessary from designing a bike to an interplanetary spacecraft, from typewriters to computers, from a horse and wagon to a sports car, and from a radio to a flat screen TV. It is the understanding of matter at the atomic level, developed by discovering physics laws, that led to the semiconductor and microchip industry, and subsequently to computing, which led to the Internet revolution and the information age. Information technology has already become the fiber of all aspects of life in the modern world.

Although understanding the microworld at the atomic level, which was developed by physics, has been used in many fields, such as engineering and technology, we could not directly control microobjects, such as molecules and atoms. A big part of the reason is although physics studies in the twentieth century reached the levels of subatomic particles, such as quarks, other fields have been taking their time to catch up with molecules and atoms. Toward the end of the twentieth century, however, studies in most science fields had reached the level of molecules and gave rise to nanoscience and nanotechnology. This was first envisioned by a great physicist Richard Feynman, and expressed on December 29, 1959 during his talk at the meeting of the American Physical Society at California Institute of Technology titled: *There Is Plenty of Room at the Bottom*. This event is now considered to be the beginning of nanotechnology. Using nanotechnology, which is the application of nanoscience, enables us to control matter at the level of molecules and atoms: Mind the difference between understanding and controlling.

So, welcome to the twenty-first century, the age of molecules.

P.4 QUANTUM PHYSICS: THE SCIENCE OF THE MOLECULAR AGE

As mentioned earlier, all things macro are composed of things micro. It is the same truth that is being told through different stories in different disciplines.

A chemist will tell us that everything is made of atoms. A physicist will even go farther and tell us that the universe started off from a few fundamental particles, such as quarks and electrons: Quarks combined to make neutrons and protons, which combine to make nuclei. A nucleus combined with electrons to make atoms, which in turn bonded together to make molecules, and so on. A biologist will tell us that life evolved from certain molecules, which combined into the basic unit of life called a cell. Cells evolved into a whole variety of unicellular and multicellular organisms due to mutations in the genes, the source of evolution. These stories are telling us that nature has designed everything in the universe, living and nonliving, at a microscopic level from the bottom up.

Being human, our greatest blessing and our greatest curse has been that we are macrobeings, whereas the universe was built from the bottom up, we started investigating it from the top down. Obviously, the first things we saw and tried to understand around us were the macro items: our eyes laid upon stars, not atoms. This is why the history of physics started from planets, or other celestial and terrestrial bodies, and eventually made its way to atoms and subatomic particles. Similarly, the discipline of biology started from the study of living organisms, such as plants and animals, and made its way up to cells and the molecules in the cell. So, we had it backwards, and in this sense we have been very good reverse engineers.

We can boast as much as we like about our superiority to other organisms, but the fact remains that we as humans have our own limitations in understanding this universe. First, we can only investigate and understand so many things in our life time. Second, we can only investigate, gather information, obtain knowledge, and develop our understanding piecemeal. Some pieces may look different from each other at the surface. However, they may just be different facets of the same thing at some other level. Having only a partial knowledge, which are in different pieces, helps to obscure the big picture. Nature works the way it does without caring about our limitations. For example, nature did not create and does not need the borders and interfaces between different disciplines of science such as physics, chemistry, biology, and so on. These borders are the creations of our own limitations.

Here is another problem. Our common sense is built upon our experience with the macroscopic world. Therefore, the physics built from interacting with the macroscopic world, called Newtonian or classical physics (also called Newtonian and classical mechanics) makes sense to us. However, when our scientific studies entered the microscopic world, the laws of classical physics collapsed right before our eyes. Quantum physics came to our rescue and to this date has successfully explained microscopic structures and phenomena. Due to our macroscopic common sense and intuition, even the pioneers of quantum physics had, and we continue to have, trouble with digesting quantum physics. Common sense tells us, it only makes sense that it does not make sense to our common sense, because common sense is built from our experience with the macroscopic world. As you will see in this book, quantum phenomena brutally violate our common sense. You will also see that classical physics actually is a special case, an approximation, of quantum physics. This means quantum physics is more fundamental, which is consistent with the fact that the universe is designed at the microscopic level, and is built from the bottom up. Therefore the reality of the macroworld that we experience on a daily basis has its roots in the microworld of atoms and molecules. This is also reflected by how a human is developed even today: Two half-cells (sperm and egg) join to make a zygote, which divides and develops into an embryo. Embryonic cells differentiate to make different parts of the body, and subsequently we have a full human being within months.

Even though quantum physics is more fundamental and general and classical physics is only an approximation of quantum physics, classical physics, which is an approximate science, works well in the macroworld. The reason is because those approximations are at the microscale; that is, they are so small they easily go unnoticed. So, while dealing with the macroworld, we can study and use classical physics and leave quantum physics for the microworld.

However, there is a problem with this approach, and hence an opportunity to be captured. As already mentioned, physics and its direct applications, as well as biology, chemistry, materials science, and other fields, have already entered the microworld. This situation has two aspects: individual and collective. Individual fields are developing their own subfields corresponding to the microscale, such as microbiology and molecular biology. All these fields are collectively organizing their efforts into what is termed as nanoscience and nanotechnology. For example, the entire field of molecular biology is biology at the microscale, or nanoscale level to be precise: The diameter of a DNA molecule is ~2 nm, for example.

As nanoscience and, consequently, nanotechnology are progressing, we are learning how to handle and control molecules and atoms at individual levels. This opens the door to an endless array of opportunities, so much so that many of us are already calling the twenty-first century the nanoage or the molecular age. Quantum physics is the law of the land in the molecular age of the twenty-first century.

P.5 WHY THIS BOOK

Because everything is made of atoms, the macroscopic reality of our everyday experience in the macroworld has its roots in the microscopic world. So, here is the key to generating new ideas in making progress in all sciences in modern times: There are many things that atoms and molecules do, which cannot be understood in terms of what objects in the macroworld can do; whereas there is nothing that macroobjects do that cannot be understood in terms of what atoms and molecules can do. Quantum physics is the science that helps us to understand what atoms and molecules can do.

With this realization and with the fact that much of modern science, engineering, and technology are already dealing with molecules and atoms, comes the increased need to learn quantum mechanics even for nonphysics majors, such as chemistry, biology, engineering, materials science, computer science, nanotechnology, and related fields. This book, designed as a complete course in quantum mechanics for these nonphysics majors, can be used as a textbook for a stand-alone course in quantum mechanics. It can be part of another course or a series of courses, such as modern physics and physical chemistry, or as a part of a short program, such as nanotechnology. This book presents a rich, self-contained, cohesive, concise, yet comprehensive picture of quantum mechanics for senior undergraduate and first-year graduate student, nonphysics majors, and for those professionals at the forefront of biology, chemistry, engineering, computer science, materials science, nanotechnology, and related fields.

Most, if not all, books on quantum physics written for science students use abstract mathematical formulation of quantum mechanics and leave its implications and connections to the real world often nonintuitive. This kind of framework may be necessary for physics students, but often is a learning hurdle for nonphysics majors. However, the centerpiece of this mathematical system and of quantum mechanics is the wave function. This book begins with the wave function, which the students have already learned about during their introductory course in classical (Newtonian) physics while learning about waves, and uses simple calculus, which most science major students learn in their junior undergraduate years, to develop and explain the concepts and principles of quantum mechanics. Comprehensive coverage of quantum theory, supported by experimental results and explained through applications and examples, is presented without the use of abstract and complex mathematical tools and formalisms, such as bra–ket vectors, Hilbert space, matrix algebra, or group theory. Other features of this book that help take the mystery out and bring quantum mechanics home for nonphysics students, researchers, scientists, and technologists include the following:

1. It is shown throughout the book how quantum principles generate the same results as classical physics when we move from the micro- to the macroworld. Due to its universal importance across different fields, another principle emphasized throughout the book is the natural tendency of systems to be in a state of equilibrium, that is, in a state of minimum energy.
2. Throughout the book, concepts and principles of quantum physics are explained in the language of nonphysics majors by presenting examples and applications from nonphysics fields including chemistry, biology, nanotechnology, and related fields.
3. Living in the Quantum World, a section at the end of each chapter, features real-world applications of one or more quantum mechanics principles discussed in the chapter.

4. Enough study checkpoints and problems with solutions are presented throughout the book to make difficult concepts easy to understand.
5. The interfaces and connections between quantum physics and nonphysics fields, such as biology, chemistry, computing, and nanotechnology, are identified or introduced in an easy to understand fashion.
6. All concepts and principles of quantum physics are explained from scratch the first time they appear and no prior knowledge of quantum physics is assumed.
7. The material is presented in a logical learning sequence: a section builds upon previous sections and a chapter builds upon previous chapters. There is no hopping from topic to topic and no jargon without explanation.
8. Pictures, illustrations, tables, notes, and cautions are used to help us to understand the difficult and tricky concepts.
9. To ease the reader smoothly into the chapter, each chapter begins with an interesting picture with a light amusing note that is in a direct or subtle way related to the main issue in the chapter.
10. To inspire the students, the quantum mechanics related individual achievements within the communal environment are emphasized by presenting brief biographies throughout the book. It also contributes to keeping the presentation interesting.

Furthermore, this book takes advantage of the amazing story of how quantum mechanics was developed. The concepts and principles that make the foundation of quantum theory are developed in the context of the history of the gradual development of quantum mechanics, which some of us find as amazing as quantum mechanics itself. This facilitates the introduction of the key concepts and principles of quantum physics as explanations for the results of those historic experiments that could not be explained with classical physics. In doing so, the book illustrates in an interesting way the process of scientific discoveries and advances.

After introducing the foundational concepts and principles of quantum mechanics, such as quantization, the uncertainty principle, quantum mechanical tunneling, and wave–particle duality in the context of explaining the results of historical experiments in the first few chapters, they are woven together in the formalism of quantum mechanics. Subsequently, quantum theory is presented in a single wave equation called the Schrödinger wave equation. The forthcoming chapters demonstrate how the principles of quantum mechanics naturally arise when we apply this wave equation to solve physical problems. The relevance of these types of problems to nonphysics fields is demonstrated throughout the book.

The application of quantum physics to simple systems, such as the hydrogen atom and the periodic table of the elements, is extended in a logical way to complex systems, such as molecules and the assemblies of molecules. This covers the topics of molecular vibrations, molecular rotations, and quantum statistics of assemblies of molecules. Also covered are the quantum mechanical foundations of modern techniques used across several fields, such as atomic and molecular spectroscopy, lasers, NMR imaging, and electron microscopy.

P.6 IN THIS BOOK

The laws of classical mechanics, or classical physics, were developed over the centuries by many scientists. Astronomers made contributions due to their interest in the heavenly (extraterrestrial or celestial) bodies. For example, the experiments, observations, and their explanation by Galileo Galilei (1564–1642) laid the foundations for the important discoveries that were made in the following two centuries. Issac Newton (1642–1727), another giant figure of those times, clearly understood the relationship among different variables involved in the motion of objects and put the work of centuries into three elegantly simple laws known as Newton's laws of motion. Newton's greatest contribution was the unification of terrestrial and extraterrestrial (celestial) forces when he pointed out that the motion of planets around the sun can be understood by the same laws that govern the motion on Earth. Newton's laws make the foundation of classical physics, which includes Maxwell's equation of EM forces, unifying the concept of electric and magnetic forces into one force: the EM force. All these concepts that make up the core of classical physics are reviewed in Chapter 1 in a concise, but cohesive fashion. The core assumption of classical physics is that physical entities exist in two different types: particles and waves. This assumption is clearly reflected in the fact that classical mechanics (mechanics of particles) and optics (as waves) are two independent disciplines in physics with their own set of principles and laws and with their own series of experiments and course works. So, Chapter 1 is oriented toward making a clear distinction between waves and particles: either you are a particle or you are a wave. This treatment by classical physics of particles and waves as two different types of entities mirrors the physical reality that we experience as macrobeings. It makes sense to us to treat material bodies or objects as particles and to treat some other phenomena as waves that do not fit into the particle picture, such as a ripple that spreads out from a single point of impact when you throw a little rock into a body of still water.

However, as mentioned earlier, because the objects of the macroscopic world are constituted of the objects of the microscopic world, the physical reality experienced by us in the macroworld has its roots in the microworld. Just before (or during the early days of) probing the microworld, the laws of classical physics were working so well that some scientists toward the end of the nineteenth century were becoming rather smug and making statements such as the one by William Thomson Kelvin in 1900: "There is nothing new to be discovered in physics now. All that remains is more and more precise measurement."

But even Kelvin knew that there were challenges coming from the microworld, which he referred to in his speech in 1900 to the Royal Institution as "two clouds on the horizon". The two clouds that Kelvin was referring to were the failure to find a medium through which EM waves would travel and the explanation of black-body radiation. The first cloud gave rise to Einstein's theory of relativity, which is very briefly mentioned in Chapter 1. The second cloud eventually led to the development of quantum physics. So much for Kelvin's prediction of the end of physics.

Chapter 2 describes how the experimental results from experiments, such as black-body radiation and the photoelectric effect, challenged the notion that EM waves (light) were just waves. Black-body radiation was one of the first experiments where classical physics broke down in explaining the results from the microworld. It is shown how quantum concepts were introduced to explain experimental results.

Chapter 3 discusses how another set of experiments challenged the classical notion that particles, such as electrons, are just particles and not waves. It explains how the core assumption of classical physics that particles and waves are two separate components of the physical reality, breaks down in front of experimental results from the microworld. These results can only be explained with the particle–wave duality of nature: A physical entity acts as a particle under certain conditions and as waves under a different set of conditions.

Wave–particle duality is at the heart of quantum physics. It gives rise to two very important concepts in quantum physics: probability (or uncertainty) and quantization. An example of probability (or uncertainty) is that you cannot predict with certainty the position and momentum of an entity at a given instant; you can only assign probabilities to different possible values of position and momentum. Quantization means that a confined entity, such as an electron in an atom, is only allowed to have certain discreet values of energy. Both of these concepts go against the grain of classical physics, where you can predict the position and momentum of a particle at a given moment and the particle can have any energy, not only certain discrete values of energy. However, as Chapter 4 explains, this quantum picture explains the experimental results well from the study of atoms in terms of Bohr's model. His model explains in a somewhat ad hoc way how quantization keeps the electron in the atom from collapsing into the nucleolus as predicted by classical physics. Chapter 4 also describes how Bohr's model explains the spectral lines emitted by heated gases, another puzzle that the physicists were facing toward the end of the nineteenth century. The Bohr model, although a semiclassical or semi-quantum mechanical model, demonstrates the power of quantization.

So, Chapters 2–4 introduce in a logical way some fundamentals of quantum mechanics in terms of the hypotheses and laws put forward by different scientists in the beginning of the twentieth century to explain some physical phenomena that classical physics failed to explain. Chapter 5 puts these fundamentals together into a few postulates and develops a minimal formalism of quantum mechanics for the rest of the book. This chapter also shows how these postulates can be summed together into one equation: The Schrödinger equation, the fundamental equation of quantum physics.

Chapter 6 is devoted to taking the mystery out of the Schrödinger equation by examining its solutions through its application to simple, but abundantly practical problems, such as confinement and vibration (or oscillation). Most of the systems in the universe are confined or bonded to their environment, such as in electrons in an atom, atoms in a molecule, and so on. It is a natural tendency of a system to be in equilibrium, and when disturbed it will oscillate (or vibrate) about its equilibrium position, such as atoms in a molecule or a crystal. These countless examples in nature can be modeled in various ways, such as a particle in a box (confinement) and a harmonic oscillator (vibrations about an equilibrium position). While applying the Schrödinger equation to these two situations in Chapter 6, we demonstrate how solving the Schrödinger equation naturally gives rise to the principles and laws of quantum mechanics discussed in the previous chapters.

All matter (living and nonliving) is made up of atoms. Hydrogen, which makes up 9.5% of the human body mass, consists of the simplest atom: A confined system of one electron bound to one proton in the nucleus. Chapter 7 demonstrates how applying the Schrödinger equation to this simple system naturally gives rise to some quantum mechanical concepts and principles, such as quantization, principle quantum number, orbital quantum number, magnetic quantum number, and the rules associated with these quantum numbers.

The structure and properties of atoms show a pattern displayed in the periodic table of elements. One of the greatest triumphs of quantum mechanics is to present an explanation for this pattern. This topic is discussed in Chapter 8, which also introduces the quantum mechanical concept of spin and spin quantum number along with Pauli's exclusion principle, regarding the occupation of quantum states. The exclusion principle also plays an important role in determining the electronic configuration of atoms.

Atoms bound together to make molecules including the molecules of life: carbohydrates, DNA, lipids, and proteins. The quantum states of molecules in terms of rotation and vibration of diatomic molecules are discussed in the framework of molecular spectroscopy in Chapter 9. There are many physical, chemical, and biological situations in which a large number of atoms or molecules are involved. The behavior of such collections or assemblies and the relationships of individual constituents to the collection are studied in statistical mechanics: statistics fused with mechanical laws. Some basic concepts and principles of statistical mechanics are developed in Chapter 10, where we clearly explore the interface between classical and quantum statistical mechanics. The vibrations and rotations of molecules are revisited in the collective statistical environment.

Now that almost all the sciences, including biology, chemistry, materials science, and computer science, that began with macroscopic studies are meeting physics in the microscopic, including the nano, world of molecules and atoms, the twenty-first century is shaping up to be the molecular age. Due to these common issues and problems that scientists are dealing with on this size scale, this area in all sciences is collectively developing as nanoscience. The developing field of nanotechnology, the application of nanoscience, is increasingly allowing us to handle molecules and atoms at individual levels. Chapter 11 discusses quantum mechanics as a common thread through different fields of nanoscience and nanotechnology.

P.7 BACK TO THE FUTURE

History repeats itself in interesting ways at various levels. As already mentioned, all science disciplines that directly or indirectly emerged from one discipline, which was natural philosophy, are now converging in many ways into one field called nanoscience and its application, which is nanotechnology. Quantum physics, being the science of the small, is the common thread that runs through all these fields at the nano and smaller scales.

Finally, let us borrow a metaphor and its elaboration from Richard Feynman who, once referring to an unknown poet, said "The whole universe is in a glass of wine". We do not know what the poet really meant by this, but a number of scientists from different disciplines sitting around a wine glass can together actually see the entire universe in it. A biologist, for example, sees the fermentation, the bacteria, and hence life. A chemist sees an interesting array of chemicals, ferments, enzymes, product, whereas a physicist sees the reflections and refraction in the glass, and the swirling evaporating liquid. Physicists and chemists both see the molecules of alcohol and atoms, such as carbon, hydrogen, and oxygen; the physicist sees protons and neutrons in the nucleus of the atoms and the quarks from which the atoms are composed. These are the same quarks from which the whole universe is made. Where did the glass come from? Rocks ... fossils ... secrets of life and the universe ... evolution of stars ... we can go on and on. I do not want to spoil the party, but remember, it is our small minds (our limitations), that divide this glass of wine, the universe, into physics, chemistry, biology, geology, and so on; nature sees no such divisions.

So, again about this glass of wine in the words of Feynman:

"Let's put it all back together, not forgetting ultimately what it is for. Let it give us one more final pleasure: drink it and forget it all!" The party of life and the dance of the universe will continue. Take this book as a micro or quantum celebration of this party.

April 2010
Silicon Valley,
California

PAUL SANGHERA
Senior Research Scientists
Infonential, Inc.
Former Research Scientist
at CERN and Cornell

FIRST, THERE WAS CLASSICAL PHYSICS 1

If I have seen further it is by standing on the shoulders of Giants.
Issac Newton, Letter to Robert Hooke, February 1676

Figure 1.0. "Nicolas Poussin (1594–1665): Blind Orion Searching for the Rising Sun (24.45.1)". Date: 1658. In *Heilbrunn Timeline of Art History*. New York: Courtesy of The Metropolitan Museum of Art, 2000.

In this picture, based on Greek mythology, blind Orion a hunter has set Cadelion, a servant, on his shoulders as a Guide to the East where the rays of the Sun would restore his eyesight. *Dwarfs standing on the shoulders of giants (Latin: nanos gigantium humeris insidentes)* is a Western metaphor with a modern-time interpretation: "One who develops future intellectual pursuits by understanding the research and works created by

Quantum Physics for Scientists and Technologists, by Paul Sanghera

notable thinkers of the past." This metaphor, first recorded in the twelfth century and attributed to Bernard of Chartres, a twelfth century French Platonist philosopher, was famously used by seventeenth-century scientist Isaac Newton. Newton himself was rather modest about his own achievements, when in his famous letter to Robert Hooke in February 1676, he wrote "If I have seen further it is by standing on the shoulders of Giants".

In This Chapter

1.1 INTRODUCTION

Physics is a discipline in natural science, the branch of science that relies the most on mathematics to create an explanation of the universe we live in. The word *science* has its origin in a Latin word that means "to know". Science is the body of knowledge of the natural world organized in a rational and verifiable way. The word *physics* has its origin in the Greek word that means nature.

Physics is that branch, or discipline, of science that deals with understanding the universe and the systems in the universe to all levels of depth from planets to fundamental constituents of matter, such as atoms, electrons, and quarks. The core part of physics is to understand the universe and everything in it in terms of the fundamental constituents of matter and the interactions between those constituents. The interactions are commonly called forces.

As human, we are macroorganisms unable to observe the microobjects and phenomenon with our naked eyes. However, what worked for us as a species that we are cognizant of phenomena in and around the range we live in. In other words, we have the capability of studying and understanding something that is beyond our intuition. The distance between two points that we resolve with our eyes is on the order of one-tenth of a millimeter (mm), and the smallest time between two instances that we can measure without the help of sophisticated tools is on the order of one-tenth of a second(s). Therefore, the journey of physics, and hence science, began by studying the macroobjects and systems. The physics of these macrosystems is called classical physics. Therefore it is important to understand and appreciate classical physics before we can understand quantum physics. This chapter presents a high-level review of the important concepts in classical physics in a concise and cohesive fashion. If you are not sure of any concept covered in this chapter, consult an introductory physics book for help from the list presented at the end of this book.

Classical physics divides the physical world into two types of physical entities: particles and waves. Your main goal in this chapter is to grasp the classical approach of physics in terms of particles and waves being different kinds of entities. To that end, we will explore three avenues: particles, waves, and forces.

1.2 PHYSICS AND CLASSICAL PHYSICS

As mentioned earlier, at its core physics is that branch of science that deals with understanding the universe and the systems in the universe in terms of fundamental constituents of matter, such as atoms, electrons, and the interactions among those constituents.

> *Note*: When physicists use the word micro, they usually mean all sizes nonmacro, including micro (10^{-6}), nano (10^{-9}), pico (10^{-12}), and smaller. In this sense, the microscale word includes nanoscale. We also use the micro word in this book in this sense unless stated otherwise.

Physics, the most fundamental science, deals with (discovering and exploring) the fundamental principles that are subsequently applied to many other disciplines of science and technology, such as biology, chemistry, material science, electronics, engineering, and nanotechnology. Think of basic physics principles being used in building practical devices and systems, such as radio, television, cellular phone, or an radio frequency identification (RFID) system, and think of the whole field of physical chemistry and biophysics. Physics, undoubtedly, has been the foundation of all engineering and technology. Understanding and application of physics laws is necessary from designing a mousestrap to designing and building a flat screen TV, a sports car, and a spacecraft. Depending on the history of their development, some fields, such as chemistry and engineering, have been ahead of other fields, such as biology, in making use of physics. It is expected that in the coming years, physics, especially quantum physics, will be used enormously in the process of understanding entities and phenomena in various fields of biosciences, including biochemistry, molecular biology, genetics, and even evolution.

Despite its sophisticated theories, physics, at the end of the day, is an experimental science. Physicists observe the phenomena of nature, find the patterns and relationships among those phenomena, and try to explain this in terms of models and theories, which after rigorous experimental tests are established as physical laws or principles.

Caution: The notion that a theory is just a theory, just a random thought, an abstract idea, or an unproven concept, is an incorrect notion. Scientists use this word as follows: A theory is an explanation of a natural phenomenon or a set of phenomena based on observation. By definition, a theory is falsifiable and faces the tests of experiments for its acceptance among the scientific community. The theory of gravitation, the theory of unification of forces, and the theory of biological evolution, known as Darwin's theory of evolution, are all examples of such theories, also called scientific theories.

The development of physical theories and principles is an endless process of back-and-forth between ideas and experiments. Scientists never regard any theory as final or as the ultimate truth. On the contrary, they are always looking for new observations that will require us to revise a theory or to discard it for a better one. By its very nature, a scientific theory can be disproved by finding phenomena and behavior that are inconsistent with it, but a theory can never be proved to be always correct. As you will see in this book, the development of quantum physics is a good example of this process of developing scientific theories.

Before quantum physics, we had classical physics, which is reviewed or overviewed in this chapter. In order to get through this chapter smoothly, let us review some basic concepts related to physics:

Particle. A particle is a small object that behaves as a whole unit in terms of its motion, properties, and behavior. Although one usually thinks of a particle as a very small object, there is no size limit on what can be treated as a particle. A point particle is an idealized and simplified representation of an entity. It can be fully described as having a spatial extent of zero (size zero), and therefore its position is completely defined by one set of coordinates (x, y, z), called Cartesian coordinates. Given it is a point, it obviously has no internal structure. While its geometry is simple, it can, however, have properties associated with it, such as mass and electrical charge. In many cases, this approximation works well in understanding the overall behavior of a system and getting out reasonable quantitative results. As seen later in this book, this concept can also be used to represent the center of mass of a system, a reference frame.

So, if one point in an object can be located to determine its position in space, in order to simplify the discussion of its motion we can treat this object as a point particle, also referred to as a particle for brevity. Any object can be considered a particle as long as we are not interested in its internal structure and rotational motion. In some cases, large objects, such as the Sun, can be modeled as point particles for the determination of certain characteristics or quantities. For example, to determine the orbit of the Earth around the Sun it is reasonable to assume that all of the Sun's mass is concentrated at one point, its geometric center. It is a good approximation for determining the force exerted on the Earth and thus determining the Earth's orbit. As another example, the molecules in the kinetic molecular theory of gases are considered as particles. Once you start considering the rotation of molecules and the fact that that they are made of atoms (internal structure), you can no more treat them as particles (i.e., point particles).

When we want to study the internal structure and dynamics of an object, the object is no more a particle, it is a system.

> *Note*: Over centuries, physicists have invented several coordinate systems to describe the position and motion of a physical entity in space. Depending on the nature of the entity and the problem, one coordinate system may be more convenient than others. For example, you know from your introductory physics course that translational motion is usually better described in a Cartesian coordinate system, and circular or rotational motion is usually better described in polar coordinates.

System. A system is a set of distinct entities often interacting with one another. A system has a structure defined by its constituents, which have a structural as well as functional relationship with one another. A system as a whole has some characteristics and a certain behavior. For example, an atom is a system constituted of electrons, protons, and possible neutrons. Similarly, planets revolving around the Sun make up a system called the solar system. An organism, such as a human being, is also a system composed of organs; and a deoxyribonucleic acid (DNA) molecule inside an organism is a system composed of smaller components called nucleotides.

Physical Quantity. A measurable observable, for example energy of an entity, such as a particle or a system, is called a physical quantity. A physical quantity describes an aspect of an object or a phenomenon quantitatively. In physics, we understand the universe, the systems in the universe, and their behavior in terms of physical quantities, as well as the relationships among these physical quantities. In other words, laws of physics are usually expressed in terms of relationships among the physical quantities. Mass, electric charge, length, time, speed, force, energy, and temperature are some examples of physical quantities. This book will use the term observable and physical quantity interchangeably.

Unit. A physical quantity is measured in numbers of a basic amount called a unit. The measurement of a quantity contains a number and a unit, for example, in 15 miles, the mile is a unit of distance (or length). Similarly, a kilogram (kg) is a unit of mass.

Force. The influence that an object exerts on another object to cause some change is the force. Where there is change, there is some force behind it. Motion of an object is an example of change. The change, for example, could be in the physical properties, such as the speed or position (location) of the object. The exact nature of this change will be determined by some physical principles, such as Newton's laws of motion, which is described later in this chapter.

Interaction. A mutual force between two objects through which they influence each other is an interaction. For example, two particles attract each other due to an attractive force or repel each other due to a repulsive force between them. Sometimes the words *interaction* and *force* are used synonymously. There are four known fundamental interactions or forces: (1) gravitational force; (2) electromagnetic force; (3) strong nuclear force; and (4) weak nuclear force.

For example, most of the forces that chemistry students learn about, such as covalent bonds, ionic bonds, hydrogen bonds, and London dispersion force, are different manifestations of the electromagnetic (EM) force.

Where there is a force there is energy, or the potential for energy, called potential energy. Force acting on a particle results in a change in the kinetic or potential energy of the particle.

Speed. The speed describes the motion of a particle. To be precise it is the change in position with time. For example, your car is moving at a speed of 70 miles per hour (mph). When you specify the direction of the speed, it becomes velocity. For example, the velocity of your car at a given moment could be 70 mph toward the North.

Energy. The measure of the ability of a force to do work is energy. There are different kinds of energies corresponding to different forces, such as EM energy corresponding to EM force, nuclear energy corresponding to nuclear forces, and so on. All of these energies have the same units. Flow of energy, that is, energy transfer, is a key to maintaining order in the universe and in our bodies. For example, energy flows from the Sun to the plants, and from the plants to our bodies in the form of food. Energy can be converted from one form to another, but it can neither be created nor destroyed. This is called the principle of conservation of energy. For example, in your microwave oven, the EM energy that you buy from your power company is converted into heat energy that warms (or cooks) your food; the form is changed, not the total content of the energy.

Work. A measure of the amount of change produced by a force acting on an object is work. For example, gravitational force between the Sun and the Earth is making the Earth revolve around the Sun. In doing so, Earth is doing some work. But how is it possible that two objects separated from each other can exert force on each other? This is where the concept of field comes into the picture.

Power. The energy transferred per unit time or the amount of work done by a force per unit time is power.

Field. The fundamental forces of nature can work between two objects without the objects physically touching each other. For example, Sun and Earth attract each other through gravitational force without touching each other. Two charged particles attract or repel each other through EM force without touching each other. This effect is called *action at a distance* and is explained in physics by the concept of a *field*. Each of the two objects that, for example, attract or repel each other from a distance, create a field in the space. This is the field that exerts the force on the other object. For example, there is a gravitational field corresponding to the gravitational force and an EM field corresponding to the EM force, and so on.

For the purpose of visualization, the force applied through a field is often represented in terms of field lines. For example, Figure 1.1 illustrates electric field lines around positively and negatively charged particles. As shown in this figure, electric field lines always point away from a positively charged particle and point toward a negatively charged particle. The symbol $\vec{E}$ stands for an electric field, and the arrow on E means the electric field is a vector.

A physical quantity can be a scalar or a vector.

Scalars and Vectors. A scalar is a quantity that has a magnitude, but no direction. For example, a speed of 70 mph is a scalar. A vector is a quantity that has both magnitude and direction. For example, a speed of

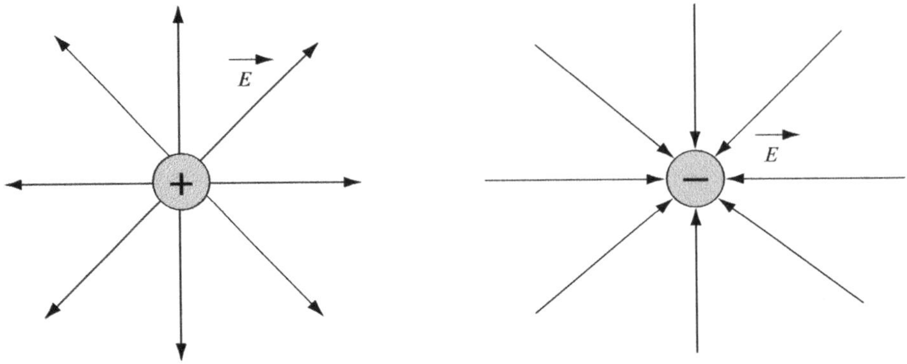

Figure 1.1. Electric field lines around a particle with electric charge.

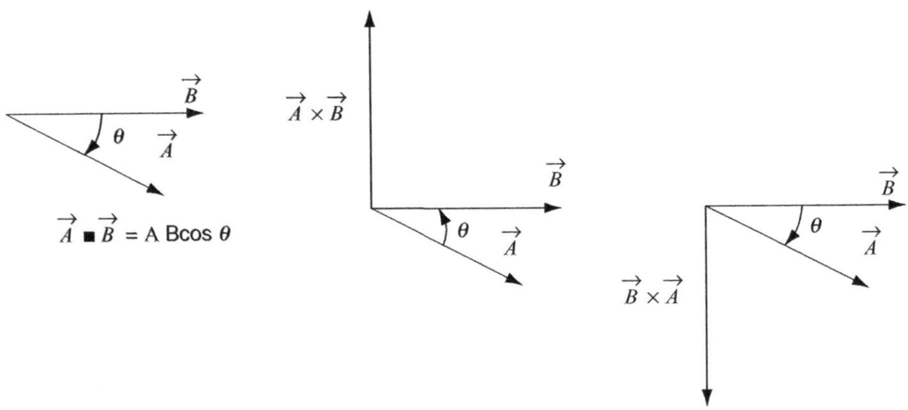

Figure 1.2. Illustration of a scalar and a vector product.

70 mph toward the North is a vector. Speed in a specified direction is called velocity.

Scalar quantities can be multiplied just like numbers, whereas vector quantities can undergo any of two kinds of multiplication called scalar product and vector product. The scalar product of two vectors $\vec{A}$ and $\vec{B}$ is denoted as $\vec{A}.\vec{B}$, and due to the dot in this notation it is also called a dot product. As the name suggests, the scalar product of two vectors is a scalar obtained by multiplying the magnitude of a vector with the magnitude of the component of the other vector in the direction of this vector:

$$\vec{A}.\vec{B} = \mathrm{AB}\cos\theta \tag{1.1}$$

where θ is the angle between the two vectors $\vec{A}$ and $\vec{B}$, as shown in Figure 1.2, which also illustrates the vector product.

A vector product of two vectors $\vec{A}$ and $\vec{B}$ is denoted as $\vec{A}\times\vec{B}$, and due to this notation it is also called a cross product. As the name suggests, the vector product of two vectors $\vec{A}$ and $\vec{B}$ is also a vector, say $\vec{C}$ perpendicular to the plane of $\vec{A}$ and $\vec{B}$, as illustrated in Figure 1.2. The magnitude of $\vec{C}$ is given by

$$\left|\vec{C}\right| = \left|\vec{A}\times\vec{B}\right| = \text{AB}\sin\theta \tag{1.2}$$

Also,

$$\vec{A}\times\vec{B} = -\vec{B}\times\vec{A}$$

PROBLEM 1.1

Forces $\vec{F}_A$ and $\vec{F}_B$ with magnitudes of 3 and 5 Ns, respectively, are acting on a particle. The force $\vec{F}_A$ is acting along the x-axis, and the force $\vec{F}_B$ is making an angle of 30° with the x-axis in the x–y plane.

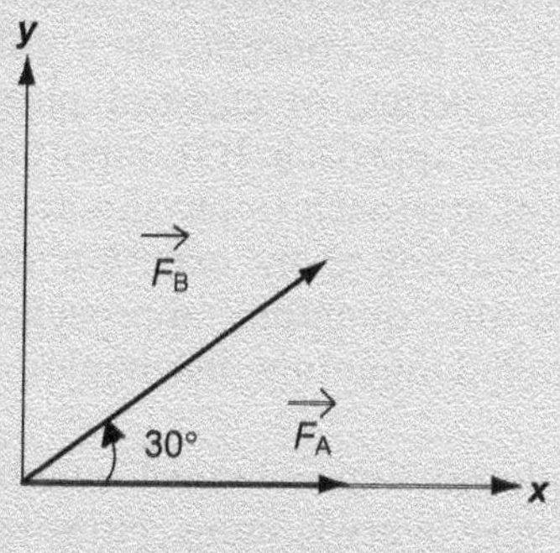

A. Calculate the scalar product of these two forces.
B. Calculate the vector product of these two forces.

Solution:

A.

$$\vec{F}_A.\vec{F}_B = F_A F_B \cos\theta = 3\times5\times\cos 30° = 12.99\,\text{N}$$

B.

$$\vec{F}_A\times\vec{F}_B = F_A F_B \sin\theta = 3\times5\times\sin 30° = 7.5\,\text{N}$$

The force as a result of the cross product will be along the z-axis

Note: A material body or particle is a physical entity that contains mass and occupies space.

Classical physics looks at the universe in terms of material bodies and waves. These material bodies, their structures, and their behavior are understood in terms of what are called physical quantities, which were explained earlier. Many physical quantities of material bodies can be studied by treating them as point particles. So, you can say that classical physics divides the physical world into particles and waves.

1.3 THE CLASSICAL WORLD OF PARTICLES

The Greek philosopher Aristotle is considered by many to be the first physicist. What is known as classical physics or Newtonian physics today is a result of the scientific work of an enormous number of scientists across several centuries, starting from Aristotle (384–322 BC) to Galileo Galilei (1564–1642), to Isaac Newton (1643–1727), to James Clerk Maxwell (1831–1879). The goal of this chapter is not to present a full review of the classical physics, but rather briefly explore a few concepts in a cohesive way in order to establish the big cohesive picture of classical physics to capture the key characteristics of the classical approach. It is important for you to understand and appreciate the classical approach to physics in order to fully understand how this approach was challenged by experimental results and how this challenge was met with the quantum approach, which developed into quantum physics, also called quantum mechanics, as opposed to classical mechanics.

In Section 1.2, we explored some physical quantities; many more can be listed, and it can become very overwhelming. However, here is the good news: Recall all introductory physics courses that you might have taken, and it will not be hard to realize that most of the physics is based on the following three fundamental concepts:

Existence of an Entity. A physical entity, such as a material particle, exists with some defining physical characteristics, such as mass and charge, pertaining to the existence of the entity. Physical entities can also be waves, which we will introduce later in this chapter.

Position in Space. At a given moment, the particle exists at some point in a three-dimensional (3D) space. This point is represented by using some coordinate system, such as a Cartesian coordinate system with rectangular coordinates x, y, and z, as shown in Figure 1.3.

Time. The particle exists at a specific point in space and at some specific time.

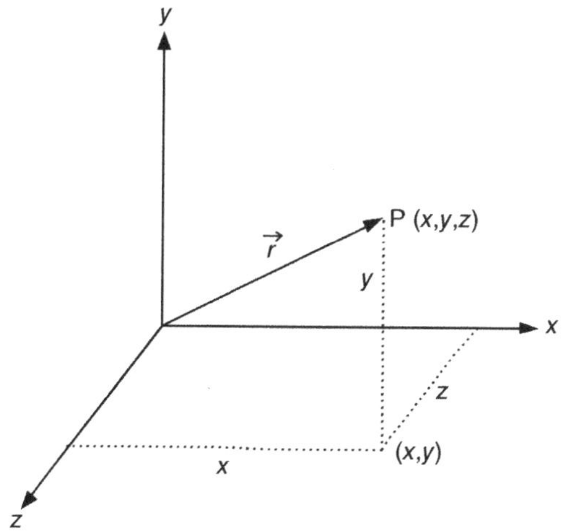

Figure 1.3. The Cartesian coordinate system used to represent the position of a particle in space.

It is important to realize that before describing the motion of a particle in space, we first must be able to determine the particle's position at a given time.

As mentioned earlier, physics relies mostly on mathematics to develop scientific explanations for entities and their behavior: the phenomena. It is time to demonstrate this phenomena by putting the physical quantities described earlier in their mathematical form. In addition, we will also derive these quantities from the three fundamental concepts just described.

One of these concepts, the position, (see Fig. 1.3) of a particle at a point P at a certain moment, is described by a vector $\vec{r}$ that goes from the origin of the coordinate system to point P. The position vector, as it is called, can be resolved into three component vectors along the x-, y-, and z-axis:

$$\vec{r} = \vec{x} + \vec{y} + \vec{z} = \hat{\imath}x + \hat{\jmath}y + \hat{k}z \tag{1.3}$$

where x, y, and z are the magnitudes of the position vector in the x, y, and z direction; and $\hat{\imath}$, $\hat{\jmath}$, $\hat{k}$ are the unit vectors along the x, y, and z direction. The magnitude of the position vector is given by

$$r = \sqrt{x^2 + y^2 + z^2} \tag{1.4}$$

If you are considering motion in a straight line or in just one dimension, you can treat the vector quantities as scalar, because the direction is known and fixed. This is exactly what we are doing for most of the quantities in the rest of this section for simplicity.

So, the most fundamental issue in physics is the existence of a physical entity at a specific point in space and time, and the rest of physics can be derived by throwing in a time-related concept called change, for example, the change of position of the entity in time. This change creates motion, which in turn gives rise to quantities like momentum and energy.

Note: Change may also apply to other fundamental properties of an entity. For example, later in this chapter you will see, that change in electric charge at a point with respect to time generates electric current.

The classical world of particles is explained in terms of some physical quantities derived from the three fundamental concepts discussed here.

1.4 PHYSICAL QUANTITIES

Physical entities (material bodies and waves) exist in space defined, for example, by the spatial coordinates (x, y, z) in the Cartesian coordinate system and by temporal coordinate, t. Figure 1.4 illustrates an example of how other physical quantities can be derived from three basic quantities: mass m of the particle, position x of the particle, and time t corresponding to three fundamental concepts discussed earlier:

- *Distance.* The distance (displacement) d that the particle has traveled during a time interval can be measured as the change in position (x) in this time duration:

$$d = x_2 - x_1 \tag{1.5}$$

  Meter (m) is a unit of distance; so is a mile.
- *Velocity.* Velocity, v, of a particle at a given moment is simply the rate of change of position of the particle with time represented by

$$v = \frac{dx}{dt} \tag{1.6}$$

  Velocity is a vector quantity, that is, it has a magnitude and a direction. The magnitude alone is called speed. Meters per second (m/s) is a unit of velocity or speed; so is mile/s.
- *Momentum.* The momentum p of a particle is the product of its mass and velocity:

$$p = mv = m\frac{dx}{dt} \tag{1.7}$$

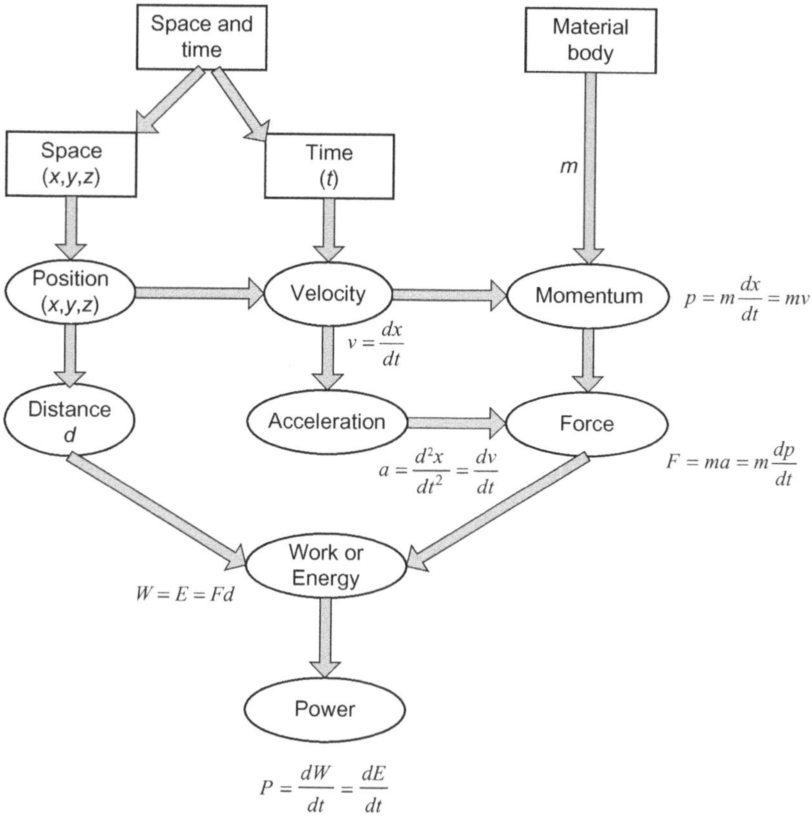

Figure 1.4. Physical quantities of material particles derived from the three fundamental concepts: existence (mass), space, and time.

- *Acceleration.* The acceleration of a particle at a given time is simply the rate of change in velocity with time:

$$a = \frac{dv}{dt} = \frac{d^2x}{dt^2} \tag{1.8}$$

- *Force.* The force applied on a particle is measured as the product of its mass and acceleration it is experiencing as a result of the force:

$$F = ma = m\frac{dv}{dt} = m\frac{d^2x}{dt^2} \tag{1.9}$$

The newton (N) is a unit of force, which is given by

$$1\,\text{N} = 1\,\text{kg}\frac{\text{m}}{\text{s}^2}$$

- *Work.* The work W is measured as a product of force and distance:

$$W = Fd = F(x_2 - x_1) \tag{1.10}$$

In general, the work performed by a force F on an object moving along a path S is given by

$$W = \int_{s1}^{s2} \vec{F}.d\vec{S} \tag{1.11}$$

where s_1 and s_2 are the start and end points of the curve. The joule (J) is a unit of work, which is given by

$$1\,\text{J} = 1\,\text{N m} = 1\,\text{kg}\frac{\text{m}^2}{\text{s}^2}$$

Energy has the same unit as work.

Energy. As mentioned earlier, the ability of a force to do work on an object is called energy. The work done by a force on an object is stored in the object in some form of energy. For example, assume you lift a body of mass m from the ground straight up and put it on a platform that is at a height h from the ground. In doing so, you have done work, W_g, against the gravitational force, $F_g = mg$, where g is the acceleration due to gravity. The work done by you has been stored into the body as its potential energy, E_p:

$$E_p = W_g = F_g h = mgh \tag{1.12}$$

When an external force works on a body at rest and gives it a speed of v, the work done on the body is equal to the kinetic energy, E_k, achieved by the body given by

$$E_k = \frac{1}{2}mv^2 \tag{1.13}$$

In general, the work done on a body is equal to the energy gained by the body and the work done by the body is equal to the energy lost by the body. For example, when you lifted up an object the energy was lost by your body and gained by the object that was lifted up. Overall, energy stayed conserved.

Power. Power is the rate of work performed or the rate of change of energy with time:

$$P = \frac{dW}{dt} = \frac{dE}{dt} \tag{1.14}$$

The watt (W) is a unit of power, which is equivalent to Joules per second (J/s).

Caution: Many quantities discussed here, such as velocity, acceleration, force, and momentum, are vector quantities; that is, they have magnitude as well as direction. For simplicity, in our treatment of these quantities, for the purpose at hand, we are only considering the magnitude in writing the equations, or alternatively considering these quantities in only one dimension, along the x-axis.

STUDY CHECKPOINT 1.1

Demonstrate how the physical quantity, called kinetic energy, can be derived from the three fundamental concepts of physics discussed in this section.

Solution:

Kinetic energy E_k is given by

$$E_k = \frac{1}{2}mv^2 = \frac{1}{2}m\left(\frac{dx}{dt}\right)^2$$

So, observe that the kinetic energy can be derived from m, x, and t.

In a nutshell, motion is caused by the change in position of an entity. But, what causes the change and exactly how does it occur? The work of several scientists over centuries elegantly put the answers to this question into three laws of motion by Issac Newton.

1.5 NEWTON'S LAWS OF MOTION

In an introductory physics course, you learn about these physical quantities by using point particles, that is, material entities with their mass concentrated at one point. Most of the physical quantities discussed so far describe the motion, or mechanics, of particles, and therefore belong to the branch of classical physics called classical mechanics. Classical mechanics is based on Newton's three laws of motion discussed in the following.

First Law. A physical body, or object, will continue in its state of rest or of uniform motion in a straight line until acted upon by an external force.

> *Note*: Newton's first law answers this question: If the position and momentum of a particle is known at time $t = 0$, what will be the position and momentum of the particle at some later time t in the absence of any external force?

In other words, the first law states that in the absence of any external force the motion of a body does not change: The body stays in a state of equilibrium. However, it does not necessarily mean that no force is acting on the body. For the body to be in equilibrium, the net result, that is, the vector sum, of all the forces acting on a body should be zero:

$$\sum \vec{F} = 0 \tag{1.15}$$

What if there is some nonzero net force acting on the body? This is when the second law of motion comes into the picture.

Second Law. If a net external force acts on a body, the body accelerates in the same direction as the force. The magnitude of the acceleration a is directly proportional to the force F that causes it:

$$a \infty F$$

This means

$$a = CF$$

where C is the proportionality constant and is equal to the inverse of the mass of the object: $C = 1/m$.

Therefore,

$$F = ma \tag{1.16}$$

It can also be written by using Eqs. 1.7 and 1.8 as:

$$\vec{F} = m\vec{a} = m\frac{d\vec{v}}{dt} = \frac{d\vec{p}}{dt} \tag{1.17}$$

Equation 1.17 is the fundamental equation of Newtonian mechanics, where p, the product of mass and speed of an entity, is called its momentum.

Note: Newton's first and second law answer this question: If the position and momentum of a particle is known at time $t = 0$, what will be the position and momentum of the particle at some later time t under the influence of a net force force F. If $F = 0$, we are talking about the first law.

The second law states how an object's motion changes when an external force is applied on it. What else does an object do in response to an external force? This question is answered by the third law of motion.

Third Law. When a body A exerts a force on a body B, then as a reaction, body B exerts back an equal and opposite force on body A. Equal and opposite means that the two forces are equal in magnitude and opposite in direction. It is also stated as: Action and reaction are equal and opposite. Therefore, this law is also known as the law of action and reaction, which means forces always occur in pairs. It can mathematically be represented as:

$$\vec{F}_{21} = -\vec{F}_{21} \tag{1.18}$$

Caution: Do not think that because these forces are equal and opposite, they will cancel out. Remember that these forces are being applied on two different bodies. They would have canceled out if they were working on the same body at the same point.

PROBLEM 1.2

Derive the first law of motion from the second law of motion.

Solution:

From the second law of motion:

$$F = ma = m\frac{dv}{dt} = \frac{dp}{dt}$$

This means that if F is zero, $a = 0$, which in turn means if no force is applied on the object, the object will keep moving with the uniform (constant) velocity v with which it is currently moving, or it will stay at rest if it is currently at rest.

It can be proved that the second law is the fundamental law from which the first and third laws can be derived.

STUDY CHECKPOINT 1.2

Specify if each quantity below is a scalar or a vector:

A. Velocity
B. Energy
C. Momentum
D. Acceleration
E. Mass
F. Temperature

Solution:

Scalars: B, E, F
Vectors: A, C, D

Note: Newton's three laws of motion based on the work of several scientists, including Copernicus, Kepler, Galileo Galilei, and Newton himself, represent a great intellectual feat of humans on this planet. These three rather simple laws can be applied to any system until certain limits are hit with respect to size and speed. For very small sizes, Newtonian physics is replaced with quantum physics, and for very high speeds (closer to the speed of light), Newtonian physics is replaced with relativistic physics based on Einstein's theory of relativity. More on this later in this chapter.

So far, we have defined the concepts and some basic physical quantities related to motion in a straight line. However, Newton's laws also apply to motion that is not in a straight line, rotational motion.

1.6 ROTATIONAL MOTION

Motion can also occur in a curvy or circular fashion called rotation. As illustrated in Figure 1.5, the position of a particle in circular motion (in two dimensions, 2D) can be determined by the radius, r, of the circle, and the angle θ that it makes with a fixed direction by extending an arc of length s. The relationship between these three quantities is given by

$$\theta = \frac{s}{r} \tag{1.19}$$

The units of angle θ determined this way are in radians (rad) and are related to degrees as follows:

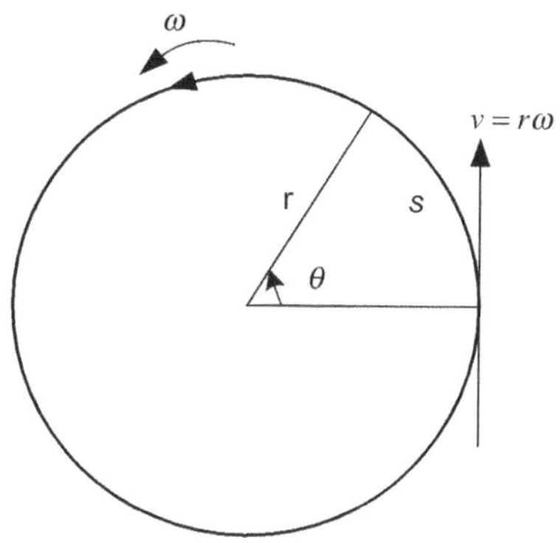

Figure 1.5. Circular motion for an arc length s corresponding to an angle θ: $s = r\,\theta$.

$$1\text{ rad} = \frac{360^\circ}{2\pi} = 57.3^\circ$$

This is an example of a polar coordinate system in 2D. The basic quantities related to rotation are derived in the following:

Angular Velocity. Because the angle θ specifies the position of a particle in polar coordinates at a given moment, the rotational motion (velocity) of the particle can be defined in terms of the rate of change of θ. This means the angular velocity, ω, of a particle moving in a circle of radius, r, is defined as the rate of change in the angle θ with time:

$$\omega = \frac{d\theta}{dt} = \frac{1}{r}\frac{ds}{dt} = \frac{v}{r} \tag{1.20}$$

Therefore,

$$v = r\omega \tag{1.21}$$

To derive Eq. 1.20, we have used the definition of $\theta = s/r$. Equation 1.21 gives the relationship between the linear velocity, v, along the length of the curve and the angular velocity ω along the curve. The direction of the linear velocity v is tangent to the curve. Note that even if the linear velocity v is constant in magnitude, its direction is continuously changing, so it is under acceleration in that sense. Hence, $\vec{v}$ is changing. The acceleration of a particle moving in a circle can be presented in terms of centripetal and tangent components.

Centripetal Acceleration. The component of the acceleration along the radius of the circle and directed toward the center of the circle, called centripetal acceleration, is given by

$$a_r = \frac{v^2}{r} \tag{1.22}$$

For example, in the case of a satellite revolving around the earth, this acceleration means that the satellite is being attracted by the Earth with a force F_s given by

$$F_s = \frac{mv^2}{r} \tag{1.23}$$

where m is the mass of the satellite. The satellite is pulling the Earth toward itself with the gravitational force F_g given by

$$F_g = G\frac{Mm}{r^2} \tag{1.24}$$

where M is the mass of the Earth and the satellite is revolving around the Earth in a circular orbit with radius r. If the two forces are equal, they will balance and the satellite will stay in the circular orbit. Therefore, the condition for the satellite to stay in orbit is

$$G\frac{Mm}{r^2} = \frac{mv^2}{r}$$

which means

$$v = \sqrt{\frac{GM}{r}} \tag{1.25}$$

this value of v is called the escape velocity.

Tangential Acceleration. This is the angular acceleration, that is, the change in angular velocity with time, and can be written as:

$$\alpha = \frac{d\omega}{dt} = \frac{d^2\theta}{dt^2} = \frac{1}{r}\frac{d^2s}{dt^2} = \frac{a_t}{r} \tag{1.26}$$

where a_t is the component of acceleration along the tangent to the circle.

Angular Momentum. The angular momentum, L, of a particle in a circular motion is given by

$$\vec{L} = \vec{r} \times \vec{p} = \vec{r} \times m\vec{v} = I\vec{w} \tag{1.27}$$

where

$$I = mr^2 \tag{1.28}$$

is a physical quantity called the moment of inertia, which is related to the geometric structure of the entity.

Therefore, the magnitude of angular momentum can be written as:

$$L = I\omega = mr^2\omega \tag{1.29}$$

The rotational kinetic energy can be described as:

$$E_k = \frac{1}{2} I\omega^2 \tag{1.30}$$

which can be shown as equivalent to the definition of kinetic energy for linear motion:

$$E_k = \frac{1}{2} I\omega^2 = \frac{1}{2}(mr^2)\left(\frac{v^2}{r^2}\right) = \frac{1}{2}mv^2 \tag{1.31}$$

Isaac Newton (1643–1727)

Isaac Newton, the great physicist and mathematician, was born on 4 January 1643 in Woolsthorpe-by-Colsterworth, England. He studied and did most of his work at Cambridge, where he became professor of mathematics. His 1687 publication of the *Philosophiæ Naturalis Principia Mathematica* (usually called the *Principia*) in 1687 is among the most influential books in the history of science. It laid the groundwork for most of classical mechanics. In this work, Newton presented three laws of motion that make up the core of classical mechanics. In addition, his work includes the universal law of gravitation, principles of conservation of momentum and angular momentum, and theory of colors. He developed the first reflecting telescope.

Newton shares the credit with Gottfried Leibniz for the development of differential and integral calculus. Newton died in his sleep in London on 31 March 1727.

In a nutshell, using three laws of motion and the physical quantities that we have derived from the three fundamental concepts (mass, position, and time), we can describe complex motion of physical objects or particles.

So far, we have treated particles pretty much separated from each other. However, it is an interesting world and interesting things happen when particles (like people) interact with each other. The physical interaction with direct touch is called collision.

1.7 SUPERPOSITION AND COLLISION OF PARTICLES

It is convenient to illustrate some physical concepts by using just one particle. However, in the real world particles hardly ever exist alone. They tend to exist as colonies or communities, and therefore they interact with each other. They collide with each other and scatter, their properties superpose. Scattering and superposition are important characteristics of particle interactions.

1.7.1 Superposition

To understand superposition, you first must understand the resolution of a physical vector quantity, such as velocity, into its components. As demonstrated in Figure 1.6 by using velocity as an example, the component of a vector in a given direction is the projection of the vector on a line in that direction. Figure 1.6 shows the projections of a velocity vector $\vec{v}$ on the axes

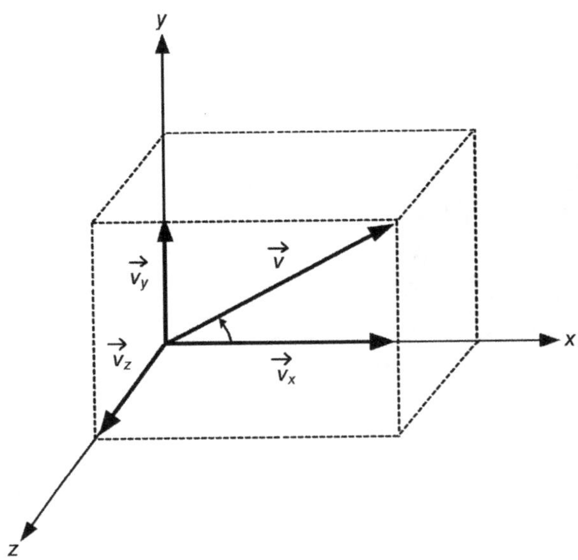

Figure 1.6. Resolving components of a vector as projections to the *x*-, *y*-, and *z*-axis.

of the rectangular (Cartesian) coordinate system: x, y, z. These projections are called rectangular components of velocity and are denoted as v_x, v_y, v_z:

$$\vec{v} = \vec{v_x} + \vec{v_y} + \vec{v_z} = \hat{\imath} v_x + \hat{\jmath} v_y + \hat{k} v_z \tag{1.32}$$

where v_x, v_y, v_z are the magnitudes of the vector $\vec{v}$ in the x, y, and z directions, respectively. The magnitude of the vector $\vec{v}$ can be written in terms of the magnitudes of its components:

$$v^2 = v_x^2 + v_y^2 + v_z^2 \tag{1.33}$$

or

$$v = \sqrt{v_x^2 + v_y^2 + v_z^2} \tag{1.34}$$

Note that we have combined the magnitudes of the components to obtain the magnitude of the original vector. Similarly, you can also combine the components of two vectors to get a single vector called the resultant vector. This property is called superposition of vectors.

As you learned in your introductory physics course, superposition of vector physical quantities, such as velocity and force, is of utmost importance. For example, any number of forces applied at a particle (or at a point of a body) has the same effect as a single force that is equal to the vector sum of all the individual forces. In other words, the forces combine or superpose. Figure 1.7a illustrates an example of two forces $\vec{F_1}$ *and* $\vec{F_2}$ acting at a point on a box in an x–y plane. The effect of these forces is the same as the force $\vec{F_R}$, which is the vector sum of the two forces. As shown in Figure 1.7b, the two individual forces can be vectorially added by first resolving into their components.

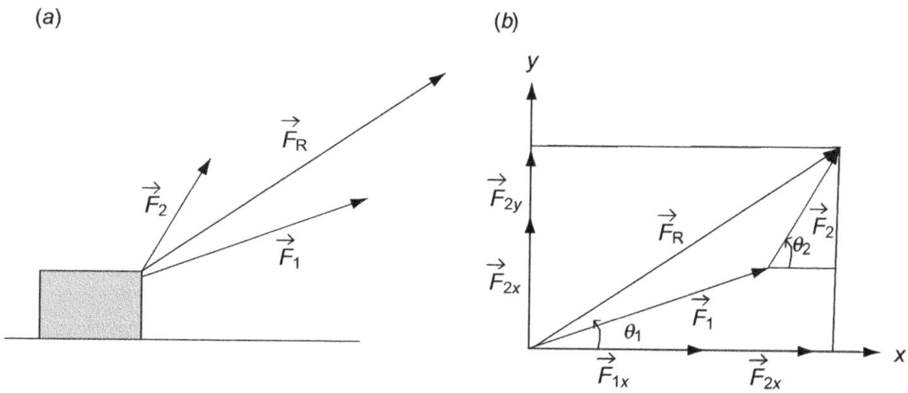

Figure 1.7. Superposition of two forces $\vec{F_1}$ *and* $\vec{F_2}$ resulting in one force $\vec{F_R}$.

$$\vec{F_1} = \hat{\imath} F_{1_x} + \hat{\jmath} F_{1_y}$$
$$\vec{F_2} = \hat{\imath} F_{2_x} + \hat{\jmath} F_{2_y}$$
$$\vec{F_R} = \vec{F_1} + \vec{F_2} = \hat{\imath} F_{1_x} + \hat{\jmath} F_{1_y} + \hat{\imath} F_{2_x} + \hat{\jmath} F_{2_y} = \hat{\imath}(F_{1_x} + F_{2_x}) + \hat{\jmath}(F_{1_y} + F_{2_y})$$

where

$$\vec{F_{R_x}} = \hat{\imath}(F_{1_x} + F_{2_x})$$
$$\vec{F_{R_y}} = \hat{\jmath}(F_{1_y} + F_{2_y})$$

Therefore,

$$F_R^2 = F_{R_x}^2 + F_{R_y}^2$$

or

$$F_R = \sqrt{F_{R_x}^2 + F_{R_y}^2} \tag{1.35}$$

We have demonstrated superposition in a 2D plane, but the extension into 3D space, and thereby using all the three coordinates (x, y, and z) is straightforward. Then, we will also have z components of each force, which we add to obtain F_{R_z}, and the magnitude of the resultant force becomes

$$F_R = \sqrt{F_{R_x}^2 + F_{R_y}^2 + F_{R_z}^2} \tag{1.36}$$

Also, there could be more than two forces acting on a particle. You decompose each of the forces acting on a particle into three coordinates, and then recombine their components along each coordinate to obtain the resultant force. By now, you should be able to realize that a number of forces acting on a particle may result in a net force of zero, that is, the components of all the forces along each of the three coordinates add up to zero. A particle (or a body) with zero resultant force acting on it is a particle in equilibrium. Zero force means zero acceleration, not necessarily zero speed; it may have a constant speed or a uniform velocity. Equilibrium of a particle with zero velocity is called static equilibrium and the equilibrium of a particle with uniform velocity is called dynamic equilibrium.

Of course, you can apply superposition to other vector physical quantities as well, such as velocity.

> *Note*: In the context of equilibrium, you can restate Newton's first law of motion as: A particle in equilibrium will stay in equilibrium until and unless acted upon by an external force.

1.7.2 Collision and Scattering

A car accident on a freeway and balls colliding on a billiard table are both examples of collisions. So are the two beams of protons colliding at a gigantic particle collider at CERN (a particle physics lab near Geneva, Switzerland).

However, for our purpose, consider a simple system of two particles A and B, as shown in Figure 1.8, moving with momentum $\vec{P}_A$, *and* $\vec{P}_B$; kinetic energy T_A, *and* T_B; and total energy E_A, and E_B. So the momentum and energy of the system before collision are given by

$$\vec{P} = \vec{P}_A + \vec{P}_B$$

$$T = T_A + T_B$$

$$E = E_A + E_B$$

Now, the two particles collide and in general may have a different velocity and direction of motion after the collision, and therefore different energy and momentum. Assume that after the collision, the two particles are moving with momentum $\vec{P}'_A$, and $\vec{P}'_B$; kinetic energy T'_A, *and* T'_B; and total energy E'_A, *and* E'_B. So the momentum and energy of the system after collision are given by

$$\vec{P}' = \vec{P}'_A + \vec{P}'_B$$

$$T' = T'_A + T'_B$$

$$E' = E'_A + E'_B$$

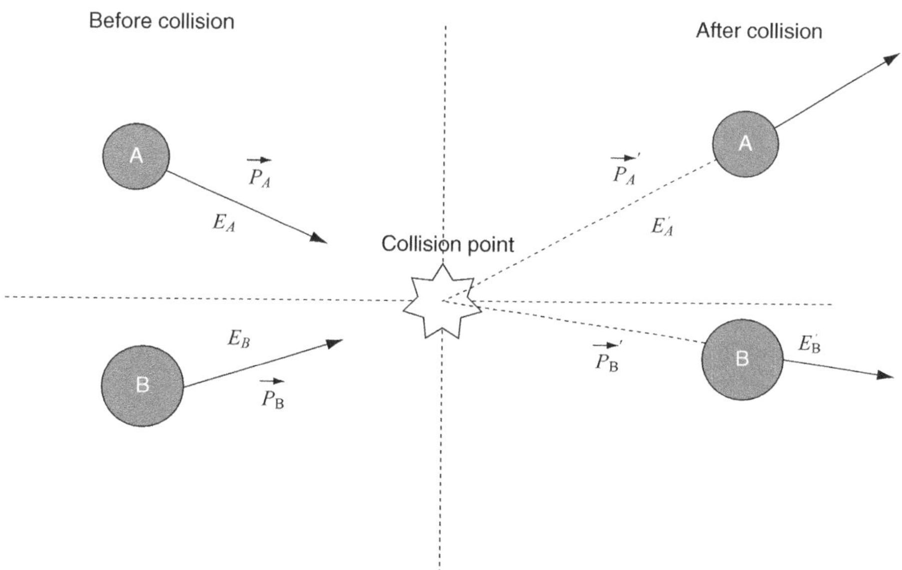

Figure 1.8. Illustration of collision between two particles.

Assume there is no external force acting on the system. Then, according to the principle of momentum conservation, the momentum of the system after the collision will be equal to the momentum of the system before the collision:

$$\vec{P}' = \vec{P}$$

Also, according to the principle of energy conservation, the total energy of the system after the collision will be equal to the total energy of the system before the collision:

$$E' = E$$

Even though the energy is conserved, its form can change, and it is the toal energy that is conserved. Therefore, one specific form of energy, such as kinetic energy, does not have to be conserved. If the kinetic energy of the system after the collision is the same as before the collision, the collision is called elastic. If the total kinetic energy of the system changes as a result of the collision, the collision is called an inelastic collision.

Because the particles as a result of collision deviate from their original path, they can be considered as scattered, and the process is called scattering. You can imagine that when there are a large number of particles, such as molecules, colliding with each other in an elastic fashion, they will be deflected from their path quite frequently. As a result, they will be moving in a zigzag path. Such a motion is called Brownian motion. Yes, the billiard ball model of collision discussed in this section works well even at the microscopic scale of molecules and subatomic particles, such as protons and electrons; well enough to predict the behavior of participants to some degree of accuracy. Chapter 2 discusses an example of collision (and hence scattering) called Compton scattering, one of the few phenomenon that pushed physicists toward quantum mechanics.

In a nutshell, collision, scattering, and superposition are important characteristics of particles. You will see that the concept of scattering and superposition also apply to waves, but in a different way.

The motion of particles can be considered as transportation of energy and momentum from one point to another in space. However, this leaves out a class of natural phenomena and physical entities, which can only be explained by using a concept called a wave.

1.8 CLASSICAL WORLD OF WAVES

A wave is a disturbance of some sort that propagates through space and transfers some kind of energy from one point to another without transporting mass, which occurs with particles. For example, when you speak to a person face to face, the sound wave travels from your mouth to the ear of the listener. The propagating disturbance here is the change of pressure in the air. As long as the wave is traveling through a point, the air pressure at that point does not stay constant over time, and your ear and brain perceives the pressure change as a sound. You must also be familiar with the water waves (ripples) that you can see by throwing a rock in a pond with still water. As another familiar example, the light that you need to see anything is a result of an EM wave. The disturbance in an EM wave is the change in the electric and magnetic field. The wave can be looked upon as propagation of this disturbance. The EM wave is a wave that carries EM energy. You will learn more about these waves later in this chapter. One of the interesting things about EM waves is that they can travel through a vacuum. This means that they do not need a material medium to travel through, whereas sound waves and water waves are mechanical disturbances that require a physical medium such as air or water, to travel through. The EM radiation is an example of periodic waves.

1.8.1 Periodic Waves

Figure 1.9 illustrates a traveling sinusoidal wave by presenting snapshots of a wave at different times $t = 0.5\,T$, $1\,T$, $1.5\,T$, and $2\,T$, where T is the period of the wave, which is the time in which the wave travels a distance equal to its wavelength.

Wavelength is one of the defining characteristics of a wave.

The wave illustrated in Figure 1.9 is called a periodic wave because the shape of the wave is the repetition of a pattern stretched from one crest to the next or from one trough to the next.

We can use the periodic wave to define some wave characteristics.

1.8.2 Defining Wave Characteristics

As shown in Figure 1.10, you can describe a wave in terms of some parameters, such as amplitude, frequency, and wavelength, as described in the following:

Wavelength. Denoted by the symbol λ, this is the distance between two consecutive crests or two consecutive troughs of a periodic wave. The distance equal to the wavelength makes one cycle of change.

Amplitude. Amplitude is the maximum amount of disturbance during one wave cycle.

Velocity. The velocity or speed, v, of a wave is the speed with which the disturbance defining the wave travels in space. As illustrated in Figure 1.11, the wave pattern with length λ travels with velocity v and covers the distance λ in time T, the period of the wave. Therefore,

$$v = \frac{\lambda}{T} = f\lambda \tag{1.37}$$

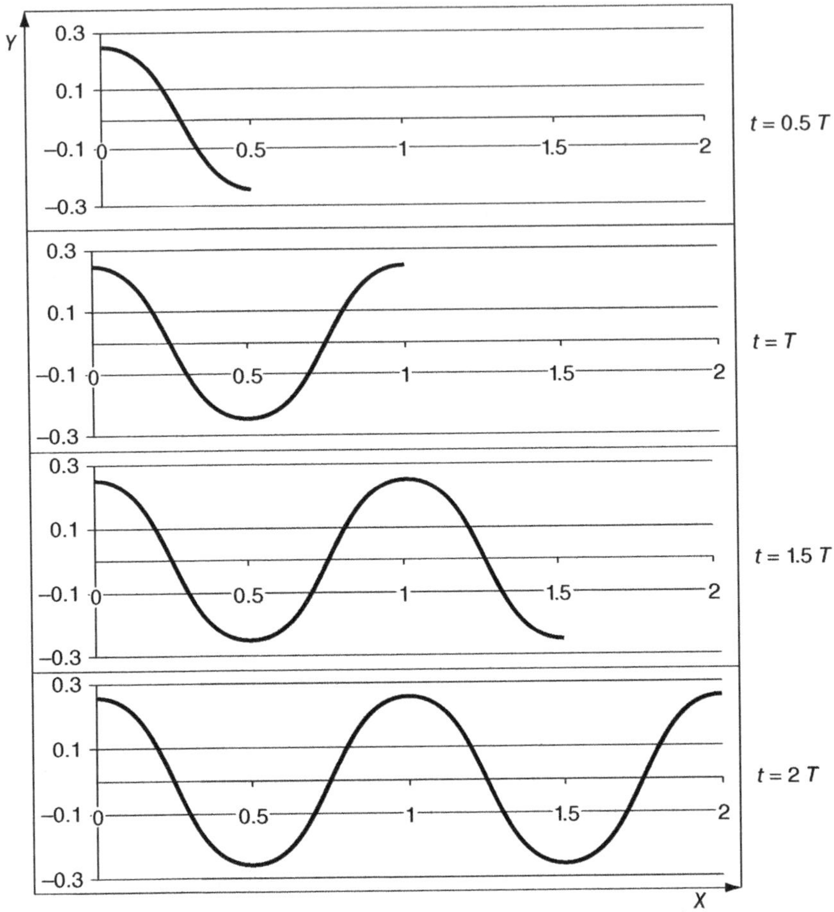

Figure 1.9. Propagation of a sinusoidal wave in time.

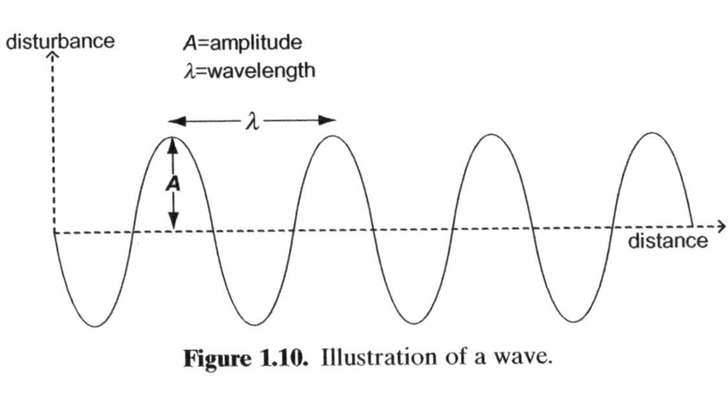

Figure 1.10. Illustration of a wave.

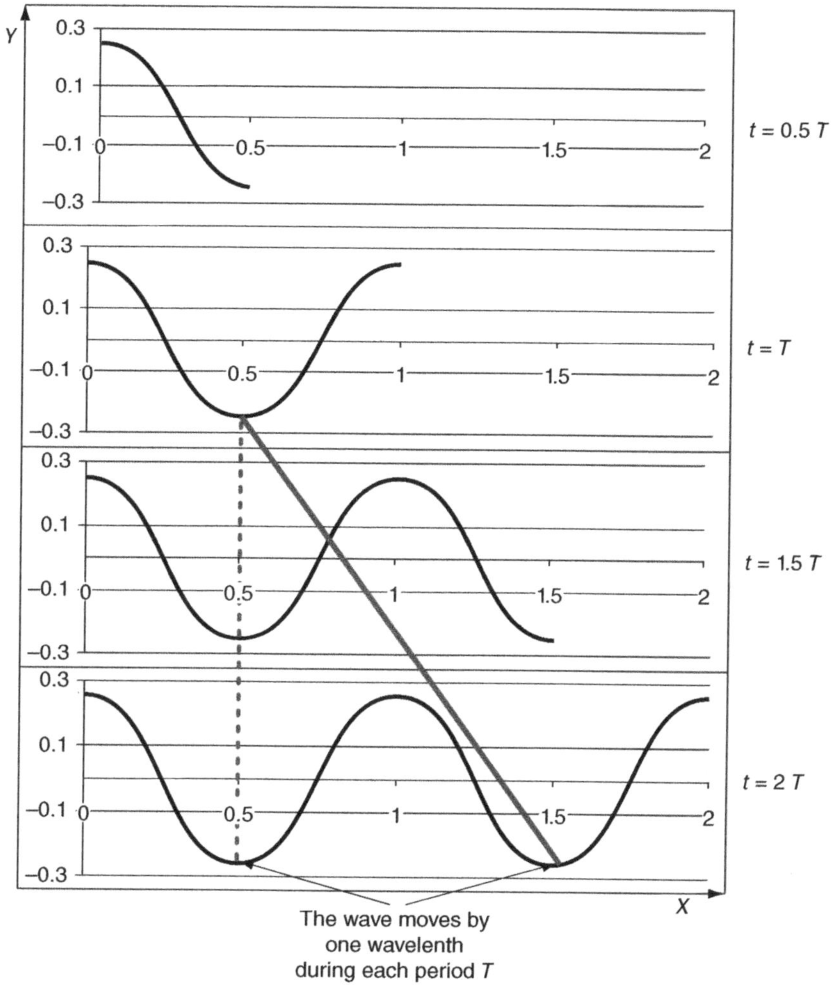

Figure 1.11. An illustration of a wave propagation.

The speed of a mechanical wave depends on the properties of the medium in which it propagates. The speed of EM waves in a vacuum is constant, which is the speed of light, c, given by

$$c = 2.998 \times 10^8 \ m/s$$

Frequency. The frequency, f, is the number of cycles per unit time a wave repeats at a given point:

$$f = \frac{1}{T} \tag{1.38}$$

where T is the time to complete one cycle and is called a period. This translates to

$$f = \frac{v}{\lambda} \tag{1.39}$$

Therefore, the frequency of an EM wave, f, propagating through free space (vacuum) is calculated by using the following equation:

$$f = \frac{c}{\lambda} \tag{1.40}$$

The frequency is measured in units of the hertz (Hz) and 1 cycle per second (cps) is 1 Hz.

Phase. A phase is the current position in the cycle of change in a wave.

Wavefront. The concept of wavefront is often used to describe the propagation of a wave. A wavefront is the locus, such as a line for a light ray in one dimension (ID) or a surface for a wave propagating in 3D, of points having the same phase. The locus is a mathematical entity (Latin for "place") that represents a collection of points with a shared property. For example, in 2D, a circle is the locus of points that are at an equal distance from a single point, the center of the circle. In 3D, the surface of a sphere is a locus of points that are at an equal distance from one point, the center of the sphere. A sphere makes a spherical wavefront for the wave propagation in 3D space.

Some concepts, such as velocity and energy, are common to both particles and waves. Other concepts, such as frequency and wavelength, are specific to waves, that is, in the framework of classical physics. Reflection of waves is very similar to the elastic collision of a material object in which the object strikes a wall and bounces back. Some other properties of waves, such as refraction and diffraction, are specific to waves. In classical physics, the terms of diffraction and refraction are not applied to particles. Although reflection and scattering actually do apply to particle interactions. These and some other wave characteristics are discussed next.

1.9 REFLECTION, REFRACTION, AND SCATTERING

As a wave, such as a light wave, travels through a medium, it can scatter. Also, when a wave, such as a light wave, strikes an interface between two transparent materials, such as air and glass, it can go under reflection and refraction. These phenomena are described in the following.

Reflection. Reflection is defined as the abrupt change in direction of a wavefront at an interface between two dissimilar media, so that the wavefront returns into the medium from which it hits the interface. For example, radio frequency (rf) waves, also called radio waves, are reflected when they strike the objects much larger than their wavelength, such as the floor, ceiling, and support beam. Metals are an obstruction to the signal because they are good at reflecting radio waves.

Refraction. Refraction is defined as the change in direction of a wavefront at an interface between two dissimilar media, but the wavefront does not return back to the medium from which it hit the interface. For example, the radio waves bend when they pass from one medium into another. Figure 1.12 illustrates reflection and refraction.

Scattering. Scattering is the phenomenon of absorbing a wave and reradiating it in various directions. Thereby it changes its direction of propagation from the original direction. Note that no absorption happens in particle scattering. As an example of wave scattering, consider the reflection of an EM wave, which is actually a scattering. For example, when a rf wave is scattered, it results in the loss of the signal or dispersion of the wave, as shown in Figure 1.13. It happens due to the interaction of the wave with the medium at the molecular level.

An application of scattering is a radar that uses scattering to detect metallic objects, such as airplanes or speeding cars.

Two other important characteristics of waves are diffraction and interference, discussed next.

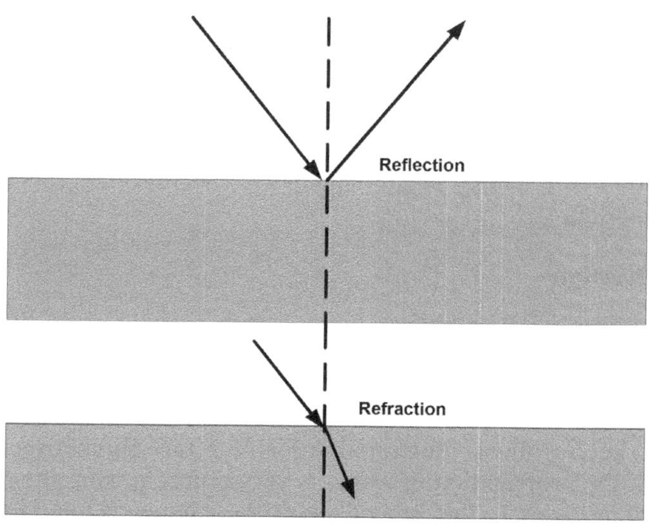

Figure 1.12. Reflection and refraction of waves; only the direction of propagation is shown.

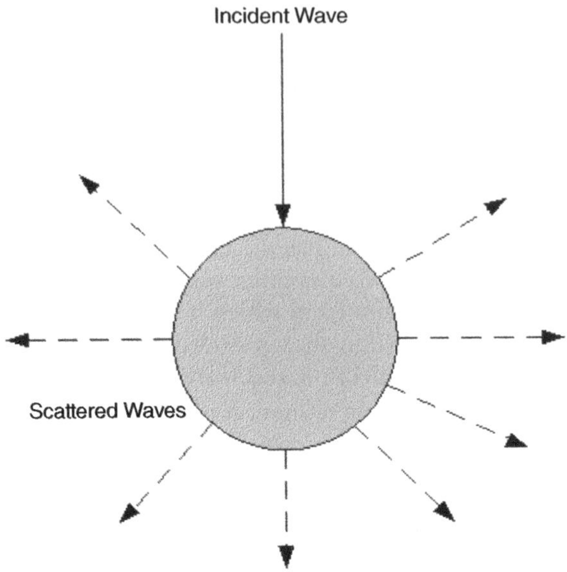

Figure 1.13. An illustration of wave scattering; only the direction of propagation is shown.

1.10 DIFFRACTION AND INTERFERENCE

It will come as no surprise if I tell you that even if the person who is talking to you turns his back on you, you can still listen, because sound is a wave, and waves exhibit a characteristic (or a phenomenon) called diffraction. If sound was carried by a particle, you would never be able to listen to a person whose back is turned to you.

Diffraction and a related property called interference are described in the following sections.

1.10.1 Diffraction

Diffraction refers to the bending of a wave when it strikes sharp edges or when it passes through narrow gaps. This makes propagation of waves quite different from the propagation of particles. The barrier will completely stop a particle coming at it as illustrated in Figure 1.14*a*, whereas waves will bend around it to keep going as illustrated in Figure 1.14*b*. This shows that in classical physics, the propagation of waves is very different from the propagation of a beam of particles. In the case of waves, the narrow opening in the barrier acts as a source.

Diffraction is related to a phenomenon called interference.

Note: One of the main objectives of this chapter is to help you understand how waves and particles are treated as different types of physical entities in classical physics. If you do not feel comfortable with any of the concepts discussed here, you should review them by using a basic physics book before you proceed; a list of books is provided in the Bibliography section at the end of this book.

STUDY CHECKPOINT 1.3

Diffraction is the process in which a wave bends around an edge or a narrow opening. Why does this bending happen in waves and not in particles?

Solution:

The slit acts as a source of waves, whereas it does not act as a source of a beam of particles. This is a unique characteristic of waves that each wavefront act as a source; this is how waves propagate.

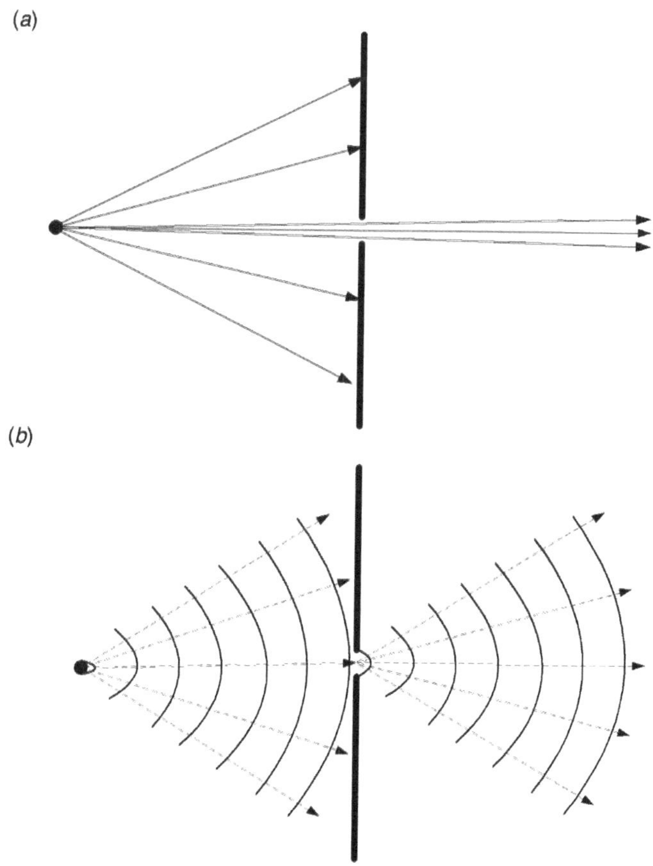

Figure 1.14. Illustration of (*a*) a beam of particles passing through a narrow opening in a barrier and (*b*) a wave transmitting through a narrow opening in a barrier.

1.10.2 Interference

Interference is defined as the interaction between two waves. Thomas Young in 1802 formulated the interference of light waves in terms of the principle of superposition: If two light waves arrive simultaneously at the same place, their effect is identical to that of the third wave, whose amplitude is the algebraic sum of the individual amplitudes of the component waves. An example of interference as a result of light from a source passing through a double slit is illustrated in Figure 1.15, where light waves passing through two different slits interfere with each other. The interference can be constructive, in which case the resultant wave has a larger amplitude, that is, the individual amplitudes are added to each other; or destructive, in which case the resultant wave has a smaller amplitude than the original wave, that is, the individual amplitudes are subtracted from each other. This constructive and destructive interference creates a pattern on the screen with high-intensity spots corresponding to constructive interference and very low intensity (dark) spots corresponding to destructive interference.

Thomas Young (1773–1829)

Thomas Young was born on June 13, 1773 in Milverton, Somerset, England into a large Quaker family. He began studying medicine in 1792, and also became interested in pure science. Young was a true polymath for his time, with interests ranging from physics to physiology and medicine, to music, to languages, and egyptology; he studied and contributed to all these fields. He obtained the degree of doctor of physics in 1796 from the University of Göttingen, Lower Saxony, Germany.

In 1801, Young was appointed as a lecturer at the recently formed Royal Institution in London, where he gave a series of lectures on a variety of topics. At that time, there was evidence for both pictures of light: waves and particles, and Newton was the proponent of the particle theory. In 1801, Thomas Young was intellectually bold enough to present a serious challenge to Newton's ideas on the topic, even though Newton was so greatly revered as a scientist that it was nearly impossible for anyone to dispute his theory.

Thomas Young died on May 10, 1829 at the age of 55 in London, England.

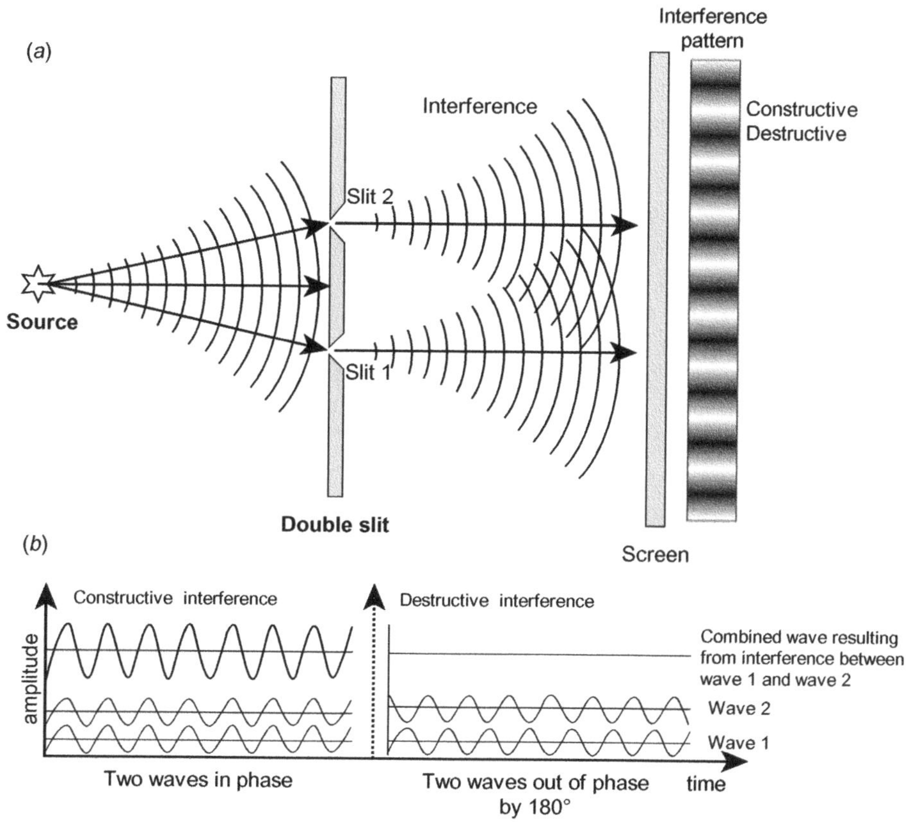

Figure 1.15. Interference of two waves produced by a double slit.

In order to avoid confusion, it is important to understand the relationship among diffraction, scattering, and interference. Diffraction is bending of a wave around small obstacles or the spreading out of waves through small openings. Consider an obstacle providing multiple closely spaced openings. A wave passing through it will split into components coming from different openings, which will interfere with each other through a process called interference or superposition. This interference gives rise to a pattern with maxima and minima of wave intensity, and is called an interference or diffraction pattern. One may also look at the bending of waves as scattering with a particle view in mind.

Newton's second law describes the motion of a particle. What describes the motion of a wave?

1.11 EQUATION OF WAVE MOTION

Newton's second law of motion, represented by Eq 1.17, provides the fundamental equation of motion for Newtonian mechanics to describe the motion of a particle. Is there any fundamental equation of motion for a wave? The answer is yes, and here it is in 1D.

$$\frac{\partial^2 \Psi}{\partial x^2} = \frac{1}{v^2}\frac{\partial^2 \Psi}{\partial t^2} \tag{1.41}$$

In general, this equation describes a wave that has a variable quantity Ψ, called a wave function, which propagates in the x direction with velocity v. For example, for the wave on a string, Ψ represents the displacement of the string from the x-axis; for a sound wave, Ψ represents the pressure difference; and for an EM wave, Ψ is the magnitude of an electric or magnetic field. For mechanical waves, Eq. 1.41 can be derived from Newton's second law of motion, and for EM waves, it can be derived from Maxwell's equations.

> *Note*: As demonstrated in Problem 1.3, the wave equation 1.41 is obeyed by any wave in 1D that is propagated without dispersion or change in shape. It is satisfied by any function of $(x - vt)$ or of $(x + vt)$.

It can be proved that a general solution of the wave equation 1.41 is given as:

$$\Psi(x,t) = \Psi_0 \sin(kx - \omega t) \tag{1.42}$$

where Ψ_0 is the maximum displacement called the wave amplitude and ω is the angular velocity given by

$$\omega = kv = \frac{2\pi}{\lambda} v \tag{1.43}$$

where k is called wavenumber.

PROBLEM 1.3

Prove that any function of $x - vt$ or of $x + vt$ will satisfy the wave equation.

Solution:

Substitute:

$$f = x - vt$$

which implies

$$\frac{\partial f}{\partial x} = 1 \tag{1}$$

$$\frac{\partial f}{\partial t} = -v \tag{2}$$

Therefore,

$$\Psi = \Psi(x - vt) = \Psi(f)$$

$$\frac{\partial \Psi}{\partial x} = \frac{\partial \Psi}{\partial f}\frac{\partial f}{\partial x} = \frac{\partial \Psi}{\partial f} \tag{3}$$

We made use of Eq. 1.

$$\frac{\partial \Psi}{\partial t} = \frac{\partial \Psi}{\partial f}\frac{\partial f}{\partial t} = -v\frac{\partial \Psi}{\partial f} \tag{4}$$

We made use of Eq. 2.
From Eqs. 3 and 4:

$$\frac{\partial^2 \Psi}{\partial x^2} = \frac{\partial^2 \Psi}{\partial f^2} \tag{5}$$

$$\frac{\partial^2 \Psi}{\partial t^2} = -v\frac{\partial^2 \Psi}{\partial f^2}\frac{\partial f}{\partial t} = v^2\frac{\partial^2 \Psi}{\partial f^2} \tag{6}$$

This means:

$$\frac{\partial^2 \Psi}{\partial f^2} = \frac{1}{v^2}\frac{\partial^2 \Psi}{\partial t^2}$$

Substitute this in Eq. 5:

$$\frac{\partial^2 \Psi}{\partial x^2} = \frac{1}{v^2}\frac{\partial^2 \Psi}{\partial t^2}$$

This is the wave equation.
Hence, the proof.

Just like particle motion, wave motion can be considered as the transportation of energy and momentum from one point to another in space, but without the transportation of matter. In mechanical waves, such as sound, water, and waves on a string, the momentum and energy are transported through a disturbance that is propagated through the medium. Some waves do not need any medium to travel through. For example, EM waves transport momentum and energy through electric and magnetic fields, which propagate through a vacuum.

In this book, we will be dealing with the EM waves in order to introduce quantum physics. So, it is relevant to review some concepts of EM, which is also an important part of classical physics. While motion of mass (material objects) can be described with Newtonian mechanics, the motion of electric charge (also referred to as just charge for brevity) is described by EM.

Again, while reviewing the basic concepts of EM, pay attention to how all these concepts are based on (or can be derived from) three fundamental concepts: charge instead of mass in Newtonian mechanics, position of charge in space, and change of position of charge with time. So the three fundamental concepts are again generally the same: entity, space, and time. An example of an EM wave visible to us is visible light.

1.12 LIGHT: PARTICLE OR WAVE?

Light has intrigued human beings for centuries. The struggle to understand the nature of light and the centuries-long controversy over it is not only one of the most interesting chapters in the history of science, but it also provides grounds to develop quantum physics. Early theories considered light as a stream of particles that emanated from a source and created a sensation of vision by entering the eyes. Newton during his time was the most influential proponent of the particle (also called corpuscular) theory of light, whereas Christian Huygens and Robert Hooks were among the major proponents of the wave theory of light. The phenomena of reflection and refraction of light known at the time were easily explained using the particle theory. Although Huygens was able to explain reflection and refraction by using wave theory, Newton rejected the wave theory of light based on the lack of evidence for diffraction. Even when the evidence for light diffraction became available, Newton's followers successfully argued that diffraction was due to the scattering of light particles from the edges of slits.

The wave theory of light made a comeback in 1801 when Thomas Young showed that sound waves and light both exhibit interference; showing that it is not possible for particles to create interference. Although Young's work went largely unnoticed by the scientific community for about a decade, Augustin Fresnel put wave theory on a firm mathematical basis, while performing extensive experiments on diffraction and interference. A big blow to Newton's particle theory came in 1850, when Jean Foucault showed by his measurements that the speed of light in water is less than that in air. In 1860, James Clerk Maxwell published his theory of EM, which predicted the speed of EM waves to be 3×10^8 m/s. He suggested that this agreement with the speed of light was not coincidental, that light was in fact an EM wave.

Maxwell's theory was experimentally confirmed by experiments performed by Hertz in 1887, who successfully generated and detected EM waves in the laboratory by using electrical means, such as a spark gap in a tuned circuit. He used the spark gap to in a tuned circuit to generate the wave and another circuit to detect it. The irony of the situation is that in the same year when Hertz's experiment confirmed Maxwell's wave theory, Hertz also discovered a phenomenon called the photoelectric effect, which could only be explained by Einstein using the particle model of light. However, we are getting ahead of ourselves now. These latter two points will be discussed in the forthcoming chapters. For now, understand that toward the end of the nineteenth century, when some experiments challenged classical physics, light was largely viewed as a wave: the EM wave.

To understand EM waves, it is important to understand electricity.

1.13 UNDERSTANDING ELECTRICITY

Electricity is the property of matter related to electric charge. Historically, the word electricity has been used by several scientists to mean electric charge. This property (electricity) is responsible for several natural phenomena, such as lightning, and is used in numerous industrial applications, such as electric power and the whole field of electronics. To understand electricity, you must understand the related concepts discussed in the following:

Electric Charge. Electric charge, also referred to as just charge, is a basic property of some particles of matter. There are two types of charge: positive and negative. For example, an electron has a negative charge and a proton has a positive charge, equal in magnitude to that of the electron. The charge of an electron (or proton) is used as a natural unit of charge. The standard symbol used to represent charge is q or Q. Two particles (or objects) with the same type of charge repel each other and two objects with the opposite types of charge attract each other. The charge is measured in units of the coulomb, denoted by C.

Charge is conserved: the algebraic sum of all the electric charges in a closed system is a constant. Note, it does not mean that charge cannot flow. Nor does it mean that charges cannot be created or destroyed. For example, consider a system of one particle called a photon, which can decay into a pair of electrons with a negative charge and a positron with a positive charge. Nevertheless, the algebraic sum of charges is still zero as before. Also, the electrically neutral objects are not neutral because they have no charge, but because they have an equal number of positive and negative charges. For example, all matter is made up of atoms and all atoms in their natural form contain an equal number of protons (positively charged particles) and electrons (negatively charged particles).

Electric Potential–Voltage. The electric potential difference V_{ab} between two points *a* and b is the work required to move one unit, *C*, called a coulomb, of charge from one point *a* to the other point *b* in space. This is commonly called an electric potential or voltage because it is measured in units of the volt, denoted by V.

Capacitance. Capacitance is the amount of charge stored in a system, called a capacitor, per unit of electric potential. In other words, the capacitance, *C*, is defined by the following equation:

$$C = \frac{Q}{V} \tag{1.44}$$

A capacitor is a device that stores charge and the resulting electric potential energy. Therefore, capacitance may also be looked upon as a measure of the ability of a device to store energy. For example, consider a parallel plate conductor shown in Figure 1.16, which contains two parallel plates as conductors. Assume that initially each conductor has a zero charge and therefore has a zero net charge. Now assume that electrons are transferred from conductor *a* to conductor *b*, so that conductor *b* acquires a $-Q$ charge and conductor *a* acquires a $+Q$ charge. The total

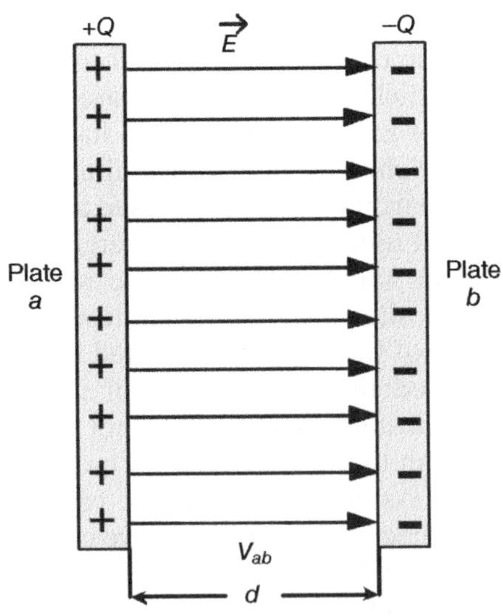

Figure 1.16. A parallel plate capacitor.

work performed for this transfer is stored as the potential energy in the capacitor. The voltage (potential difference) between the plates can be written as:

$$V_{ab} = Ed \tag{1.45}$$

where E is the electric field between the two plates (conductors), and d is the distance between the plates. Therefore, from Eqs. 1.44 and 1.45, the capacitance of the capacitor can be written as:

$$C = \frac{Q}{V_{ab}} = \frac{Q}{Ed} \tag{1.46}$$

This example of a capacitor is the so-called parallel plates capacitor: Two metallic plates separated from each other and each plate carrying equal and opposite charge, Q, with a potential difference between them, V. Capacitance is measured in units of the farad, denoted by F. For example, if the charge on each plate of a parallel plate capacitor is 1 C, and the voltage between them is 1 V, the capacitance of the capacitor will be 1 F.

Electric Current. Electric current is the rate of flow of an electric charge per unit time, and can be defined by the following equation:

$$I = \frac{Q}{t} \tag{1.47}$$

In this equation, I is the current and Q is the amount of charge that flowed past a point in time t. More generally, it can be represented in terms of the rate of the charge change with time:

$$I = \frac{dQ}{dt} \tag{1.48}$$

Current is measured in ampere units, (A). For example, one C of charge flowing past a point in 1 s represents 1 A of current. The materials, such as metals, that permit relatively free flow of charge are called conductors, whereas the materials, such as glass, that do not allow free flow of charge are called insulators.

Resistance. Resistance is a measure of opposition offered by a material to the flow of charge through it. The resistance can be measured by Eq. 1.49.

$$I = \frac{V}{R} \tag{1.49}$$

This means that the larger the resistance, the smaller the current. Resistance is measured in units of the ohm (Ω). For example, if the voltage of 1 V creates 1 A of current in a conductor, the resistance of the conductor is 1 Ω.

Electric Energy. The amount of work that can be done by an amount of electric charge across a potential difference is the electric energy. For example, the energy, E, of a charge Q across a voltage V is given by Eq. 1.50.

$$E = QV \tag{1.50}$$

Electric Power. The rate of work performed by an electric current is the electric power. In other words, it is the electric energy produced or consumed per unit time, and is given by Eq. 1.51.

$$P = \frac{E}{t} = \frac{QV}{t} = IV \tag{1.51}$$

The power is measured in units of the watt (W). For example, the power consumed to maintain a current of 1 A across a voltage of 1 V is 1 W.

PROBLEM 1.4

Show that electric power can also be expressed by the following equations:

$$P = I^2 R$$

$$P = V^2/R$$

Solution:

We know that

$$P = IV$$

We also know that

$$I = V/R$$

Therefore,

$$P = IV = (V/R)V = V^2/R$$

But,

$$I = V/R \quad \text{means} \quad V = IR$$

Therefore,

$$P = IV = I \times IR = I^2R$$

Electric Field. The electric field is a field used by the charges at a distance to exert a force on each other. In other words, the charges at a distance interact with each other through their fields, called electric fields. Here is how it works. A particle as a result of its charge modifies the properties of the space around it. Any other charged particle, due to the charge it carries, experience that modification in terms of an electric force. As explained earlier, the fields are usually represented by field lines, as illustrated in Figure 1.17 for two charged particles interacting with each other through their electric fields.

Note: An electric force on a charged particle is exerted by the electric field created by other charged particles.

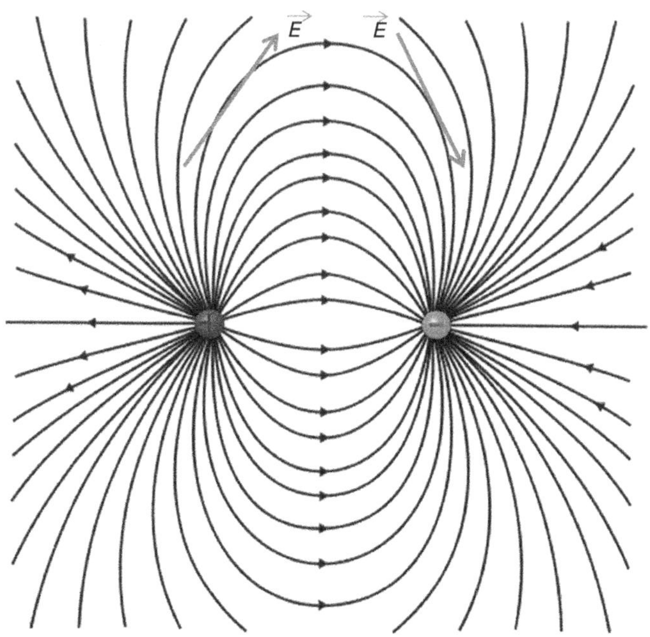

Figure 1.17. A particle with a positive and a negative electric charge applying force on each other through electric field lines. Courtesy of Sharayanan.

An electric field due to a particle with charge q at a distance r is defined as the force exerted on a unit charge at that point and is given as:

$$\vec{E} = k\frac{q}{r^2}\hat{r} \tag{1.52}$$

where k is called the Coulomb constant with a value of 8.988×10^9 N.m^2/C^2, and $\hat{r}$ is the unit vector along the line from the particle (considered to be at the origin) to the point where the electric field is being considered. The unit vector $\hat{r}$ is obviously the direction of the electric field and, as shown in Figure 1.17, it is always tangent to the field lines. The number of field lines passing through an area represent the strength (magnitude) of the field in that area. A quantity related to the number of electric field lines passing through an area is called an electric flux, denoted as Φ_E. Now, recall that an electric field at a point is a tangent to the field line (or curve) at that point. From differential calculus, you immediately know that the magnitude of the electric field will be given by the slope of the tangent line to the flux (as a function of area), that is, $\partial\Phi_E/\partial A$:

$$E = \frac{\partial \Phi_E}{\partial A} \tag{1.53}$$

Therefore,

$$\Phi_E = \int \vec{E}.d\vec{A} \tag{1.54}$$

It turns out that the flux through a closed surface is the same as the flux through the sphere around that surface with the source charge at the center of the sphere:

$$\Phi_E = \oint \vec{E}.d\vec{A} = \oint \vec{E}.\hat{n}dA = EA = k\frac{q}{r^2}4\pi r^2 = 4\pi kq = \frac{q}{\varepsilon_0} \tag{1.55}$$

Note that the radius of the sphere does not matter. We used the definition of E from Eq. 1.52 and

$$k = \frac{1}{4\pi\varepsilon_0} \tag{1.56}$$

Equation 1.55 is the mathematical representation of Gauss's law, which states that the total electric flux through a closed surface is equal to the total charge inside the surface divided by ε_0, a constant called the free permittivity of space.

Two charges of the same type exert a repulsive force on each other, and two charges of opposite types exert an attractive force on each other. This force is called an electric force. A charge in motion creates another kind of force called a magnetic force applied through a magnetic field.

1.14 UNDERSTANDING MAGNETISM

Magnetism is the property of a material that enables two objects to exert a specific kind of force on each other called a magnetic force, which is created by an electric charge in motion. To understand magnetism, you must understand the related concepts discussed in this section.

1.14.1 Magnetic Field

A magnetic field is a field produced by a moving charge that it used to exert a magnetic force on another moving charge. A magnetic field is also produced by some magnetic materials, such as permanent magnets. It turns out that a magnetic field in such materials is also produced due to moving charges, the circulating current on the atomic scale. So, in both cases a magnetic field is generated by circulating currents: In conventional electric circuits, the current circulates at the macroscale (the scale of the circuit), whereas in magnetic materials the current circulates at the atomic scale.

> *Note*: A moving charge (i.e., current) creates a magnetic field in the space around it. Note that a moving charge is still a charge; so it will generate an electric field as well.

What is common among a doorbell, an electric motor, an electromagnet, and a transformer? All contain coils of wires with a large number of turns spaced closely enough so that each turn closely approximates a planar circular loop of current. Each loop of current I generates a magnetic field B according to a rule called the right-hand rule, as illustrated in Figure 1.18. Imagine the fist made with your right hand with the thumb sticking straight up. If the fingers curl in the direction of $\vec{I}$, the thumb points in the direction of $\vec{B}$.

To reverse this situation, now consider a long straight current carrying conductor. The magnitude of the magnetic field at a distance r from this wire carrying current I is defined as:

$$B = \mu_0 \frac{I}{2\pi r} \tag{1.57}$$

where the direction of the magnetic field B is again defined by the right-hand rule depicted in Figure 1.19. As illustrated by this figure, the magnetic flux

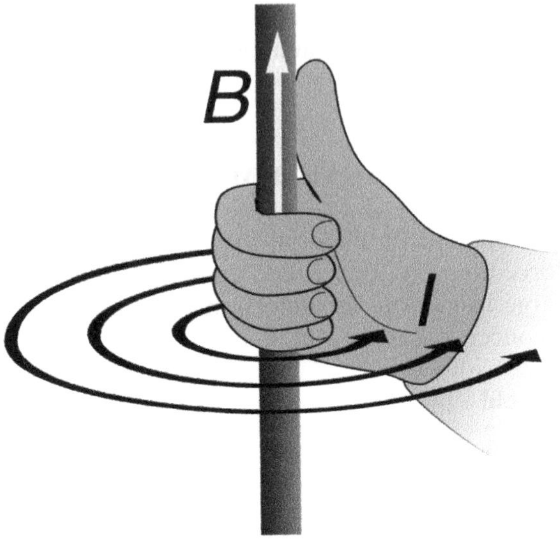

Figure 1.18. The right-hand rule for the magnetic field B generated by a current in a loop. Based on images by José Ferando. Courtesy of Wikipedia Commons.

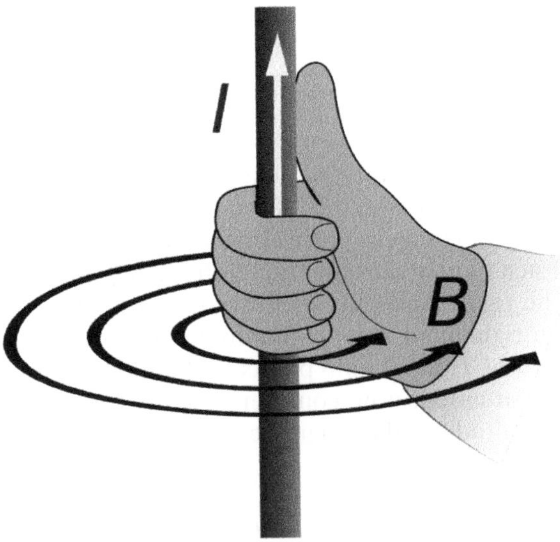

Figure 1.19. The right-hand rule for the magnetic field B generated by a current running through a long straight conductor. Based on images by José Ferando. Courtesy of Wikipedia Commons.

lines are coaxial circles with a radius r, and thereby with a circumference of $2\pi r$. The line integral of B along this closed path can be taken as;

$$\oint \vec{B}.d\vec{l} = \frac{\mu_0 I}{2\pi r}\oint d\vec{l} = \frac{\mu_0 I}{2\pi r}2\pi r = \mu_0 I \tag{1.58}$$

As you can see, the end result is independent of the radius of the circle (the closed path); the equation is true for any closed path that encloses a steady current I. This equation is the mathematical form of Ampere's law, which states that the line integral of a magnetic field along any closed path is equal to μ_0 times the net current I that crosses the area bounded by the path.

Just like an electric field, a magnetic field is also represented by field lines, the magnetic field lines, which give rise to the concept of a magnetic flux.

1.14.2 Magnetic Flux

Analogous to the electric flux described earlier, a magnetic flux is a measure of the quantity of the magnetic field lines through a certain area. It is proportional to the strength of the magnetic field and the surface area under consideration. For example, the current running through a wire in a circuit will create the magnetic field, and hence the magnetic flux in the area around it. Flux is a general property of any field, and is not limited to only magnetic and electric fields.

In a somewhat analogous way to an electric flux, we can explore the mathematical representation of a magnetic flux. The magnitude of a magnetic field at a point is equal to the magnetic flux per unit area across the area perpendicular to the field. This means that the relationship between the magnetic flux and the magnetic field $\vec{B}$ is given by

$$\frac{\partial \Phi_B}{\partial A_\perp} = B \tag{1.59}$$

Therefore,

$$\Phi_B = \oint \vec{B}.d\vec{A}_\perp = \oint \vec{B}_\perp.d\vec{A} = 0 \tag{1.60}$$

Note: The direction of an area is given by the unit vector $\hat{n}$ perpendicular to the surface of the area. Therefore $\vec{B}.\vec{A} = BA\cos\theta$, where θ is the angle that $\vec{B}$ makes with $\hat{n}$.

Equation 1.60 is the mathematical form of Gauss's law for magnetism. This integral is zero because any flux (field) line that enters a closed area exits that area too. So the net flux through the surface is zero, because unlike an electric charges, there is no such thing as isolated magnetic charges. Theoretically they are called magnetic monopoles. In other words, unlike electric field lines that begin and end on electric charges, magnetic field lines never have end points, they always form closed loops.

Faraday's law for a magnetic flux states that the change in magnetic flux creates an electromotive force (emf), which is practically a voltage. In other words, the changing magnetic flux through a circuit will induce a current in the circuit. Recall that the magnetic flux can be created by the current in a circuit according to Amperes law. Faraday's law says the reverse: The change in magnetic flux can create a current.

Analogous to the electric flux, magnetic flux Φ_B through a surface with area A is defined as:

$$\Phi_B = \int \vec{B}.d\vec{A}$$

$$\varepsilon = -\frac{\partial \Phi_B}{\partial t} \tag{1.61}$$

As mentioned earlier the change in this flux induces an emf ε:

This is Faraday's law of induction, which states that the induced emf in a closed loop equals the negative of the rate of change of a magnetic flux through the loop. The induced emf is actually the electric force represented by the electric field E:

$$\varepsilon = \oint \vec{E}.d\vec{l} \tag{1.62}$$

the line integral of $\vec{E}$ around the closed path. Therefore, Faraday's law can also be written as:

$$\oint \vec{E}.d\vec{l} + \frac{\partial \Phi_B}{\partial t} = 0 \tag{1.63}$$

To sum up, consider two electric circuits next to each other. There will be a magnetic flux through the second circuit due to the current in the first circuit: Ampere's law. If you change the current in the first circuit, it will change the magnetic flux through the second circuit, and the change in magnetic flux will create the current through the second circuit according to Faraday's law. So, you see that electricity and magnetism are related to each other and can be looked upon as two facets of what is called electromagnetism.

1.15 UNDERSTANDING ELECTROMAGNETISM

Electromagnetism is the unified framework to understand electricity, magnetism, and the relationship between them. In other words, to understand the electric field, magnetic field, and the relationship between them. To see this relationship, first recall that a charge creates an electric field, and the same charge, when it starts moving, creates a magnetic field. The electric field exerts an electric force, whereas a magnetic field exerts a magnetic force: Both originate from the electric charge. Therefore, they are intimately related: A changing electric field produces a magnetic field and a changing magnetic field produces an electric field. Due to this intimacy, the electric and magnetic force are considered two different manifestations of the same unified force called an EM force. The unified form of an electric and magnetic field is called an EM field, and an electric and magnetic field are considered its components. In other words, an EM force is exerted by an EM field.

Where there is a force, there is energy. The energy corresponding to an EM force is called EM energy or radiation. This energy is transferred from one point in space to another through what are called EM waves.

So, what is the frequency of EM waves? Electromagnetic waves cover a wide spectrum of frequencies, and the ranges of these frequencies are one way to define different types of EM waves.

1.15.1 Types of Electromagnetic and Other Waves

Waves can be grouped according to the direction of disturbance in them and according to the range of their frequency. Recall that a wave transfers energy from one point to another in space. That means there are two things going on: The disturbance that defines a wave, and the propagation of a wave. In this context, the waves are grouped into the following two categories:

Longitudinal Waves. A wave is called a longitudinal wave when the disturbances in the wave are parallel to the direction of propagation of the wave. For example, sound waves are longitudinal waves because the change of pressure occurs parallel to the direction of wave propagation.

Transverse Waves. A wave is called a transverse wave when the disturbances in the wave are perpendicular (at right angles) to the direction of propagation of the wave.

Electromagnetic waves are transverse waves, which means that the electric and magnetic fields change (oscillate) in a plane that is perpendicular to the direction of propagation of the wave. Also, note that the electric and magnetic field in an EM wave are perpendicular to each other.

Note: Electric and magnetic fields (E and B) in an EM wave are perpendicular to each other and are also perpendicular to the direction of propagation of the wave.

STUDY CHECKPOINT 1.4

Are the waves shown in Figures 1.9 and 1.10 longitudinal or transverse waves?

Solution:

Transverse waves.

Because electric and magnetic fields change in a plane that is perpendicular to the direction of wave propagation, the direction of change still has some freedom. Different ways of using this freedom provide another criterion for classifying EM waves into the following:

Linearly Polarized Waves. If the electric field (and hence the magnetic field) changes in such a way that its direction remains parallel to a line in space as the wave travels, the wave is said to be linearly polarized.

Circularly Polarized Waves. If the change in the electric field occurs in a circle or in an ellipse, the wave is said to be circularly or elliptically polarized. Therefore the polarization of a transverse wave determines the direction of disturbance (oscillation) in a plane perpendicular to the direction of wave propagation.

Caution: Only transverse waves can be polarized, because in a longitudinal wave the disturbance is always parallel to the direction of wave propagation.

So, you can classify EM waves based on the direction of disturbance in them (polarization). The other criterion to classify EM waves is the frequency. This classification gives rise to what is called the EM spectrum.

1.15.2 Electromagnetic Spectrum

Have you ever seen EM waves with your naked eye? The answer, of course, is yes! Visible light is an example of EM waves. In addition to visible light, EM waves include radio waves, ultraviolet (UV) radiation, and X-rays; which, of course, are not visible to the naked eye. These different kinds of EM waves only differ in their frequency, and therefore the wavelength. The whole frequency range of EM waves is called the EM spectrum, which is shown in Figure 1.20 and Table 1.1 along with the names associated with different frequency ranges or regions within the spectrum. The human eye has evolved to detect visible light that happens to be in the range where the Sun puts out most of its light energy that hits Earth.

PROBLEM 1.5

Consider two EM waves represented by A and B:

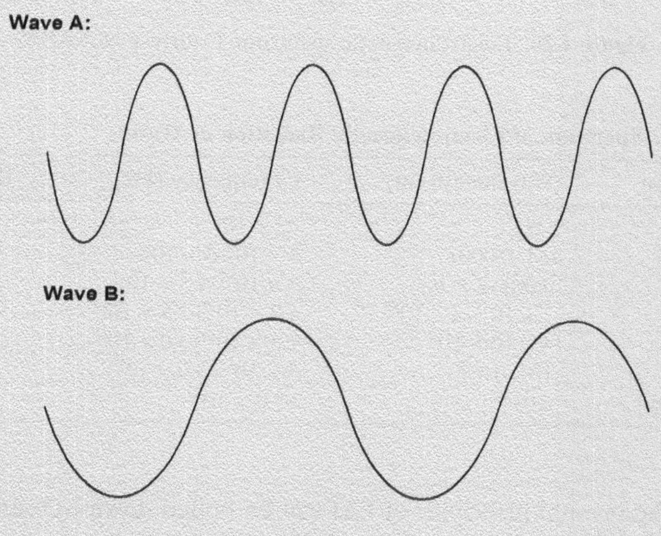

A. Which wave has the higher wavelength?
B. Which wave has the higher frequency?
C. If one of these waves represents visible light and the other one UV radiation, which wave is which?

Solution:

A. Wave B has longer or higher wavelength.
B. Wave A has higher frequency because it has smaller wavelength.
C. Wave B represents visible light and wave A represents UV radiation, because UV radiation has a higher frequency than that of a wave of visible light.

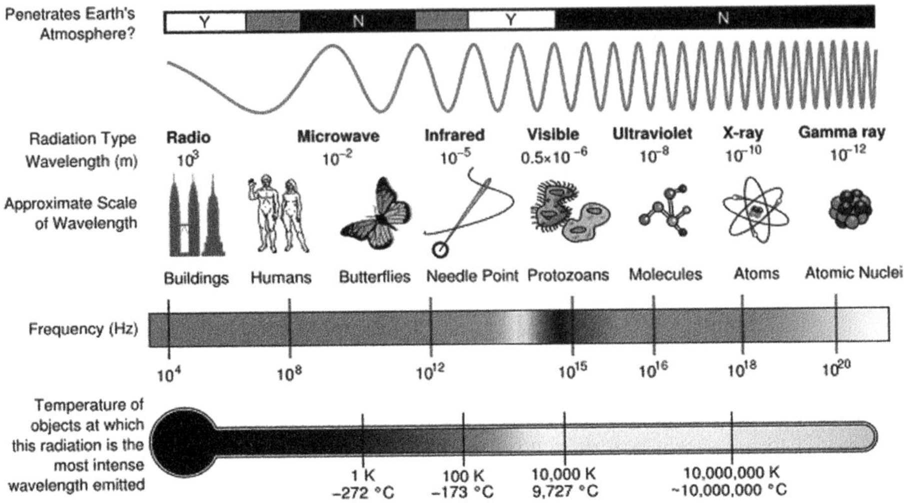

Figure 1.20. Electromagnetic spectrum. Courtesy of NASA.

TABLE 1.1. Spectrum of Electromagnetic Radiation or Waves

Wave Region	Wavelength (m)	Frequency (Hz)	Energy (eV)
Radio	>0.1	$<3 \times 10^9$	$<10^{-5}$
Microwave	0.1–0.0001	3×10^9–3×10^{12}	10^{-5}–0.01
Infrared	0.001–7×10^{-7}	3×10^{12}–4.3×10^{14}	0.01–2
Visible	7×10^{-7}–4×10^{-7}	4.3×10^{14}–7.5×10^{14}	2–3
Ultraviolet	4×10^{-7}–10^{-9}	7.5×10^{14}–3.0×10^{17}	3–10^3
X-Rays	10^{-9}–10^{-11}	3×10^{17}–3.0×10^{19}	10^3–10^5
Gamma Rays	$<10^{-11}$	$>3.0 \times 10^{19}$	$>10^5$

All the fundamental principles of EM can be boiled down to four equations called Maxwell's equations.

1.16 MAXWELL'S EQUATIONS

In the early nineteenth century, when electric and magnetic forces were thought to be two different forces, the physical quantity charge was organized into two different units in electrostatic phenomena and in magnetic phenomena resulting in two different physical dimensions. To their surprise, physicists noticed that the ratio of these two different units had the unit of velocity and its value was measured to be precisely equal to the velocity of light. At that time, physicists had no idea how to explain this and took it as a remarkable coincidence. As often happens in science, the questions put to bed by one generation of scientists are often confronted sooner or later by a future generation, if not the same.

When the electric and the magnetic fields do not vary with time, we can analyze them independent of each other. However, when they do vary with time, they are interrelated. For example, Faraday's law states that a magnetic field varying with time generates an electric field, for example, as demonstrated by induced emfs in inductors and transformers. In the reverse, Ampere's law states that an electric field varying with time generates a magnetic field. James Clerk Maxwell (1831–1879) realized the disturbance element in the fact that the variation in either an electric or magnetic field in time induces the other in the adjacent regions of space. So, he considered the possibility of an EM disturbance caused by time varying electric and magnetic fields propagating through space as a wave. The mystery of the ratio of electric and magnetic charges being equal to the speed of light was solved in 1865, when Maxwell showed that an EM disturbance (wave) would propagate in space with the speed of light, and hence light was a wave and was electromagnetic in nature. Furthermore, Maxwell realized that the basic principles of electromagnetism can be expressed in terms of four equations, which display the mutual interaction between electric and magnetic fields discussed above.

Although Maxwell developed the concept of displacement current, he did not discover Maxwell's equations single handedly. His genius was to put them together and recognize their importance especially in terms of EM waves. Maxwell's equations are presented in Table 1.2 in both differential and integrals forms, where ρ is the electric charge density, J is the electric current density, Q is the charge enclosed within the surface, I_c is the conducting current, Φ_E is the electric flux, and Φ_B is the magnetic flux. Furthermore, $\vec{E}$ is the electric field, $\vec{B}$ is the magnetic field, and ε_0 and μ_0 are called vacuum electric and vacuum magnetic permeability, respectively.

TABLE 1.2. Maxwell's Equations in Differential and Integral Forms

Differential Form	Integral Form	Origin
$\vec{\nabla}.\vec{E} = \frac{\rho}{\varepsilon_0}$	$\oint \vec{E}.d\vec{A} = \frac{Q}{\varepsilon_0}$	Gauss's law for an electric field
$\vec{\nabla}.\vec{B} = 0$	$\oint \vec{B}.d\vec{A} = 0$	Gauss's law for a magnetic field
$\vec{\nabla} \times \vec{E} + \frac{\partial \vec{B}}{\partial t} = 0$	$\oint \vec{E}.d\vec{l} + \frac{\partial \Phi_B}{\partial t} = 0$	Faraday's law of induction
$\vec{\nabla} \times \vec{B} = \mu_0 \vec{J} + \varepsilon_0 \mu_0 \frac{\partial \vec{E}}{\partial t}$	$\oint \vec{B}.d\vec{l} = \mu_0 \left(I_c + \frac{\partial \Phi_E}{\partial t} \right)$	Ampere's law of circuit

The differential forms of Maxwell's equations presented in Table 1.2 involve a vector $\vec{\nabla}$, which is given by

$$\vec{\nabla} = \frac{\partial}{\partial x}\hat{\imath} + \frac{\partial}{\partial y}\hat{\jmath} + \frac{\partial}{\partial z}\hat{k} \tag{1.64}$$

which means, for example,

$$\vec{\nabla}.\vec{E} = \frac{\partial E_x}{\partial x} + \frac{\partial E_y}{\partial y} + \frac{\partial E_z}{\partial z} \tag{1.65}$$

and

$$\nabla^2 E = \frac{\partial^2 E_x}{\partial x^2} + \frac{\partial^2 E_y}{\partial y^2} + \frac{\partial^2 E_z}{\partial z^2} \tag{1.66}$$

To start from the top in Table 1.2, the integral forms of the first two Maxwell equations involve the integrals of electric and magnetic fields over a closed surface. The first equation represents Gauss's law for electric fields, which states that he surface integral of an electric field $E_\perp$ over any closed surface is equal to the charge enclosed by the surface divided by ε_0. Analogously, Gauss's law for magnetic fields expressed in the second equation states that the surface integral of the magnetic field $B_\perp$ over a closed surface is always zero. The third Maxwell's equation in the table expresses Faraday's law of EM induction, which shows how a time varying magnetic field generates circulating electric field. Finally, the fourth Maxwell equation expresses Amperes law by showing how an electric current can generate a magnetic field. The second term on the right-hand side of the equation was added by Maxwell, in analogy with the Faraday's law, to account for the fact that a time varying electric field also generates a circulating magnetic field.

In the absence of charges and currents, Maxwell's equations can be transformed into the following 3D wave equations:

$$\nabla^2 \vec{E} - \frac{1}{c^2}\frac{\partial^2 \vec{E}}{\partial t^2} = 0 \tag{1.67}$$

and

$$\nabla^2 \vec{B} - \frac{1}{c^2}\frac{\partial^2 \vec{B}}{\partial t^2} = 0 \tag{1.68}$$

where c is a constant given by

$$c = \frac{1}{\sqrt{\varepsilon_0 \mu_0}} = 2.9979 \times 10^8 \text{ m/s} \tag{1.69}$$

Note: The permittivity of vacuum ε_0 is also called the electric constant and permeability of vacuum μ_0 is also called the magnetic constant. The relationship between the speed of light, the electric constant, and the magnetic constant are given by

$$c = \frac{1}{\sqrt{\varepsilon_0 \mu_0}}$$

which symbolizes the unification of electric and magnetic forces into an EM force.

This turns out to be the speed of light. As a matter of fact, solutions to these differential equations represent transverse electric and magnetic waves traveling with the speed of light.

To ease your way into quantum mechanics, you need to be comfortable with a few more concepts related to waves, namely, confinement, standing waves, and wavegroups.

1.17 CONFINEMENT, STANDING WAVES, AND WAVEGROUPS

This chapter established that most of the physics is based on three fundamental concepts: physical entity, space, and time. Time represents the interval over which there is a change, such as motion of the physical entity. As you know, Newton's first law states that an object (or a particle) that is moving with a velocity v, will continue moving with this velocity until an external force is applied to it. The world and the universe is full of external forces. Under the influence of these forces, the motion of objects is restricted or constrained in a certain fashion. For example, the motion of earth around the Sun is constrained in an orbit. The motion of a pendulum is constrained as well. Such constraints are called confinement.

1.17.1 Confinement

Both particles and waves travel, that is, move in time. Confinement refers to the situation in which the motion of a particle or a wave is constrained (confined) to a certain region in space. The confinement on motion of a physical entity understandably results in constraints on the values of some observables of the entity, such as energy and momentum. For example, imagine a particle confined to a box: Its position is always contrained to be within the range of values that represent the dimension of the box.

It is important for you to understand standing waves, an example of confinement, because it is a good ice breaker into quantum mechanics.

1.17.2 Standing Waves

A standing wave is a wave that remains in a constant position. For that reason, it is also called a stationary wave. The first time I learned about standing waves was in terms of a string firmly attached to rigid walls on two ends, I wondered how I was going to use that knowledge. So, I stoked it in the backburner of my brain. However, I found its usefulness during my student life some time later when I started learning quantum mechanics. Furthermore, standing waves have several applications in about all fields of science and technology, and, of course, in music.

So, if you already have some knowledge of standing waves, recall it. Nevertheless, I will give you a brief refresher in this section. A standing wave will help you a great deal in understanding the Schrödinger equation, the defining heart of quantum mechanics, which you will deal with in this book. A standing wave is a wave that results from the interference of several waves confined in space. Waves on a piano string and in an organ pipe are examples of the situations, which corresponding to certain wave frequencies, can give rise to a stationary wave pattern called standing waves.

We will explore the concept of standing waves by considering a simple situation: A string tied to a tuning fork at one end A, and fixed at the other end B. Assume that the length of the string, stretched from A to B, is L. Further assume that the tuning fork is vibrating with frequency f, wavelength λ, and a small amplitude. Now, consider a wave traveling from end A (the fork) to end B. At end B, the wave is reflected back in an inverted form. After traveling a distance of $2L$, it is reflected back in an inverted form again from end A at the fork. Because it has been inverted twice, it will only differ in phase from the wave that the fork generates at this moment. However, if the wavelength is exactly equal to the length of the string, the wave that is reflected twice is now in phase with the wave created by the fork at the moment of second reflection. Therefore, the two waves interfere constructively. The amplitude of the wave resulting from interference is twice that of any of the two interfering waves. Following the same argument, you can envision another event of constructive interference after traveling a distance of the next $2L$, and so on. This way, the amplitude of the wave at the string continues to grow until it reaches a maximum value at which the energy provided by the fork to the string equals the energy lost to damping effects, such as imperfect flexibility of the string and the reflection phenomenon. At this point, the wave on the string has a constant position and energy and is called the standing wave.

> *Note*: Because the tuning form is vibrating with a small amplitude, the end A can be considered a fixed point for the purpose of reflection and interference.

When the wavelength λ of the wave generated by the tuning fork and transported to the string is equal to the length of the string L, the tuning fork is called in resonance with the string and the frequency of the fork (and the wave) is called the resonance frequency. Note that the constructive interference will happen if $2L$ is any integer times λ. In other words, the resonance frequency may have any of the values that result in the following values for the wavelength:

$$2L = n\lambda_n \tag{1.70}$$

Therefore,

$$\lambda_n = \frac{2L}{n} \tag{1.71}$$

where n is any positive integer. Accordingly, the resonance frequency can assume the following values:

$$f_n = \frac{v}{\lambda} = n\frac{v}{2L} = nf_1 \tag{1.72}$$

where f_1 is called the fundamental frequency. The different frequencies corresponding to the different n values are called harmonics. Therefore, a harmonic of a wave is a component frequency of the wave that is an integer multiple of the fundamental frequency.

The amplitude of a standing wave, like any other wave, is the amplitude of the maximum vertical displacement of the wave, and the displacement is represented by its wave function Ψ. Let us derive this wave function from the wave functions of the individual waves that the standing wave is composed of. The displacement of the wave traveling from A to B, is given by

$$\Psi_R = \Psi_0 \sin(kx - \omega t) \tag{1.73}$$

where

$$k = \frac{2\pi}{\lambda} \tag{1.74}$$

is called the wavenumber and

$$\omega = 2\pi f \tag{1.75}$$

is called the angular frequency. Similarly, the displacement of the second wave on the string traveling from B to A is given by

$$\Psi_L = \Psi_0 \sin(kx + \omega t) \tag{1.76}$$

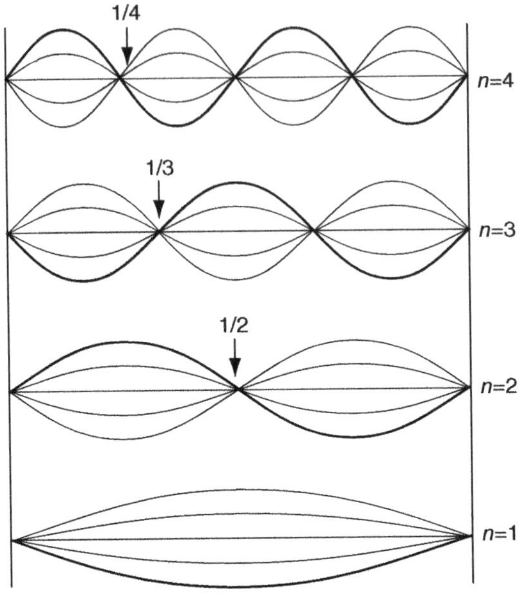

Figure 1.21. Standing waves on a string fixed at both ends.

The displacement of the standing wave, that is, the resultant of superposition of the two waves, is obtained by adding the corresponding wave functions:

$$\Psi = \Psi_R + \Psi_L = \Psi_0 \sin(kx - \omega t) + \Psi_0 \sin(kx + \omega t) \tag{1.77}$$

which can be simplified to be written as:

$$\Psi = 2\Psi_0 \cos\omega t \sin kx \tag{1.78}$$

The wave function for standing waves is illustrated in Figure 1.21 for the first four n values.

STUDY CHECKPOINT 1.5

From the expressions for Ψ_R and Ψ_L, prove that

$$\Psi = 2\Psi_0 \cos\omega t \sin kx$$

Solution:

$$\Psi = \Psi_R + \Psi_L = \Psi_0 \sin(kx - \omega t) + \Psi_0 \sin(kx + \omega t)$$

Use the trigonometric identity:

$$\sin A + \sin B = 2\sin\frac{A+B}{2}\cos\frac{A-B}{2}$$

$$A+B=2kx; \quad A-B=2\omega t$$

which implies

$$\Psi = 2\Psi_0 \cos\omega t \sin kx$$

Note that the wave function of the standing wave not only has the information about the displacement of the wave, but also about other characteristics, such as the resonance condition that we derived earlier without considering the wave function. Problem 1.9 demonstrates how the resonance condition can be derived from the wave function, which is

$$\lambda_n = \frac{2L}{n} \tag{1.79}$$

and therefore,

$$f_n = \frac{v}{\lambda_n} = n\frac{v}{2L} \tag{1.80}$$

where v is the phase velocity and, also in this case, is the group velocity because we are assuming that the string is a perfectly flexible string.

In general, a single wave made of several different waves is called a wavepacket or a wavegroup. The phase velocity is the velocity of the individual component waves and the group velocity is the velocity of the wavepacket. Note that the two are not the same.

1.17.3 Wavegroups

When waves of slightly different frequencies interfere with each other, they form a pattern of wavegroups or wavepackets due to the superposition of the individual waves. Because the individual waves have different frequencies, they will be traveling with different speeds, but the group as a whole will be traveling with a speed called a group velocity. Figure 1.22 illustrates a simple example of the superposition of two waves A and B with frequencies f_1 and f_2 generating a third wave with frequency f that consists of wavegroups.

Assume that the two individual waves A and B have the same amplitude Ψ_0, have wave numbers k_1 and k_2, and angular frequencies ω_1 and ω_2. From Eq. 1.73, we can write the wave functions for these two waves:

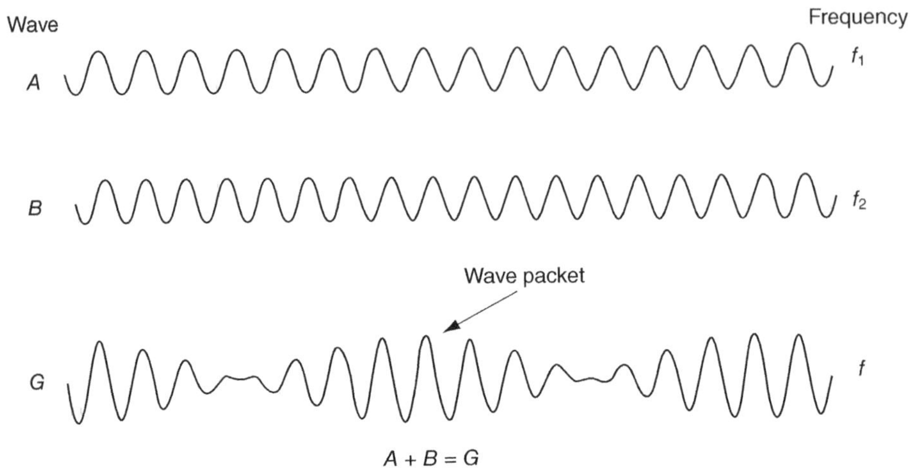

Figure 1.22. Superposition of two waves A and B with frequencies f_1 and f_2 generating a third wave with frequency f that has wavegroups.

$$\Psi_A = \Psi_0 \sin(k_1 x - \omega_1 t) \tag{1.81}$$

$$\Psi_B = \Psi_0 \sin(k_2 x - \omega_2 t) \tag{1.82}$$

The wave function, Ψ, of the resultant wave G in Figure 1.22 is given by

$$\Psi = \Psi_A + \Psi_B = \Psi_0 [\sin(k_1 x - \omega_1 t) + \sin(k_2 x - \omega_2 t)] \tag{1.83}$$

We use the trigonometric identity

$$\sin c + \sin d = 2 \sin \frac{c+d}{2} \cos \frac{c-d}{2} \tag{1.84}$$

to rewrite Eq. 1.83 as:

$$\Psi = 2\Psi_0 \sin\left[\frac{1}{2}(k_2 + k_1)x - \frac{1}{2}(\omega_2 + \omega_1)t\right] \cos\left[\frac{1}{2}(k_2 - k_1)x - \frac{1}{2}(\omega_2 - \omega_1)t\right] \tag{1.85}$$

By substituting

$$\bar{k} = \frac{1}{2}(k_2 + k_1)$$

$$\bar{\omega} = \frac{1}{2}(\omega_2 + \omega_1)$$

$$\Delta k = \frac{1}{2}(k_2 - k_1)$$

$$\Delta\omega = \frac{1}{2}(\omega_2 - \omega_1)$$

we can rewrite Eq. 1.85 in a simplified form:

$$\Psi = 2\Psi_0 \sin(\bar{k}x - \bar{\omega}t)\cos\left(\frac{\Delta k}{2}x - \frac{\Delta\omega}{2}t\right) \tag{1.86}$$

Use Eq. 1.86 to realize the relationship between Δk and Δx, where Δx is the width of the packet along the x-axis, which must be at least one-half of the wavelength. This means that the end points of the wavepacket (where the amplitude is zero), say x_1 and x_2, must have a phase difference of π. Mathematically, this means that the difference in the terms $\Delta k\, x/2$ in Eq. 1.86 at points x_1 and x_2 must be equal to π. This means

$$\frac{\Delta k\, x_2}{2} - \frac{\Delta k\, x_1}{2} = \pi$$

which means

$$\Delta k\left(\frac{x_2}{2} - \frac{x_1}{2}\right) = \frac{\Delta k\, \Delta x}{2} = \pi$$

which implies

$$\Delta k\, \Delta x = 2\pi \tag{1.87}$$

Also, Eq. 1.86 means that the group or wavepacket travels with wavenumber $\Delta k/2$ and angular frequency $\Delta\omega/2$. Therefore, the speed of the wavegroup is given by

$$v_g = \frac{\Delta\omega}{\Delta k} \approx \frac{d\omega}{dk} \tag{1.88}$$

For simplicity, we have only considered the wavegroup as a superposition of two waves. In general, a wavepacket can be thought of as a superposition of many waves with angular frequencies and wavenumbers distributed over limited ranges. As illustrated in Figure 1.23, superposition of several waves of different frequencies will give rise to a more localized interference pattern, the wavegroup.

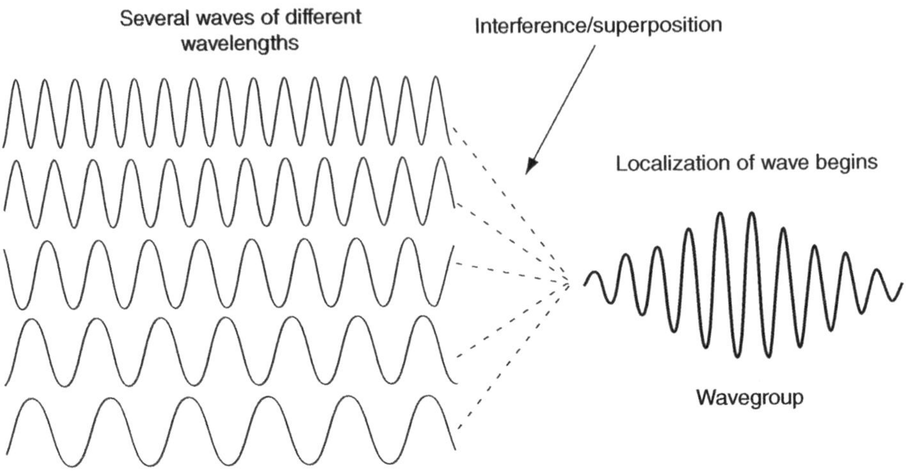

Figure 1.23. Localization of waves in the form of a wavegroup.

> **STUDY CHECKPOINT 1.6**
>
> As illustrated in Figure 1.23, a wavegroup is more localized, that is, the spread over space is less. How about the wavenumber?
>
> **Solution:**
>
> Wave number is given by
>
> $$k = \frac{2\pi}{\lambda}$$
>
> Because there are many wavelengths composing the wavegroup, therefore the wavenumber is more spread out.

So, in classical physics, a physical entity could be either a particle or a wave. We have reviewed important characteristics and the behavior of particles and waves. Now let us compare particles and waves in the context of these characteristics in order to distinguish particles from waves.

1.18 PARTICLES AND WAVES: THE BIG PICTURE

As mentioned earlier, classical physics treats particles and waves as two separate kinds of entities. We have discussed some of the characteristics that

TABLE 1.3. Comparison of Particles and Waves

Characteristic	Particles	Waves
Collision	Yes	No particle-like collision
Scattering	Yes Deviate direction after collision	Yes, but different from particle scattering: Reradiated in various directions after absorption.
Reflection	No. Although particles do deviate from the path after meeting an obstacle; scattering.	Yes
Refraction	No	Yes
Diffraction	No	Yes
Interference/ superposition	No Forces acting on particles can go under superposition, but particles themselves do not go under interference and superposition like waves do.	Yes Waves interfere and make an interference/diffraction pattern corresponding to constructive and destructive interference.
Energy	Transport energy	Transport energy

particles and waves have and some of the phenomena that they exhibit. Table 1.3 compares particles and waves in terms of phenomena they exhibit.

Classical physics is based on our perception of physical reality with our five senses. Particles are particles because they carry mass and they collide with each other as solid objects. Waves are waves because they have frequency, wavelength, and they go under phenomena, such as interference and diffraction.

Recall again the three fundamental concepts on which the physics is based: physical entity, space that specifies the position of the entity, and time that is used to specify the change in position of the entity. According to Newton's law, the change in the position of rest or in the state of uniform motion occurs due to external forces. There are countless forces all around and within us. However, all of those forces are different manifestations of the four fundamental forces of nature. Both kinds of entities, particles and waves, are subject to one or more of these interactions.

1.19 THE FOUR FUNDAMENTAL FORCES OF NATURE

Physicists study matter from elementary particles to galaxies and the universe, and all the fundamental forces that keep the universe and the systems in the universe in order. Although the fundamental forces acting on material systems are universal, the size of the material systems does matter in many ways in determining which force is significant for a given situation. For example, we as humans are macrobeings, and therefore we started our studies from the macroworld by using macrotools starting with our five senses. However, we are incapable of seeing the microworld with our naked eyes. Table 1.4 presents some examples of material systems on the size scale of macro to micro including nano and beyond nano.

Throughout the history of physics, physicists have discovered the following:

- All matter is composed of two kinds of a very few elementary particles called quarks and leptons: there are six types of quarks and six types of leptons. For example, a proton is made up of three quarks, as is a neutron. An electron is an example of a lepton. All atoms are made of protons, neutrons, and electrons.

TABLE 1.4. Examples of Material Systems on the Size Scale

Size	Scientific Notation (m)	Example (In This Neighborhood)	Observation Tools/ Techniques
1.7 m	1.7×10^{0}	Human height	Human eye
1 cm	1×10^{-2}	Wedding ring	Human eye
1 mm	1×10^{-3}	Thickness of a CD	Human eye Optical microscope
100 μm	1×10^{-4}	Plant cell	Human eye Optical microscope Electron microscope
10 μm	1×10^{-5}	Animal cell	Human eye Optical microscope Electron microscope
1 μm	1×10^{-6}	Bacterial cell	Optical microscope Electron microscope
100 nm	1×10^{-7}	Virus	Electron microscope
10 nm	1×10^{-8}	Virus, Protein molecule	Electron microscope
1 nm	1×10^{-9}	Protein molecule, aspirin molecule	Electron microscope
100 pm	1×10^{-10}	Water molecule	Electron microscope
10 pm	1×10^{-11}	Atom	Indirect observation by tools (e.g., cyclotrons)
0.001 pm	1×10^{-15}	Proton	Indirect observation by tools (e.g., cyclotrons)
<1 pm	$<1 \times 10^{-12}$	Other subatomic particles	Indirect observation by tools means (e.g., particle accelerators and particle colliders)

- Elementary particles are bound together to make atoms, molecules, and bulk matter through the presence of four fundamental forces: Strong and weak nuclear forces to form the nuclei of atoms, EM force to form atoms and molecules, and gravity to hold together large systems like planets, stars, and galaxies. These forces are expressed through particles, for example, EM force is expressed by an elementary particle called a photon. In other words, two charged particles interacting with each other via an EM force means they are exchanging photons. Photons are the mediators of EM force.
- These fundamental forces also govern the behavior and properties of particles and waves and of the bulk material. For example, chemistry students know that the stronger the intermolecular forces, which are some sort of EM forces, the higher is the melting and the boiling point of a substance.

STUDY CHECKPOINT 1.7

A. Consult with a chemistry textbook to write down the names of some types of intermolecular forces.
B. Intermolecular forces belong to which fundamental force?

Solution:

A. London dispersion forces, dipole–dipole-forces, and hydrogen bonding.
B. Electromagnetic force

Note: When we say EM force is expressed through photons, we mean that the two particles, for example, an electron and a proton, experience an EM force between them by exchanging photons. In other words, these two particles interact with each other and photons are the means of interaction; they intermediate the interaction. For this reason, an EM force and any other of the four forces are also called interactions.

In a nutshell, fundamental forces help form the material systems from basic building blocks, keep these systems intact, and hold this planet and the universe together in some order. There are four fundamental forces (gravitational, electromagnetic, weak and strong forces) that govern our universe. These forces are discussed next in Sections 1.19.1–1.19.4.

1.19.1 Gravitational Force

Also called gravity, this is the force of attraction between any two material bodies with nonzero mass. According to gravitational law, first formulated by

Newton in 1687, every material body attracts every other material body in this universe with a force that is directly proportional to the product of the masses of the two bodies and inversely proportional to the square of the distance between them. It is the gravitational force that keeps us on Earth and not floating randomly in space. Also, it is the gravitational force that keeps Earth revolving around the Sun. Gravity is the weakest of the fundamental forces.

The gravitational force in classical mechanics is given by Newton's law of gravitation, which states that two particles experience an attractive force F_g, that is directly proportional to the product of their masses m_1 and m_2, and inversely proportional to the square of the distance r between them:

$$F_g = G\frac{m_1 m_2}{r^2} \tag{1.89}$$

where G is the proportionality constant and is known as a gravitational constant. So, the gravitational field exerts a gravitational force on any particle with mass. You can see that mass is the origin of a gravitational force.

1.19.2 Electromagnetic Force

This is the force that an EM field exerts on any particle with an electric charge: Charge is the origin of an EM force. Electromagnetic force manifests itself through an electric and a magnetic force. The magnitude of an electric force between any two charged particles is given by Coulomb's law, which states that two charged particles experience a force F_e that is directly proportional to the product of their charges q_1 and q_2, and inversely proportional to the square of the distance r between them:

$$F_e = k\frac{q_1 q_2}{r^2} \tag{1.90}$$

where k is the proportionality constant and is known as the Coulomb constant. Notice the striking similarity between equations representing an electric and a gravitational force. However, note that gravitational force is only an attractive force because there are no negative mass particles, whereas an electric force is attractive between unlike charges and repulsive between like charges, because particles can have positive or negative charge. We can present Coulomb's law in terms of an electric field. The particles with charges q_1 and q_2 exert an electric force on each other through their electric fields. The electric field $\vec{E}_1$ at position $\vec{r}$ due to the particle with charge q_1 at the origin is defined as:

$$\vec{E}_1 = k\frac{q_1}{r^2}\hat{r} \tag{1.91}$$

where $\hat{r}$ is the unit vector along $\vec{r}$. This electric field is the force applied on any particle at the position $\vec{r}$ per unit charge of that particle. Therefore, the electric force applied by this electric field on a particle with charge q_2 is given by

$$\vec{F}_{1,2} = q_2\vec{E}_1 = k\frac{q_1 q_2}{r^2}\hat{r} \tag{1.92}$$

which is Coulomb's law.

The magnetic force component $\vec{F}_B$ of an EM force is given by

$$\vec{F}_B = q\vec{v}\times\vec{B} = qvB\sin\theta \tag{1.93}$$

where q is the magnitude of charge of a particle moving with a velocity $\vec{v}$ in a magnetic field with magnetic induction $\vec{B}$ and θ is the angle made by the velocity vector with the magnetic induction vector. Therefore, both electric and magnetic forces can be written as one force, the EM force:

$$\vec{F}_{\mathrm{EM}} = \vec{F}_E + \vec{F}_B = q\left(\vec{E} + \vec{v}\times\vec{B}\right) \tag{1.94}$$

This is the force experienced by a particle with a charge q and moving with a velocity $\vec{v}$ in an electric field $\vec{E}$ and a magnetic field $\vec{B}$.

Electromagnetic force is responsible for holding the electrons and protons together in an atom and also for binding atoms into molecules. This force plays an important role in all forms of chemical phenomenon, such as chemical reactions. Most of the forces we experience in our daily lives other than gravitational force during our interaction with ordinary objects, such as pushing, pulling, scrubbing, have their origin in an EM force at the molecular level, that is, the molecules in our body interact with the molecules of the body we interact with by using an EM force. Light is also associated with the EM force. Therefore, the macroscopic world is largely run by gravitational and EM forces.

1.19.3 Weak and Strong Nuclear Forces

There are two kinds of nuclear forces:

Weak Nuclear Force. This nuclear force is responsible for radioactive decay. It is 10^{13} times weaker than the strong nuclear force.

Strong Nuclear Force. This force keeps the subatomic particles intact. For example, it is responsible for keeping three quarks bound together to form a proton.

STUDY CHECKPOINT 1.8

Which fundamental force facilitates biochemical reactions and many processes occurring in our body?

Solution:

Electromagnetic force.

1.19.4 Four Fundamental Forces: The Big Picture

These four fundamental forces are also called interactions because particles and other systems of matter use these forces to interact with each other. Physicists use the terms force and interaction interchangeably. This universe starting from subatomic particles to galaxies is held together by these interactions. In other words, our universe is an expression of the greatest underlying orchestra being played by fundamental particles and the interactions among them. The basic characteristics of these forces, such as relative strength, and the most effective range in the context of the real world are presented in Table 1.5.

The strength of the electric force between two particles is inversely proportional to the square of the distance between them, that is, it decreases as the particles are separated from each other. Because theoretically the force between two charged particles never drops all the way to zero, you can say that the range of the EM force is infinite. However, practically speaking, the effective range of an EM force in matter only goes to $\sim 5 \times 10^{-10}$ m, that is, 0.5 nm. This is because of the cancellation effects of positive and negative charges in the matter, whereas the range of gravity stays infinite because it is

TABLE 1.5. Characteristics of Fundamental Forces of Nature

Force/Interaction	Intermediate Particle	Relative Strength	Most Effective Range (m)	Mostly Affected Matter or Process
Gravitational	Graviton	1	Infinitely large	Macrosystems (e.g., planets and objects on the planets)
Electromagnetic	Photon	10^{36}	10^{-9}	Microsystems and nanoparticles (e.g., molecules and subatomic particles)
Weak nuclear	W^+, W^-, Z^0	10^{25}	10^{-18}	Nuclear level: Radioactive decay
Strong nuclear	Gluon	10^{38}	10^{-15}	Nuclear level: Quarks

only an attractive force. Furthermore, it is the force that keeps protons and electrons bound together to form atoms, and atoms bound together to form molecules, crystals, and other materials, such as metals. Therefore, it is of specific interest to different fields in nanotechnology. Because quantum physics is needed to understand phenomena in the micro (and nano)-scale and the most effective force in this scale is the EM, force, we will be dealing with this force in this book.

Note: Out of the four fundamental interactions in the universe, the interaction that is relevant to fields discussed in this book (biology, chemistry, nanotechnology, and computer science) is the EM interaction, which exhibits itself in our world in many forms including electricity and magnetism. In chemistry classes, the EM force appears (or expresses itself) under different forms and names, such as electrostatic force, van der Wall's force, London dispersion force, hydrogen bonding, atomic bonding, and so on. You know from your introductory physics course that the EM force is also the force that leads to friction, lets your car move smoothly, and let you walk on earth and pick up things and hug your friends.

Although physics has made advances in exploring all of these four interactions, other sciences has mostly made use of EM forces. Physicists have learned repeatedly that the four fundamental forces appear to be a manifestation of a single unified force. As a matter of fact, the whole history of physics can be looked upon in terms of exploring unification.

1.20 UNIFICATION: A SECRET TO SCIENTIFIC AND TECHNOLOGICAL REVOLUTIONS

The whole history of the development of science can be explained in terms of one concept called unification. Most of the breakthroughs in science and technology have to do something with some kind of unification at some level. So, what is unification? Broadly speaking, unification is the process or act of uniting two or more apparently different things into one.

Nature works the way it does. We, the human, have our limitations in understanding it. We develop our understanding, gather information, and obtain knowledge in pieces. Some pieces may look different from each other at the surface. However, they may be the different facets of the same thing at some other level. For example, nature did not create different disciplines of science, such as physics, chemistry, biology, and so on. These and the borders among them are our own creations representing our limitations. To see the underlying unity behind diversity has been the key to major breakthroughs in science. The history of Physics, the mother of all sciences, demonstrates this very well. According to physics, our universe is shaped and governed by four fundamental forces: gravitational, electromagnetic, weak nuclear, and strong nuclear forces.

Here are some examples of unification in Physics. The realization that it is the same force that keeps us on Earth and makes the Earth rotate around the Sun enabled the physicist to recognize one of the four fundamental forces of nature: the gravitational force. From this realization, Newton was able to write the general formula for this force:

$$F = G\frac{m_1 m_2}{r^2}$$

where F is the force with which two bodies with mass m_1 and m_2 separated by a distance r attract each other; G is the gravitational constant. The two bodies may be you and the Earth or the Earth and the Sun. This unification was nothing less than a revolution in classical physics. For example, the airplanes and the satellites make use of this unified concept.

As another example, the electric force is the force that repels two similarly charged particles away from each other and attracts two oppositely charged particles toward each other. The magnetic force is the force that makes magnets repel or attract each other. Electric and magnetic forces were supposed to be two different forces. In the nineteenth century, it was proved that electric and magnetic forces are two different manifestations of the same force called the EM force. James Clark Maxwell is given credit for this unification of electric and magnetic forces, and the equations in physics that illustrate this unification are called Maxwell's equations of electromagnetism. So, the fields of electricity and magnetism were unified into one field called electromagnetism. The great product of this unification is the realization (or discovery) that light is an EM wave.

In the second half of the twentieth century, Abdus Salam, Steven Weinberg, and Sheldon Glashow demonstrated that the EM force and the weak nuclear force are the two different manifestations of the same force called the electroweak force. They were awarded the Nobel Prize in physics (1979) for this unification. This unification predicted the existence of some fundamental particles unknown at the time. The existence of one of these particles was confirmed at several experiments at CERN, in which the author of this book also participated.

Physicists believe that the four fundamental forces that govern our universe today are the low-energy manifestations of the single unified force, say the super force, at high energy that existed at the moment the universe was created. The work is still in progress. The unification of forces is illustrated in Figure 1.24.

Some unification events are summarized in Table 1.6.

Unification is in the very spirit of scientific thinking: looking for a common principle to explain a multitude of observations and phenomena. This thinking has its root in our being cognizant of the environment around us and around our knowledge and theories. If we divide the world into macro and micro, we are macrobeings. However, if we consider the universal scale of size, we actually fall in the mid-range: We cannot see very small, such as atoms, and we

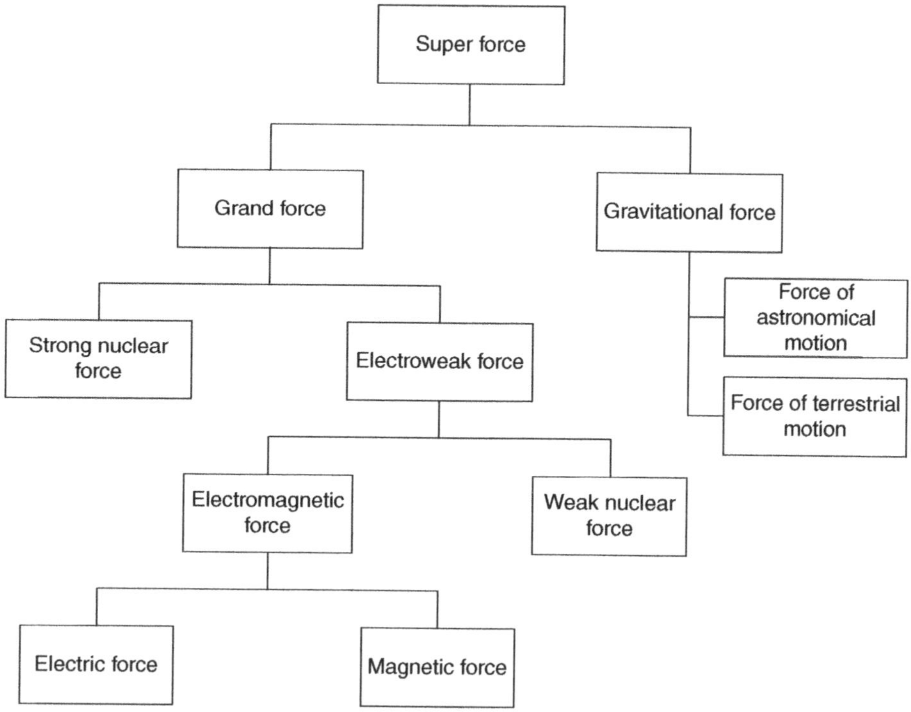

Figure 1.24. The unification of forces of nature.

TABLE 1.6. Some Examples of Unification Events

Unification Event	Time (s)	Result
Gravity	1700	Universal gravitation. Unifies the idea that the forces of attraction between material bodies that act outside of Earth are the same as those that act on Earth
Electricity and magnetism	1800	Light or photons Electromagnetic force
Electromagnetic force and weak nuclear force	1900	Electroweak force New particles W^+, W^-, Z^0
Theoretical efforts to unify electroweak force with the strong nuclear force and then with the gravity into one single force.	1900	Grand unification and super unification. Experimental evidences needed.

cannot see very large, such as the whole earth, entities at once or the universe in its entirety. Therefore our first observations and theories resulting from them also work well in the mid-range and breakdown on both ends of the spectrum. In this book, you will see how classical physics breaks down in the microworld, the world of the very small. However, keep in mind that classical physics also breaks down in some ways in the world of the very large. For example, classical physics fails to explain the behavior of objects moving at very large speed. Quantum physics is needed to explain phenomena in the microworld, whereas the special theory of relativity is needed to explain the world of very large speeds.

1.21 SPECIAL THEORY OF RELATIVITY

As we mentioned earlier, from the work of several physicists, including Young, Fresnel, and Maxwell, it was realized that light was a wave. As it often happens, with this achievement or solution, came another puzzle that eventually led to Einstein's special theory of relativity. At that time, all the known waves needed a medium to travel: sound waves travel through air, water waves travel through water, and so on. So the puzzle was What is the medium that light waves need to propagate through? It was known that light reaches us from outer space and possibly traveling through a vacuum on the way. This is an ideal example of how we (even the scientists) are prisoners of our past experience and knowledge. Great scientific discoveries are made by those who dare to break out of these self-made prisons. Continuing with the story, because all the waves discovered by then needed a medium to travel through, scientists assumed that light must need a medium too, even though it was known that light was apparently traveling through a vacuum to reach us from outer space.

So, driven by the need of a medium for light waves, scientists placed a hypothetical medium, called ether, in the vacuum, and everywhere else in the universe to be consistent. And off came the amazing properties of this mystic medium to fit in to our existing knowledge. It must have very low density so that the planets could move through it with no measurable change in their orbit positions and yet it must be so elastic that light, which was known to have an incredibly high speed, could propagate through it. However, the beauty of science is that you can theorize and fantasize as much as you can, but at the end of the day, your theory must withstand the test of experiments. It was not easy at the time to design an experiment to test the existence of ether. However, scientists tried for years by developing ever finer instruments and analysis methods. Eventually, there followed the experiment by Albert Michelson (1887) in collaboration with Edward Morley, which was designed to measure the motion of earth through ether. The experimental apparatus was sensitive enough to measure the ether drift due to the motion of the Earth, but to the surprise of the scientific world, no drift was detected.

Albert Einstein, a patent clerk at that time, interpreted the negative results of the Michelson–Morley experiment to develop what we know today as the special theory of relativity that he published in 1905. We are not covering this theory in this book, however, we present in this section some important features of this theory relevant to this book.

Energy Mass Equivalence. Mass, m, and energy, E, are equivalent and interconvertible according to the following equation:

$$E = mc^2 \tag{1.95}$$

where c is the speed of light through a vacuum, a constant; m is the relativistic mass; and E is the relativistic energy. For example, the Sun is powered by the conversion of mass into energy, which follows this formula. Given that $c = 3 \times 10^8\,\text{m}\,\text{s}^{-1}$, the factor c^2 is a very big number. This finding leads to the potential to release tremendous amounts of energy.

Relativistic Mass. The mass, m, of an object moving at speed v is related to the mass, m_0, of the object at rest by the following equation:

$$m = \frac{m_0}{\sqrt{1 - \frac{v^2}{c^2}}} \tag{1.96}$$

The mass m is called the relativistic mass and m_0 is called the rest mass.

Note that the concept of relativistic mass does not exist in Newtonian physics, in which the mass of a particle is always a constant and is equal to the rest mass. As shown in Study Checkpoint 1.9 and in Figure 1.25, this is a good approximation for an object moving substantially less than the speed of light; that is, we are dealing with speed, $v \ll c$.

STUDY CHECKPOINT 1.9

Show that the relativistic mass is equivalent to the rest mass at speeds substantially less than the speed of light.

Solution:

For

$$v \ll c$$

$$\frac{v^2}{c^2} \approx 0$$

Therefore,

$$m = \frac{m_0}{\sqrt{1 - \frac{v^2}{c^2}}} = m_0$$

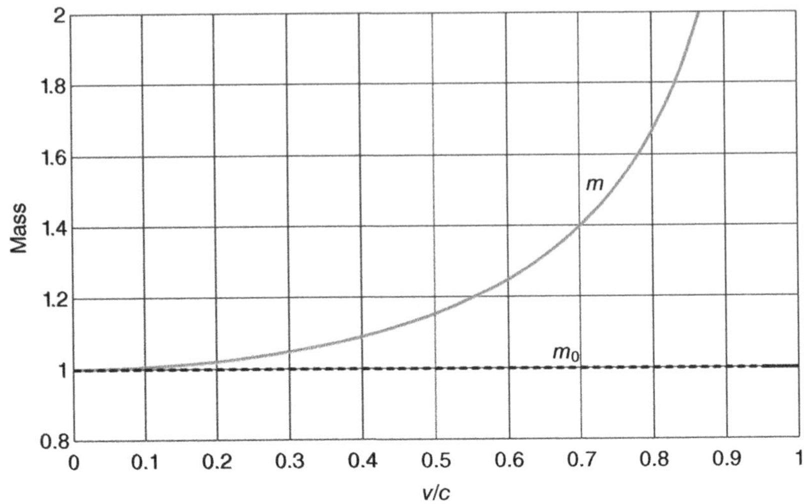

Figure 1.25. Comparison of rest mass (m_0) and relativistic mass m as a function of v/c for a particle with a rest mass of 1 unit.

Relativistic Energy and Momentum. We can write the relativistic energy in terms of the rest mass:

$$E = mc^2 = \frac{m_0}{\sqrt{1-\frac{v^2}{c^2}}} \tag{1.97}$$

Similarly

$$p = mv = \frac{m_0 v}{\sqrt{1-\frac{v^2}{c^2}}} \tag{1.98}$$

We can write the relativistic momentum in terms of the rest mass.

Let us examine how the total energy (i.e., the relativistic energy), the rest energy, and total momentum (relativistic momentum) are related. From Eqs. 1.97 and 1.98:

$$E^2 - p^2c^2 = \frac{m_0^2c^4}{1-\frac{v^2}{c^2}} - \frac{m_0^2v^2c^2}{1-\frac{v^2}{c^2}} = \frac{m_0^2c^4\left(1-\frac{v^2}{c^2}\right)}{1-\frac{v^2}{c^2}} = m_0^2c^4$$

Therefore,

$$E^2 = p^2c^2 + m_0^2c^4 \tag{1.99}$$

or

$$E = \sqrt{(p^2c^2 + m_0^2c^4)} \tag{1.100}$$

Equation 1.100 is the equation we will use in this book when we are dealing with relativistic energy or momentum. In most of the book, our treatment is nonrelativistic because this book covers nonrelativistic quantum mechanics.

We have covered a lot of material in this chapter with the intention of offering a brief review of classical physics. From this review, you should develop an understanding of the classical approach to physics. The key points of this classical approach emerging from the material covered in this chapter are summarized next in Section 1.22.

1.22 CLASSICAL APPROACH

Emerging from the review of classical physics presented here, you should see some core elements that define the classical approach of looking at the physical world. These elements are summarized in this section.

1.22.1 Separation of Particles and Waves: Either It Is a Particle or a Wave

In classical physics, particles and waves are two different types of entities. Particles are localized objects and can often be treated as point particles. Large particles and bulk matter are made of some fundamental particles and not waves. A moving particle transports energy by virtue of its moving mass. Waves, on the other hand, are spread out, a disturbance of some sort that transports energy from one point to another without transporting matter, and cannot be localized to a point. Particles can transport energy too, but by moving their mass. Waves have some unique characteristics, such as interference or superposition, in which the two waves can cancel each other out or add to each other. In other words, their amplitudes add or subtract depending on whether the interference is constructive or destructive. In the framework of classical physics, particles do not exhibit this interference phenomenon.

Other properties unique to waves include frequency, wavelength, refraction, and diffraction.

1.22.2 Either It Is Here or There: The Certainty

Knowing the current position and momentum of a particle under a given force, we can precisely predict its position and momentum at any time in the future.

Classical physics can make these predictions with full certainty; it is deterministic. Physical observables can be predicted and determined with certainty.

1.22.3 The World Is Continuous: Any Value Within a Range Is Possible

In the framework of classical physics, a particle can have any value within the acceptable range of values for its observables, such as position, momentum, and energy. In other words, the spectrum of possible values is a continuous spectrum. As an example, consider a ball of 1 kg mass. It will not be surprising if I say I can give any velocity to the ball between 0 and 25 $\mathrm{m\,s^{-1}}$. If the kinetic energy E of the ball is $\frac{1}{2}mv^2$, all values of kinetic energy between 0 J and 312.5 J are possible. In other words, as shown in Figure 1.26, the possible values of E for this ball for velocities in the range of 0 and 25 m/s are represented by a continuous curve.

1.22.4 Common Grounds Among Particles and Waves: A Red Flag

As mentioned earlier, classical physics treats particles and waves as different types of entities. However, let us make a preliminary case to challenge that notion. While particles and waves have their own unique characteristics, they also share some properties or observables. For example, a moving particle transports energy from one place to another in the form of kinetic energy. Waves transport energy too. By definition, waves are traveling disturbances that carry energy. This finding is true about standing waves, discussed in this chapter, as well because standing waves are just superpositions of traveling waves. Furthermore, most waves, such as water and sound waves, need a

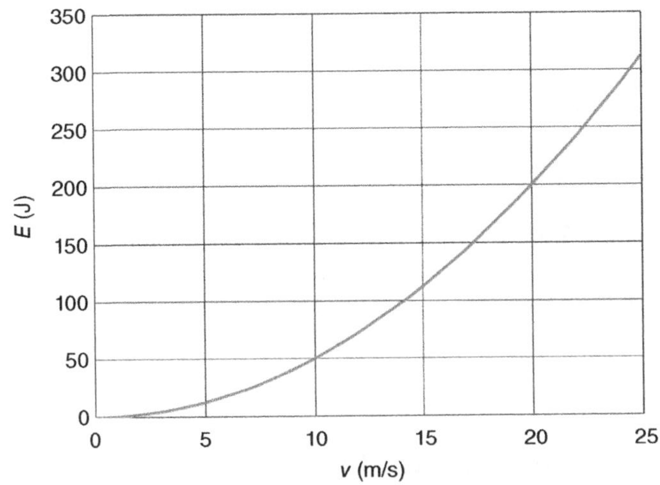

Figure 1.26. Kinetic energy as a continuous function of the velocity of a ball of 1 kg mass.

material medium to travel, and particles are material as well. Even the waves that do not need a medium to travel through, such as EM wave, need particles for their existence: You need charged particles to create EM waves.

Now here is a dilemma: Energy is transported in waves, but matter is not. However, as mentioned earlier, in the beginning of the twentieth century, Albert Einstein proposed in his special theory of relativity that matter (mass) and energy are equivalent according to Eq. 1.95:

$$E = mc^2 \tag{1.95}$$

where m is the mass of the particle and c is the speed of light.

If energy and mass are equivalent, then the waves can be considered transporting mass as well because they certainly transport energy. I will leave you with this dilemma, as an exercise to solve as you go through the forthcoming chapters. You can take it as a starting point in doubting classical physics theoretically. In Chapter 2, we will see how experimental results challenged the classical approach in a big way.

1.23 SUMMARY

Most of the concepts in classical physics can be derived from the three fundamental concepts: physical entity, space in which the entity exists, and time with which the entity changes its position in space. Classical physics, the physics of the macroworld, divides the physical world into two types of entities: particles and waves. Both particles and waves transport momentum and energy from one point to another in space. Waves do this transportation without transporting the matter. In classical physics, the motion of particles is governed by the fundamental equation of motion representing Newton's second law of motion:

$$F = ma = m\frac{dv}{dt} = m\frac{d^2x}{dt^2}$$

Similarly, the propagation of waves is determined by the following fundamental equation:

$$\frac{\partial^2\Psi}{\partial x^2} = \frac{1}{v^2}\frac{\partial^2\Psi}{\partial t^2}$$

The world and the universe is held together in an order due to the four fundamental forces of nature. These four forces encompass all the forces that cause change and motion in physical entities. These four forces are gravitational, electromagnetic, strong nuclear, and weak nuclear forces.

In a nutshell, this chapter has been a very brief review of classical physics aimed at making you realize that the classical approach divides the world of physical entities into two parts: particles and waves. This division mirrors the physical reality that we sense or experience as macrobeings. However, because the macroscopic world has been basically built from the objects of the microscopic world, such as molecules and atoms and subatomic particles, therefore the physical reality that we experience in the macroworld has its roots in the microworld. In the microscopic world, the standalone classical pictures of particles and waves collapse. This is where quantum mechanics comes into the picture with one promise: unification of two concepts: particles and waves. You will learn in Chapters 2 and 3 how this unification was historically achieved.

1.24 ADDITIONAL PROBLEMS

Problem 1.6 Two forces $\vec{F}_A$ and $\vec{F}_B$ with magnitudes 3 and 5 N, respectively, are acting on a particle. The force $\vec{F}_A$ is acting along the x-axis, and the force $\vec{F}_B$ is acting along the y-axis.

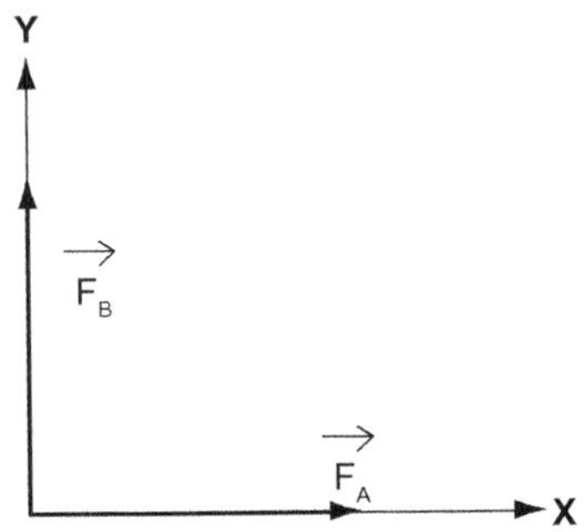

A. Calculate the scalar product of these two forces.
B. Calculate the vector product of these two forces.
C. Calculate the resultant force.

Problem 1.7 A ball is thrown up into the air with initial velocity u. The position of the ball as a function of time is given by the following equation:

$$x = ut + \frac{1}{2}gt^2$$

where g, the acceleration due to gravity, is a constant with a magnitude equal to 9.81 m/s^2. Calculate the instantaneous velocity of the ball at any time t.

Problem 1.8 If the ball in Problem 1.7 was thrown up with a velocity of 39.2 m/s, what will be its velocity at 4.0 s after the throw. Interpret your result.

Problem 1.9 The wave function of a standing wave is given by

$$\Psi = 2\Psi_0 \cos \omega t \sin kx$$

A. Derive the boundary conditions to represent the fact that the string is fixed at $x = 0$ and $x = L$.

B. Use the boundary conditions in A to derive the resonance condition:

$$\lambda_n = \frac{2L}{n}$$

Problem 1.10 Show that

$$\Psi(x,t) = \Psi_0 \sin(kx - \omega t) \tag{1}$$

is a solution of the wave equation:

$$\frac{\partial^2 \Psi}{\partial x^2} = \frac{1}{v^2} \frac{\partial^2 \Psi}{\partial t^2} \tag{2}$$

PARTICLE BEHAVIOR OF WAVES 2

The quanta really are a hopeless mess.

Albert Einstein

Figure 2.0. Images of space, captured by NASA's Hubble Space Telescope, show the earliest images of the galaxy, 600 million years after the Big Bang, released on January 5, 2010. UPI/NASA. Courtesy of NASA. ESA, R. Windhorst, S. Cohen, and M. Mechtley, (Arizona State University, Tempe, AZ), R. O'Connell (University of Virginia, VA), P. McCarthy (Carnegie Observatories), N. Hathi (University of California, Riverside, CA), R. Ryan (University of California, Davis, CA), and H. Yan (Ohio State University, OH).

This image shows a rich tapestry of 7500 galaxies stretching back through most of the universe's history. The closest galaxies observed today and seen in the foreground emitted their light ~1 billion years ago. Alternately, the farthest galaxies, a few of the very faint red specks that you would see if the figure was in color, are seen as they appeared 13 billion years ago, or roughly 650 million years after the Big Bang that, started the universe according to the Big Bang theory. This theory predicts that the cosmic microwave background fills all of observable space. Most radiation energy in the universe is in the cosmic microwave background, a faint glow almost exactly the same in all directions, that is not associated with any star, galaxy, or other object. This glow should be strongest in the microwave region of the radio spectrum. Hence, the name *cosmic microwave background radiation*, because according to the Big Bang theory, a microsecond after the Big Bang, the universe consisted of a hot plasma (a state of matter) of photons, electrons, and baryons, such as protons and neutrons. The photons were constantly interacting with the rest of the plasma through scattering. As the universe expanded, the plasma cooled enough to create favorable conditions for electrons to combine with protons and form hydrogen atoms. This event happened at ~3000 K, when the universe was ~379,000-years old. At this point, the photons scattered off the now electrically neutral atoms and began to travel freely through space, resulting in the decoupling of matter and radiation. The temperature, and hence the frequency (color) of the photons, has continued to diminish ever since; it is now down to 2.725 K corresponding to the microwave part of the electromagnetic (EM) radiation spectrum. The temperature of this radiation will continue to drop as the universe keeps expanding.

One of the greatest powers of the Big Bang theory is its prediction of almost perfect black-body (Section 2.3) spectrum in the cosmic microwave background. One of the greatest successes of the Big Bang theory is that this prediction is supported by experimental results. For example, The cosmic microwave background radiation (CMBR) has a thermal black-body spectrum at a temperature of 2.725 K, thus the spectrum peaks in the microwave range frequency of 160.2 GHz, corresponding to a 1.9-mm wavelength. The glow is almost, but not quite, uniform in all directions. It shows a very specific pattern equal to that expected if the inherent randomness of a red-hot gas is blown up to the size of the universe. These predictions are supported by the data collected by *The Far-Infrared Absolute Spectrophotometer* (*FIRAS*) instrument on the NASA Cosmic Background Explorer (COBE) satellite, which has carefully measured the spectrum of the cosmic microwave background. The FIRAS project members compared the CMBR with an internal reference black body and the spectra agreed to within the experimental error.

In This Chapter

2.1 INTRODUCTION

In Chapter 1, you learned about particles and waves and that classical physics makes a clear distinction between these entities: particles are particles and waves are waves. You also learned that light is considered to be a wave in the classical world. Particles are explained by Newtonian mechanics and waves are explained by the wave equations. To some scientists during the nineteenth century, this picture seemed complete. So much so that William Thomson Kelvin (also known as Lord Kelvin), a great physicist of the nineteenth century, thought that the basic outline of physical theory was more or less complete. Kelvin noted, however, that there remained "two small clouds": The two problems that would soon be solved and dispersed by classical physicists. Little did he (or anyone else at that time) know that these two little clouds carried the seeds of two new theories that would deconstruct the entire nineteenth century physics and define the physics of the twentieth century. One of these two theories is the theory of relativity, which came out of one problem (cloud), the challenge of explaining the results of the Michelson–Morley experiment. This experiment was set up to detect the hypothetical universal medium called ether through which light was assumed to travel. At that time, the concept of light as an EM wave had not yet been developed. Light, as understood at that time, needed a medium to travel through. However, the experimental results from the Michelson–Morley experiment showed no evidence of such a medium. Einstein's interpretation of these negative results led him to the theory of relativity. The other theory is quantum theory, whose development started with an effort to resolve the second cloud, the failure of classical theory to explain the radiation emitted by a black body.

In general, the classical wave picture of light ran into big trouble in the second half of the nineteenth century due to the observation of some phenomena that classical physics could not explain. These phenomena included blackbody radiation, the photoelectric effect, and the Compton effect. The successful explanations presented by physicists for the results from the experiments exhibiting these phenomena revealed that light consists of particles. In other words, light can exhibit particle behavior. Classical physics had no justification for such a behavior. Some radical ideas and models were proposed to explain this strange behavior, which contributed to the development of a new physics that we now know as quantum physics.

While you go through this chapter, pay attention to the three underlying streams of thought: failure of classical physics to explain a set of phenomena, evolution of the quantum picture to successfully explain these phenomena, and the role of quantum physics in understanding and using these phenomena in various applications.

2.2 THE NATURE OF LIGHT: THE BIG PICTURE

Sunlight is directly or indirectly a major source of energy for planet earth. In fact, it would be impossible to sustain life on earth without the regular supply of sunlight. As you will see in this and forthcoming chapters, light also played a significant role in the evolution of quantum physics. In our macroworld, we see the physical reality in terms of particles and waves (the classical picture) and it makes sense to our senses. All matter is considered to be made of fundamental particles, such as protons, neutrons, and electrons. Light (energy) is a wave that interacts with matter (particles) to give rise to interesting and useful phenomena. For example, it is the interaction of sunlight with molecules, called chlorophyll, contained within the cells of plants that make energy in the form of carbohydrates, such as starch and cellulose.

Note: As you know from Chapter 1, light, that is, visible light, is one narrow part of the EM radiation spectrum. Also, note that light, like any other part of EM radiation, represents pure energy and has no mass. As you will see, this view survives even quantum mechanics.

As you learned in Chapter 1, visible light is part of the EM radiation spectrum with wavelength in the range of ~400–700 nm. Your eye has receptors that can absorb this energy and convert it into images: fascinating! In addition to visible light, the EM radiation includes infrared (IR) radiation, ultraviolet (UV) radiation, and X-rays. As seen in Chapter 1, James Clerk Maxwell was among those who first postulated EM radiation in the form of waves, called EM waves. In other words, EM radiation is a self-propagating wave in space. In addition to having fundamental wave characteristics, such as intensity, frequency, and wavelength, EM radiation was found also to exhibit wave behavior, such as reflection, refraction, and diffraction. Electromagnetic radiation can travel as a self-propagating wave in space, and it can travel without the presence of matter. This is why we can see stars and galaxies millions of light years away from us. A light year is the distance that light travels in one calendar year, and is equal to 9.45×10^{12} km, as shown in Problem 2.1.

Note: Usually light means the visible part of the EM spectrum. However, sometimes in the literature the whole EM spectrum is also referred to as light. So, gather the exact meaning of the word light from the context.

PROBLEM 2.1

Calculate the distance of one light year into kilometers.

Solution:

$$\text{Speed of light} = 2.998 \times 10^8 \text{ m/s}$$

$$\text{Light year} = 2.998 \times 10^8 \text{ m s}^{-1} \times 1 \text{ year} \frac{365 \text{ days}}{1 \text{ year}} \times \frac{24 \text{ h}}{1 \text{ day}} \times \frac{60 \text{ min}}{1 \text{ h}} \times \frac{60 \text{ s}}{1 \text{ min}}$$

$$= 9.45 \times 10^{15} \text{ m} = 9.45 \times 10^{12} \text{ km}$$

From the historical perspective, it is noteworthy to mention Newton's role in understanding the behavior of light. Largely responsible for developing the particle part of classical physics, Newton saw light as a stream of particles. Recall from Chapter 1 that he put forward the Corpuscular theory of light in which light consists of corpuscles, that is, particles. The wave and particle view of light were the subjects of disagreement that goes back to as early as the seventeenth century. Although this issue was too early to be resolved during those times, we will see that neither Newton nor the proponents of the wave nature of light were wrong. In the beginning of the nineteenth century, experiments performed by scientists, such as Thomas Young, supported the wave nature of light. However, toward the end of the nineteenth century, just when Maxwell thought he had a perfect theory to explain electromagnetic radiation, further investigations of EM waves contradicted the wave view of light and challenged the classical theory. Scientists ran into trouble with the wave nature (the classical picture) of light, which could not correctly explain certain phenomena exhibited by light, such as black-body radiation, the photoelectric effect, and the Compton effect, also called Compton scattering. This chapter discusses how the classical picture of light failed to explain these phenomena and how a quantum picture of light started emerging from bold attempts to explain these phenomena.

2.3 BLACK-BODY RADIATION

Interaction between matter and light has been significant both as a tool in the advancement of science and as the role it plays in the many everyday processes needed to sustain life on our planet, as explained earlier. Light interacts with matter through various phenomena, such as reflection, refraction, diffraction, and absorbance.

In many situations, when studying the interaction of light with matter, it is useful to assume that matter is a perfect absorber of light. A black-body is defined in physics as an object that absorbs all the light that falls on it: No incident light is reflected and none is allowed to pass through it. It may sound counterintuitive to some, but a black body is also a good radiator, that is, the black body can re-emit radiation, because the abilities of a body to absorb and to radiate light are closely related to each other. The underlying principle is *It is a natural tendency of any system in the universe to be in equilibrium with its surroundings.* A body at constant temperature, by definition, is in thermal equilibrium with its surroundings. This means it absorbs light energy (from its surroundings) at the same rate as it emits. The radiation emitted by a *black body* therefore is called black-body radiation.

A black body absorbs all EM waves (light) falling on it regardless of their frequency. Therefore a black body is also an ideal source of radiation, the black-body radiation, which is a mixture of frequencies that are determined by only one thing, the temperature of the black body. In other words, which characteristic frequencies are present in black-body radiation, and what the predominant frequency is depends on the temperature, T, of the body. For example, a black body emits IR light at room temperature, and visible light (red through blue) at a temperature a few hundred degrees above room temperature. If the temperature is increased further, a black body starts emitting UV light, and so on. For example, a piece of coal is black when it is cold. It starts emitting visible light as you raise its temperature by starting the fire. As coals or stones are heated to higher and higher temperatures, they emit higher energy lights, and as a result move from being red to blue.

In the real world, it is hard to find an ideal (perfect) black body. In the laboratory, a black body can be approximated by a cavity, a hollow object that has a small hole leading to its interior. Any radiation that enters the hole is trapped inside by the back-and-forth reflection until it is completely absorbed. This concept is illustrated in Figure 2.1. Toward the end of the nineteenth century, black-body radiation was being examined with great interest, because it was one of those phenomena that were challenging the existing wisdom of physics, the classical physics.

2.3.1 The Classical Collapse

Classical physics was unable to explain the experimentally observed spectrum of black-body radiation illustrated in Figure 2.2, which shows the intensity of the emitted radiation as a function of wavelength at different temperatures. As illustrated in this figure, the experiments revealed that the peak of the radiation curve (the highest intensity) moves to higher intensities and shorter

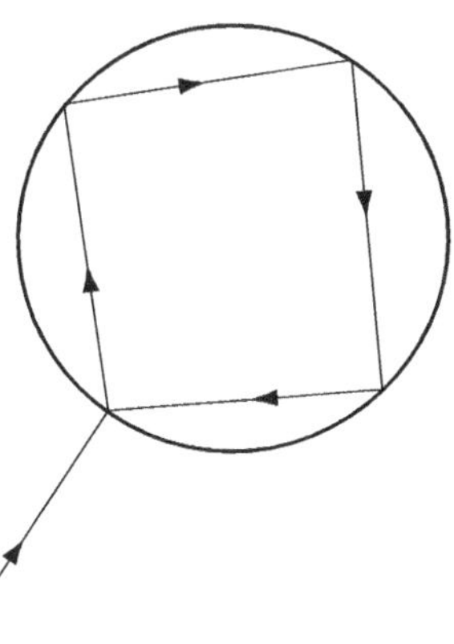

Figure 2.1. Two-dimensional illustration of light captured in a spherical cavity.

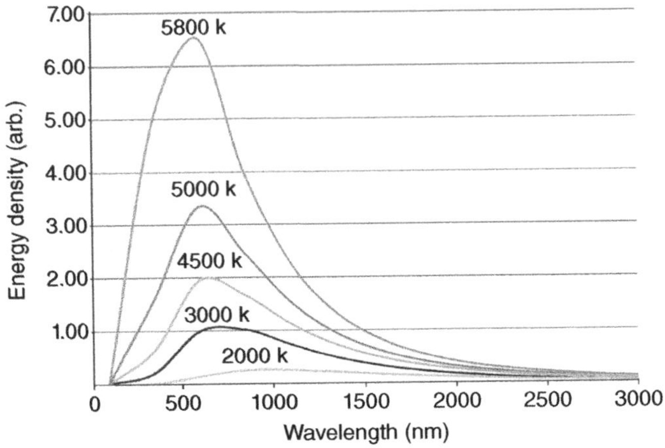

Figure 2.2. The energy density distributions of the radiations emitted by a black body at different temperatures supported by experimental data.

wavelengths as the temperature of the black body (cavity walls) is increased. Deriving these curves supported by experimental results from classical physics was a major challenge in the late nineteenth century. Lord Rayleigh and James Jeans were among those who were working on this problem. In an attempt to explain the experimental observations, they used everything classical physics could offer on this issue to derive the following formula, the Rayleigh–Jeans formula:

$$u(f,T)df = E_a \frac{8\pi}{c^3} f^2 df$$
$$= \kappa T \frac{8\pi}{c^3} f^2 df \tag{2.1}$$

where, $u(f,T)df$ is the total energy per unit volume (energy density) in the cavity in the frequency interval from f to $f + df$ at a given temperature T, c is the speed of light, T is the temperature in kelvin (K) of the black body in equilibrium with the surroundings, E_a is defined in Eq. 2.3, and κ is a constant called the Boltzmann's constant with the value given by

$$\kappa = 1.381 \times 10^{-23} \text{ J/K}$$

As shown in Problem 2.2, Eq. 2.1 can be rewritten in terms of wavelength:

$$u(\lambda,T)d\lambda = \frac{8\pi\,\kappa T}{\lambda^4} d\lambda \tag{2.2}$$

PROBLEM 2.2

Rewrite the Rayleigh–Jeans formula in terms of wavelength.

Solution:

$$f = \frac{c}{\lambda}$$

$$\frac{df}{d\lambda} = -\frac{c}{\lambda^2}$$

which implies

$$df = -\frac{c}{\lambda^2}d\lambda$$

Also,

$$u(f,T)df = -u(\lambda,T)d\lambda$$

Use these equations to rewrite Eq. 2.1 as:

$$\begin{aligned}-u(\lambda,T)d\lambda &= \kappa T\frac{8\pi}{c^3}f^2df\\ &= -\kappa T\frac{8\pi}{c^3}\left(\frac{c^2}{\lambda^2}\right)\frac{c}{\lambda^2}d\lambda\\ &= -\frac{8\pi\kappa T}{\lambda^4}d\lambda\end{aligned}$$

which implies

$$u(\lambda,T)d\lambda = \frac{8\pi\,\kappa T}{\lambda^4}d\lambda$$

Here is a simple derivation (or rather explanation) of Eq. 2.1. The term

$$\frac{8\pi f^2}{c^3}df$$

represents the number of independent standing waves in the frequency interval of f to $f + df$ per unit volume of the cavity. The term E_a represents the average energy per standing EM wave in the cavity, evaluated as following:

$$E_a = \frac{\int_0^\infty E e^{-E/kT} dE}{\int_0^\infty e^{-E/kT} dE} = kT \tag{2.3}$$

The integration comes from the underlying assumption of classical physics that the black body emits energy in a continuous fashion. You can also realize Eq. 2.3 from an important theorem of classical physics called equipartition of energy, which states that the average energy per degree of freedom of an entity that is a member of a system of such entities (think of a molecule of an ideal gas) in thermal equilibrium at temperature T is $\frac{1}{2}kT$. A standing wave in a black body has two degrees of freedom, and hence its energy is kT. We will discuss the equipartition theorem in Chapter 10 (e.g., Section 10.3), where we explore statistical quantum mechanics.

By looking carefully at Eqs. 2.1 and 2.2, you can predict a problem, that is, the energy density keeps increasing without limit with the increase in frequency or decrease in wavelength. This means for infinitely high frequency (UV frequencies), which is at a very low wavelength, the energy density will be infinity. As illustrated in Figure 2.3 for $T = 5800\,\mathrm{K}$, the experimental results do not support the absurd predictions of these calculations. This huge deviation of experimental results from theoretical predictions at low wavelengths (high frequencies) is known as the UV catastrophe of classical theory.

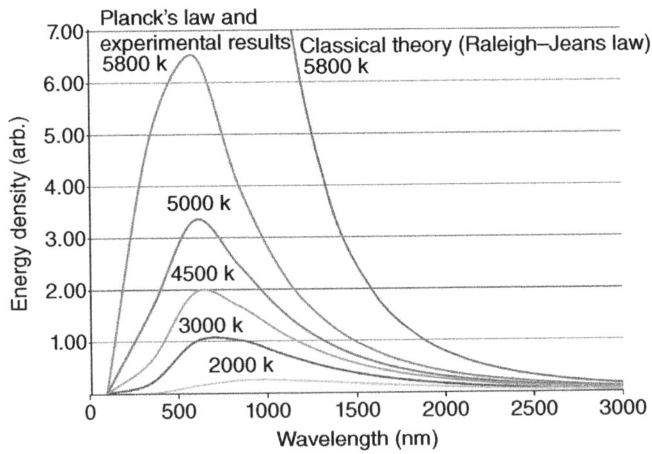

Figure 2.3. The intensity (energy density) of black-body radiation as a function of wavelength at different temperatures. Prediction of the classical theory does not match with the experimental data at $T = 5800\,\mathrm{K}$.

STUDY CHECKPOINT 2.1

True or False: The experimental results agree more with the Raleigh–Jeans law in the region of long wavelengths than in the region of very small wave lengths.

Answer: True

Hint: Compare Eq. 2.2 with Figure 2.3.

This disagreement encouraged critical thinking about the existing EM theory.

2.3.2 The Quantum Rescue

As pointed out in Section 2.3.1, the root of the problem in the classical formulae 2.1 and 2.2 lies in the underlying assumption that the energy emission is continuous, which gives rise to integration in Eq. 2.1 shown in Eq. 2.3. Obviously, this classical assumption is challenged by the experimental results, as illustrated in Figure 2.3. So, what would the alternative be? If energy is not emitted continuously, a reasonable alternative to explore is that it is emitted in discrete amounts. It was unconventional to think like that during those times, that is exactly what Max Planck, a physicist working on this problem at the time, proposed: The radiation, E, is emitted in energy steps of hf. This has two implications to modify Eq. 2.1:

1. In Eq. 2.1, E is given by

$$E_n = nhf \tag{2.4}$$

   where $n = 0, 1, 2, \ldots$
   and h is a constant with the value given by

$$h = 4.1357 \times 10^{-15} \text{ eV s} = 6.6261 \times 10^{-34} \text{ Js}$$

   As you see, h is a proportionality constant between energy and frequency, called Planck's constant. Robert Millikan made the first measurement of Planck's constant in 1916.
2. The integrals representing continuous emission in Eq. 2.3 should be replaced with sums representing discrete emission of energy:

$$\int \rightarrow \sum$$

By doing this and substituting the value of E_a from Eq. 2.4 into Eq. 2.3, we obtain

$$E_a = \frac{\sum_{n=0}^{\infty} n\varepsilon e^{-n\varepsilon/\kappa T}}{\sum_{n=0}^{\infty} e^{-n\varepsilon/\kappa T}}$$

$$= \frac{\varepsilon}{e^{\varepsilon/\kappa T} - 1}$$

$$= \frac{hf}{e^{hf/\kappa T} - 1} \tag{2.5}$$

Therefore, Eq. 8.1 becomes

$$u(f,T)df = \frac{8\pi f^2}{c^3} \frac{hf}{e^{hf/\kappa T} - 1} df \tag{2.6}$$

This formula can also be written in terms of wavelength λ, as shown in the following:

$$f = c/\lambda$$

which implies

$$df/d\lambda = -c/\lambda^2$$

which implies

$$u(f,T)df = -u(\lambda,T)d\lambda$$

with the use of these formulas, we obtain from Eq. 2.6:

$$u(\lambda,T)d\lambda = \frac{8\pi hc}{\lambda^5} \frac{1}{e^{hc/\lambda\kappa T} - 1} d\lambda \tag{2.7}$$

Equations 2.6 and 2.7 are different forms of what is called Planck's formula for black-body radiation that Planck came up with in 1900, and turned out to be in complete agreement with experimental observations at that time. This continues to be the case to date. As an example, Figure 2.4 shows the comparison of data with Planck's theory for the cosmic background described in Figure 2.0. The figure presents the cosmic microwave background spectrum measured by the *Far-Infrared Absolute Spectrophotometer* (FIRAS) instrument on the Cosmic Background Explorer (COBE) satellite operated by the National Aeronautics and Space Administration (NASA). This is the most-precisely measured black-body spectrum in nature. The solid curve shows the expected intensity from a single temperature black-body spectrum, as predicted by the

Big Bang theory. The FIRAS data, taken at 34 points equally spaced along the curve, match the curve so exactly with error uncertainties less than the width of the black-body curve, that it is impossible to distinguish the data from the theoretical curve generated by Planck's law after taking into account the predictions of the Big Bang theory as described in the Figure 2.0.

The agreement is so good that the datapoints and error bars on this graph are obscured by the theoretical curve representing the prediction of Planck's theory.

Alert: I have written the Rayleigh–Jeans and Planck's formula for spectral energy density $u(f,T)$. You will sometimes find these formulas in the literature for spectral radiance $I(f,T)$. To avoid confusion, remember that the two variables are related by the following equation:

$$u(f,T) = 4\pi/c \;\; I(f,T)$$

Max Planck (1858–1947)

Max Planck was born on April 23, 1858 in Kiel, Germany, and received his education at Munich and Berlin. While considering a career in science at Munich University in 1874, Planck discussed the prospects of physics research with Philipp von Jolly, a physics professor at the time in Munich at Munich University. Von Jolly candidly advised Planck to consider a career in some other discipline because physics was essentially complete with little prospects for further development. Planck was not persuaded.

Planck worked on black-body radiation because he thought it could be understood independent of atomic structure, which was still a mystery during those times. He first discovered (guessed) the correct formula for black-body radiation and then strived for its physical interpretation: radiation is emitted in energy steps of hf. Planck received the Nobel Prize in 1918 for this discovery that is now considered as the start of the quantum or modern physics. Although he is considered the founder of quantum physics, Planck remained skeptical of the quantum reality for a long time.

Max Planck spent most of his life in Germany, where he died on October 4, 1947 at the age of 89.

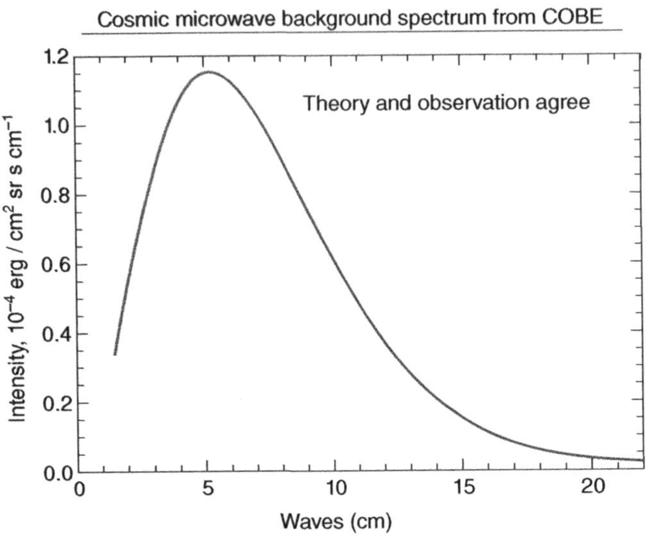

Figure 2.4. The cosmic microwave background spectrum measured by the FIRAS instrument on the COBE satellite. Courtesy of NASA.

PROBLEM 2.3

Write the Rayleigh–Jeans formula in terms of wavelength:

Solution:

$$u(f)df = \frac{8\pi\kappa T}{c^3} f^2 df$$

Use

$$f = c/\lambda$$
$$df = -c\, d\lambda/\lambda^2$$

which implies

$$u(\lambda)d(\lambda) = \frac{8\pi\kappa T}{\lambda^4} d\lambda$$

or

$$u(\lambda) = \frac{8\pi\kappa T}{\lambda^4}$$

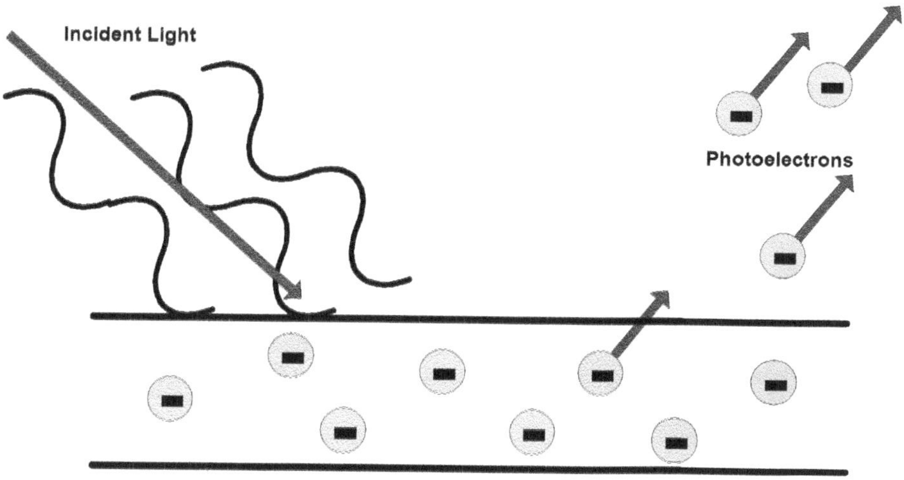

Figure 2.5. Illustration of the photoelectric effect.

During the same time, while physicists were struggling to explain the experimental observation of black-body radiation, there was another unexplained phenomenon related to light (EM waves) called the photoelectric effect.

2.4 THE PHOTOELECTRIC EFFECT

The photoelectric effect, illustrated in Figure 2.5, refers to the phenomenon in which electrons are emitted from a metal surface as a result of EM radiation incident on the surface. The ejected electrons are called photoelectrons. This effect is sometimes also called the Hertz effect, after the name of the physicist who brought it into the limelight. Further studies of this effect by other scientists contributed significantly to the understanding of the quantum nature of light, and therefore of wave–particle duality, discussed in Chapter 3.

2.4.1 The Photoelectric Effect: The Experiment

The photoelectric effect is one of the experiments that shattered classical physics at the atomic scale and gave birth to quantum physics. In Chapter 1, we learned the classical picture of light in the form of EM waves. In the late nineteenth century, researchers were performing experiments on EM waves.

Heinrich Rudolf Hertz
(1857–1894)

Heinrich Rudolf Hertz, a physicist and electronic engineer, was born on February 22, 1857 in Hamburg, Germany. The frequency unit hertz (Hz) is named after him. Hertz obtained his Ph.D. from the University of Berlin in 1880, where he continued as a postdoctoral researcher.

In 1883, Hertz accepted a post as a lecturer in theoretical physics at the University of Kiel. In 1885, Hertz became a full professor at the University of Karlsruhe, where he performed experiments on EM waves. He was the first to demonstrate the existence of EM waves in 1888 by building and using the apparatus to produce and detect very high frequency (VHF) and ultrahigh frequency (UHF) radio waves.

Hertz died of Wegener's granulomatosis at a young age of 36.

While performing experiments on producing and detecting EM waves in 1887, Heinrich Hertz observed the photoelectric effect. His experimental apparatus consisted of a spark gap in which a spark would be observed upon the detection of EM waves. He observed that sparks occurred more readily when UV light was directed at one of the metal balls in the apparatus. He reported this observation, but never followed up on it. However, other researchers, such as J.J. Thomson and Von Lenard, did follow up on this observation by studying different aspects of this phenomenon using suitable experimental setups.

Note: A spark gap refers to the gap between two conducting electrodes and is usually filled with a suitable gas, such as air. Upon applying a suitable voltage between the electrodes, a spark forms that ionizes the gas and drastically reduces its electrical resistance. As a result, an electric current flows until the path of ionized gas is broken or the current reduces below a minimum value.

Figure 2.6 presents a schematic diagram of a typical experimental setup to study the photoelectric effect. As shown in the figure, the apparatus includes an evacuated tube that contains two electrodes connected to a source of variable voltage. The light falls on the surface of a metal plate knocking some electrons, called photoelectrons, off the surface. Depending on their kinetic energy (KE), some of these photoelectrons reached the other end, the cathode, in spite of the negative polarity of the cathode. The vacuum helps the photoelectrons move at a constant speed to produce a measurable current, called

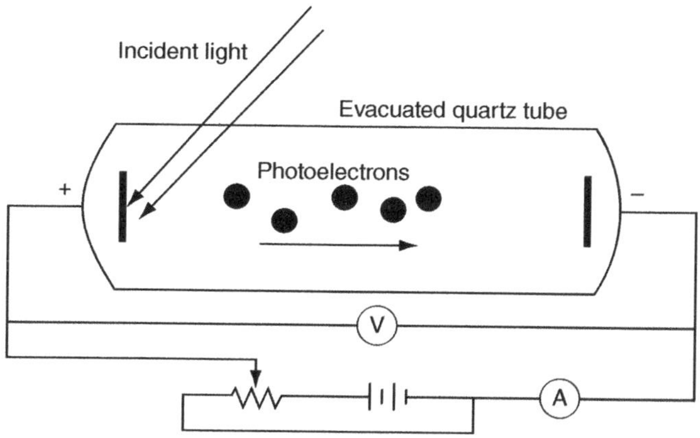

Figure 2.6. Illustration of an experimental setup for the photoelectric effect.

the photocurrent. Establishing the voltage difference, called the barrier voltage, between the two electrodes facilitates the determination of the KE of the photoelectrons. The slower moving electrons, the electrons with a low value of KE, are repelled back before they could reach the cathode. When the voltage opposing the movement of electrons toward the cathode reaches (or exceeds) the threshold value V_0, no photoelectrons arrived at the cathode. This threshold voltage, also called the *barrier* or *stopping voltage*, represents the maximum KE of any emitted electron, that is, the photoelectron.

On the surface, the photoelectric effect was not a surprise, factoring in the principle of conservation of energy. One could argue that the incident EM waves had some energy, which converted into the kinetic energy of the emitted electrons. However, the devil was in the detail discussed next in Section 2.4.2.

2.4.2 The Classical Collapse

In Chapter 1, you learned that energy is the common property of waves and particles. Furthermore, matter (mass) is a property of particles and waves are pure energy, but can manifest themselves in moving matter. So, considering the principle of conservation of energy, the photoelectric effect should not be a surprise. One can argue that the EM waves incident on the metal surface transferred their energy to electrons in the atoms, which came out as photoelectrons. However, the classical picture was challenged by the details of experimental observations collected from different experiments, for example, from an experiment by Millikan displayed in Figure 2.7, as described in the following:

1. When light strikes the metal surface, photoelectrons are emitted instantly as demonstrated by the current measured. This is true regardless of the

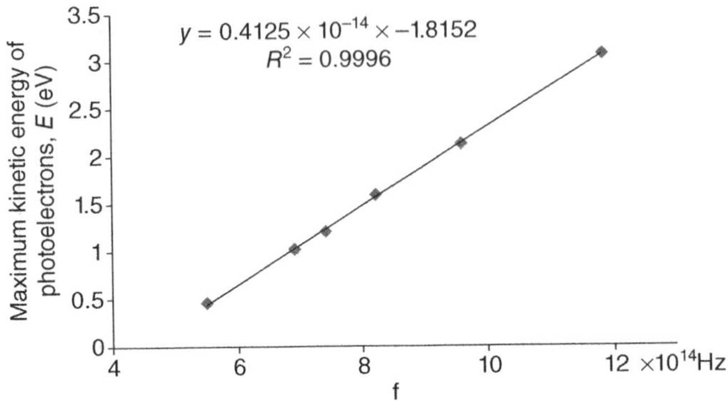

Figure 2.7. Maximum kinetic energy of photoelectrons as a function of photon frequency Based on photoelectric data for sodium (Na) from R.A. Millikan, *Phys. Rev.* 7, 355 (1916).

intensity of light once it is above a certain threshold. This instant emission of photoelectrons cannot be explained by classical physics according to which the energy of the EM wave incident on the surface is spread across the wavefronts, and therefore it will take some time for the electrons in the atom to absorb that energy before they could leave the atom in the form of photoelectrons. As shown by a simple calculation in Problem 2.4, it may take days and months.

2. At a given frequency of the incident EM waves, the strength of the current (i.e., the number of photoelectrons) is found to be directly proportional to the intensity of light. In other words, at a given frequency, the brighter light yields more photoelectrons than the dimmer light, but the average energy of a photoelectron remains the same. In contrast to this experimental observation, the classical picture of EM waves predicts the increase in the photoelectron energy with the increase in the intensity of light. This is because the light with greater intensity should have greater energy, and therefore will give a stronger push (energy) to the photoelectrons.
3. The energy of emitted photoelectrons is found directly proportional to the frequency of the incident light. As mentioned earlier, classical physics expects the photoelectron energy to be proportional to the intensity of light, and not frequency.
4. At a frequency below a certain critical (or threshold) frequency, f_0, no photoelectrons are emitted. The value of f_0 is found to be a characteristic of the specific metal used. Again, this observation cannot be explained by the EM wave theory of classical physics.

STUDY CHECKPOINT 2.2

From Figure 2.7, read the threshold frequency for photoelectric effect for Na.

Solution:

The threshold frequency will be the value of x when the value of y is zero.
From the text in the body of Figure 2.7:

$$f_0 = x = \left(\frac{1.8152}{0.4125}\right) \times 10^{14}\ \text{Hz} = 4.40 \times 10^{14}\ \text{Hz}$$

PROBLEM 2.4

Consider a one-atom thick layer of Na that is square in shape with one side of the square equal to 50 cm. Light falls uniformly on this surface and produces photoelectrons. Assume that each atom has a diameter of 0.1 nm. Further assume that each atom absorbs enough light to produce a photoelectron with a kinetic energy of 10 eV. How much time will it take for the atom to absorb this energy from the light falling on the surface with an intensity of $10^{-6}\ \text{W/m}^2$?

Solution:

Power being absorbed by the surface = $10^{-6}\ \text{W/m}^2 \times 0.25\ \text{m}^2 = 0.25 \times 10^{-6}\ \text{W}$
Diameter of an atom = 0.1 nm
Approximate number of atoms in 1 nm^2 = 100
Approximate number of atoms in 0.25 m^2 = $(100/\text{nm}^2) \times 0.25\ \text{m}^2$
$= 2.5 \times 10^{19}$
Average power being absorbed by an atom = $0.25 \times 10^{-6}\ \text{W}/(2.5 \times 10^{19})$
$= 10^{-26}\ \text{W}$
Time taken to absorb 10 eV:

$$\text{time} \times \text{power} = \text{energy}$$

which implies

$$\begin{aligned} \text{time} &= \text{energy/power} = 10\ \text{eV}/10^{-26}\ \text{W} = 10\ \text{eV}/10^{-26}\ \text{J/s} \\ &= 10\ \text{eV}/(10^{-26} \times 6.24\ 10^{18}\ \text{eV/s}) \\ &= 1.60 \times 10^{8}\ \text{s} \\ &= 1.60 \times 10^{8}\ \text{s} \times 1\ \text{day}/86{,}400\ \text{s} \\ &= 1852\ \text{days} \end{aligned}$$

This is the classical picture that obviously fails!

STUDY CHECKPOINT 2.3

In light of photoelectric experiments, which of the following increases with an increase in the intensity of the incident light?

A. Kinetic energy of the photoelectron.
B. Mass of the photoelectron.
C. Number of photoelectrons.
D. Number of photoelectrons and their kinetic energy.

Answer:
C. The number of photoelectrons.

Note: Robert Milikan received the Nobel Prize in physics in 1923 for his work on the elementary charge of electricity and on the photoelectric effect, whereas Einstein received the Nobel Prize in physics in 1921 for his services to theoretical physics, especially for his discovery of the law of the photoelectric effect.

So, you can see how classical physics falls on its face in front of the experimental observations of the photoelectric effect.

2.4.3 The Quantum Rescue

As mentioned in Section 2.4.2, the energy of photoelectrons was observed to be directly proportional to the frequency of the incident light, f, and there was no photoelectron emitted below a threshold frequency f_0. Thus, we can say that the experiments showed that the photoelectron energy, E, was directly proportional to $f - f_0$, which can mathematically be written as:

$$E = h(f - f_0) \tag{2.8}$$

where h is the proportionality constant. If you measure and plot E against $(f - f_0)$, Eq. 2.8 tells you that you can read the value of h as a slope of this linear plot.

All matter is composed of atoms, which in turn are composed of electrons, protons, and neutrons. In metals, some electrons are free to move from atom to atom: This is why they have high electrical and heat conductivity. These electrons at the metal surface can be easily removed. The minimum energy required to free an electron from a metal is called the *work function* of the metal.

It turned out that the measured value of h was the same as that of Planck's constant in the black-body radiation experiments discussed earlier. As discussed earlier, to explain black-body radiation, Planck proposed that the energy of light comes in discrete amounts. In order to explain the experimental observations of the photoelectric effect, one needs to go a step further to propose that light is made up of particles that contain these discrete amounts of energy. That is exactly what Albert Einstein proposed in essence in 1905 in his research paper that partly won him the Nobel Prize.

STUDY CHECKPOINT 2.4

From Figure 2.7, determine

A. The value of the Planck constant, h.
B. The work function for Na.

Solutions:

A.
From Figure 2.7:

$$y = 0.4125 \times 10^{-14} x - 1.8152$$

Comparing this to Eq. 2.8:

$$E = h(f - f_0) \tag{2.8}$$

We obtain the slope as:

$$\begin{aligned} h &= 0.4125 \times 10^{-14} \text{ eV s} \\ &= 0.4125 \times 10^{-14} \times 1.602 \times 10^{-19} \text{ Js} \\ &= 6.61 \times 10^{-34} \text{ Js} \end{aligned}$$

B.
From Figure 2.7,

$$y = 0.4125 \times 10^{-14} x - 1.8152$$

Comparing this to Eq. 2.8:

$$E = h(f - f_0) \tag{2.8}$$

We obtain the y-intercept as the work function:

$$W = hf_0 = 1.8152 \text{ eV} = 1.82 \text{ eV}$$

According to Einstein's proposal, the energy of light is not spread out across wavefronts as proposed by the classical picture, but is split in small packets called quanta of light, now known as photons. Planck had already proposed in his attempt to explain black-body radiations that the energy was given to the EM waves in the form of discrete amounts. Einstein further proposed that energy is carried by EM waves in discrete amounts (light quanta), and energy is transferred from light to electrons in the form of discrete amounts in terms of light quanta. So, when it comes to transferring energy, it is not between the energy spread across the wavefront and the electron, it is between the light quanta and the electron. This is what explains the instantaneous transfer of energy from light to electrons. Each light quanta that is a photon at frequency *f*, has the same energy, *hf*. So, the key point is that changing the intensity of light will change the number of light quanta, but not the energy of the quanta. This proposition that an electron acquires energy by absorbing a single photon (light quanta), explains why change in light intensity changed the number of emitted electrons, but not their energy. Furthermore, the higher the frequency, *f*, of light, the higher the energy, *hf*, of photons, and therefore the higher the energy of the emitted electrons. Below a certain frequency, the threshold frequency, no electrons will be emitted, no matter what the intensity of light.

Albert Einstein (1879–1955)

Albert Einstein, a physicist, was born on March 14, 1879 in Ulm, Wurttemberg, Germany. He is well known for discovering the theory of relativity, and especially for his mass–energy equivalence equation: $E = mc^2$. He was awarded Nobel Prize in physics in 1921 "for his services to Theoretical Physics, and especially for his discovery of the law of the photoelectric effect". Although Einstein was one of a very few who got the genie of quantum mechanics out of the bottle, he never liked the underlying idea of quantum mechanics: the probability or uncertainty. He expressed his disbelief in this concept in his famous quote: God does not play dice. Einstein died on April 18, 1955 at the age of 76, in Princeton, New Jersey, USA.

Einstein, named the person of the century by Time magazine in 1999, is the most recognized scientist in history. Across the countries and cultures, his name has become synonymous to genius.

So, all the pieces fell in place to explain all the experimental observations by the following propositions by Einstein:

1. Energy in light exists in the form of light quanta: the photons.
2. When light with frequency *f* falls on a metal surface, an electron in an atom absorbs one photon with energy *hf*.

3. Part of the absorbed energy, hf_0, is spent to remove the electron from the surface of the metal, where f_0 is the threshold frequency, and the rest of the energy turns into the kinetic energy of the photoelectron. This can be expressed in the following equation:

$$E_{EM} = hf = hf_0 + E_{ke}$$

where,

- h is Planck's constant.
- f is the frequency of the incident light, and therefore of the photons.
- f_0 is the threshold frequency of light for the photoelectric effect to happen.
- hf is the energy of the incident light, and therefore of incident photons.
- E_{ke} is the maximum kinetic energy of the photoelectrons.

The energy hf_0 is called the work function of the metal and is needed for an electron to leave the metal surface. The work function of some elements is presented in Table 2.1. Einstein's explanation of the photoelectric effect partly won him the Nobel Prize in physics in 1921.

TABLE 2.1. The Values of the Work Function for Some Elements

Element	Symbol	Work Function (eV)
Aluminum	Al	4.28
Calcium	Ca	2.87
Cesium	Cs	2.14
Copper	Cu	4.7
Germanium	Ge	4.2
Gold	Au	5.1
Iron	Fe	4.5
Lead	Pb	4.28
Lithium	Li	2.9
Magnesium	Mg	3.68
Manganese	Mn	4.1
Mercury	Hg	4.49
Nickel	Ni	5.15
Potassium	K	2.3
Silicon	Si	4.85
Sodium	Na	1.82
Silver	Ag	4.7
Zinc	Zn	4.33

Source: Herbert B. Michaelson, "The work function of the elements and its periodicity." *J. Appl. Phys.* **48**, 4729 (1977).

PROBLEM 2.5

Near-UV light of wavelength 300 nm is incident on a Na surface of area 4 cm^2. The intensity of the incident light is 2 W/m^2 and only 1% of the incident photons produce photoelectrons. Answer the following:

A. What is the maximum KE of an emitted photoelectron?
B. What is the rate of production of photoelectrons?

Solution:

A.
The energy of a photon:

$$E_P = hf = \frac{hc}{\lambda}$$
$$= \frac{4.14 \times 10^{-15} \text{ eV s} \times 3 \times 10^{8} \text{ m/s}}{300 \times 10^{-9} \text{ m}}$$
$$= 4.14 \text{ eV}$$

Work function of Na $= hf_0 = 2.3$ eV

Maximum kinetic energy of a photoelectron $= hf - hf_0$
$$= 4.14 - 2.3 \text{ eV}$$
$$= 1.8 \text{ eV}$$

B.

Total energy incident on the surface $= 2 \text{ W/m}^2 \times 4 \text{ cm}^2$
$$= 8 \times 10^{-4} \text{ J/s}$$
$$= 8 \times 10^{-4} \times 6.24 \times 10^{18} \text{ eV/s}$$
$$= 4.99 \times 10^{15} \text{ eV/s}$$

Total number of photons:

$$N_P = \frac{4.99 \times 10^{15} \text{ eV/s}}{4.14 \text{ eV/photon}}$$
$$= 1.21 \times 10^{15} \text{ photons/s}$$

Number of photoelectrons produced = 1% of the number of photons
$$= 1.21 \times 10^{13} \text{ electrons/s}$$

In a nutshell, experimental observations of black-body radiation revealed that light energy is radiated in discrete amounts. The photoelectric effect demonstrated that light energy is also absorbed in discrete amounts or light quanta.

STUDY CHECKPOINT 2.5

A. Write an equation for the energy of a photon in terms of its frequency, which summarizes Planck's explanation of black-body radiation and Einstein's explanation of the photoelectric effect.
B. Write the equation in part A in terms of wavelength of the photon.

Solution:

A. $E = hf$
B. $E = hc/\lambda$

Toward the end of the nineteenth century, physicists were also making intense efforts to understand the origin and nature of a penetrating radiation called X-rays, discovered in 1895 by Wilhelm Röntgen (1845–1923), a professor at the University of Wurzburg, Germany.

2.5 X-RAY DIFFRACTION

X-rays are another player in the development of quantum physics. After intense studies of X-rays, it became clear by 1912 that X-rays were a form of EM waves. As explained in Chapter 1, when EM (light) waves pass through a narrow slit, they are scattered in such a way that they seem to spread out, or bend around the edges of the slit, a phenomenon called diffraction. When EM waves, such as X-rays, pass through a set of evenly spaced narrow slits, called diffraction grating, the scattered waves interact to form a series of bright bands corresponding to constructive interference and dark bands corresponding to destructive interference, a pattern called a diffraction pattern. The most effective diffraction of waves occurs when the wavelength of the waves and the width of the slits are comparable in magnitudes. Scientists including Arnold Sommerfeld (1868–1951) and Wilhelm Wien (1864–1928) estimated the wavelength of X-rays to be in the range of 10–0.01 nm. Given that the distance between atoms in a crystal was typically on the order of 0.1 nm, Max von Lau (1879–1960) suggested that the X-rays should scatter from the atoms of a crystal. If X-rays were a form of EM waves, it should exhibit interference effects once scattered from the crystal atoms.

Lau was right, and making a long story short, much of our current knowledge about the internal structure of the crystals has come from the study of X-rays scattered or diffracted by the crystals, a technique called X-ray diffraction crystallography, a form of X-ray spectroscopy. The spacing of the layers of atoms in a solid crystal, which work as the width of slits, is usually in the range of ~2–20 Å, which is within the range of the wavelength of X-rays:

0.1–100 Å. X-ray crystallography uses the diffraction of X-rays through the target crystal material to determine the internal structure of the crystal.

> *Note*: A lattice is an arrangement of isolated points in space, called **lattice points**, in a regular pattern, which shows the positions of atoms, molecules, or ions in the structure of a crystal. Recall that the atoms in a crystal are arranged in a periodic array.

A crystal consists of a regular array of atoms, and each of these atoms is capable of scattering any EM wave that is incident on it. In other words, atoms in a crystal are arranged in planes that are parallel to each other. X-ray diffraction allows us to determine the structure of a crystal by determining the position of atoms in the lattice. A useful parameter in this determination is the spacing, *d*, between the parallel planes. This is how this technique works:

1. A collimated beam of X-rays is directed at the crystal.
2. The X-ray photons in the incident beam are scattered in all directions from different layers of atoms in the crystals.
3. Due to the regular arrangement of atoms in the crystal, in certain directions, the scattered waves interfere with one another constructively. These waves can be measured.
4. The scattered waves with constructive interference (that are measured) are those waves that are parallel to each other, as shown in Figure 2.8.

> *Note*: A collimated beam of light is a light beam in which the light rays are parallel to each other, and therefore have a plane wavefront.

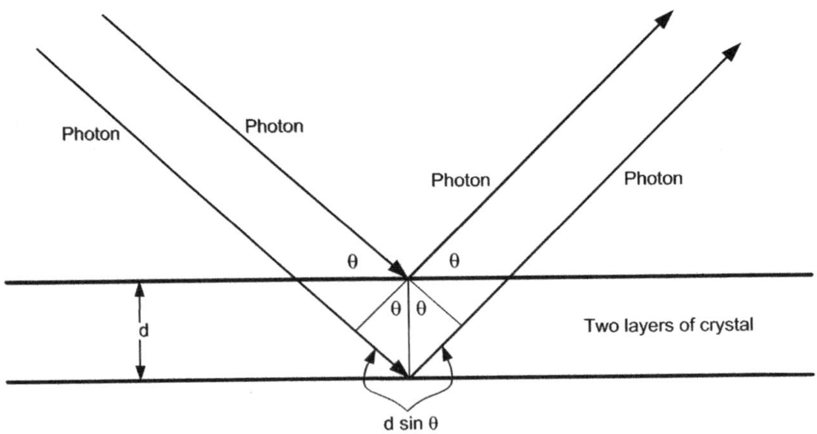

Figure 2.8. Scattering of X-rays from two parallel planes of a crystal with separation *d*.

The spacing in two layers of the crystal results in the difference in the path that the two interfering X-rays scattered from the two layers travel, called the path difference.

For the constructive interference to happen, the following conditions must be true:

1. The EM rays are parallel to each other. This, as shown in Figure 2.8, gives rise to the path difference of $2d \sin \theta$, where d is the spacing between the layers and θ is the incident angle.
2. The path difference between the arrays is an integral multiple of the wavelength of X-rays, λ. This gives rise to the path difference of $n\lambda$, where n is an integer.

From these two conditions, it implies

$$2d \sin \theta = n\lambda \tag{2.9}$$

Equation 2.9, first suggested by W.L. Bragg in 1912, is called Bragg's law. It can be used to determine either the spacing d if the wavelength λ is known or vice versa. Typically, the incident X-ray beam is fixed in a direction and the crystal is rotated through a wide range of angles, and the spectrum of the scattered X-rays is recorded. This is called an X-ray diffraction scan. To obtain a complete crystal structure, three sets of X-ray spectra are obtained, each corresponding to crystal rotations around one of the three mutually perpendicular planes of the crystal. The data from these spectra is analyzed to determine the positions of the atoms in the unit cell of the crystal. Recall that a crystal is a repetition of its unit cell in three dimensions (3D).

Readers with a biology background will be familiar with the famous X-ray diffraction photograph presented in Figure 2.9*a* originally obtained in the early 1950s from the experiment conducted by Rosalind Franklin and Maurice Wilkins. From this diffraction image and data, James Watson and Francis Crick proposed the double helix structure of the deoxyribonucleic acid (DNA) molecule, something like that shown in Figure 2.9*b*, for which they were awarded the Nobel Prize in physiology and medicine in 1962. This has been one of the most important discoveries in molecular biology.

Arthur Holly Compton (1892–1962) discovered that the wavelength of X-rays increases after scattering. He explained this phenomenon in 1923 using the quantum approach. His work helped convince the remaining doubters of the reality of photons: the quanta of EM radiation.

Speaking about doubters, although Planck won the Nobel Prize for offering the explanation for black-body radiation based on the concept that light energy is emitted in discrete amounts or quanta, he did not believe in the physical reality of light quanta for a long time. He was not the only skeptic. He was only one of many scientists who were skeptics of this emerging picture. It was the discovery and explanation of the Compton effect that convinced many of these skeptics that light quanta was a physical reality.

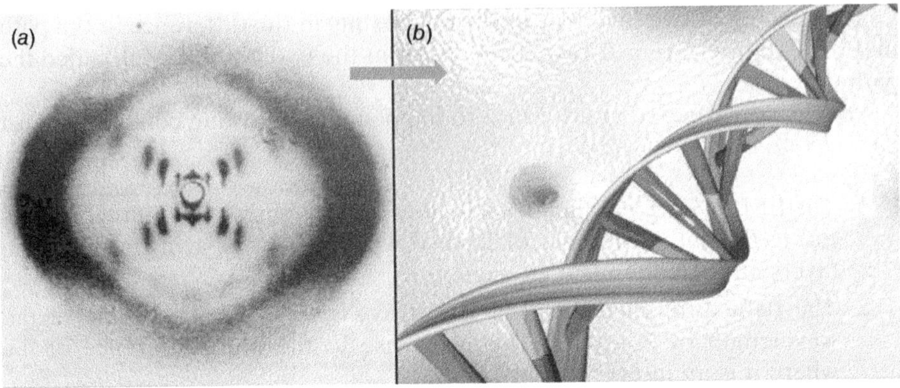

Figure 2.9. The double helix structure of the DNA molecule (*b*) derived the pattern of dark spots in the X-ray diffraction photograph of crystalline DNA (*a*). Courtesy of National Human Genome Research Institute.

2.6 THE COMPTON EFFECT

Black-body radiation, as explained by Planck's formula, demonstrated that light is emitted by a black body in discrete amounts called light (or energy) quanta. The photoelectric effect, explained by Einstein, demonstrated that light is not only emitted in quanta, but is also absorbed in quanta, and perhaps carried in quanta as well. These conclusions point to the ultimate conclusion: Light consists of quanta, individual independent particles. But there are implications. If light consists of particle-like quanta, then these quanta should behave like particles when they hit other particles, such as electrons. Think of two billiard balls: ball A hitting the ball B at rest. As a result of this collision, ball B will start moving in a certain direction determined by the momentum of ball A, and the energy and momentum of ball A will change after the collision due to the conservation of energy and momentum, as explained in Chapter 1. Because in the quantum picture, the energy is directly proportional to frequency (i.e., inversely proportional to wavelength), change in energy means change in frequency (and therefore change in wavelength). In other words, it should be possible that light hits an electron at rest, puts it into motion, and as a result changes its own frequency to a lower value. This behavior of light cannot be explained in the classical picture. But if light is composed of particle-like light quanta, then this behavior is expected. This is an example of the kinds of questions and issues with which physicists were struggling when Arthur Compton in 1918 started studying X-rays, a form of EM waves with wavelength in the range of 10–0.01 nm, scattered by free electrons. This presented the situation of particle–particle scattering (or collision) discussed in Chapter 1 if light quanta were actually particles. It was Compton who coined the term photon for the light quanta.

Arthur Holly Compton (1892–1962)

Arthur Holly Compton was born on September 10, 1892 in Wooster, Ohio. He attended the College of Wooster and Princeton University, and started studying X-ray scattering in 1918. While at Washington University in St. Louis, he observed the shift in the wavelength of scattered X-rays, now known as the Compton effect. He explained this phenomenon in 1923 in terms of the quantum picture of light, and coined the term photon for the light quanta.

The discovery of the Compton effect and its explanation, which won Compton the Nobel Prize in Physics in 1927, convinced many skeptics of the physical reality of light quanta, the photon.

Compton died on March 15, 1962 at the age of 69 in Berkeley, California, USA.

The X-ray scattering experiment is illustrated in Figure 2.10. As shown in this figure, a photon with momentum P_1, energy E_1, frequency f_1, and wavelength λ_1 collides with a free (unbound) electron at rest. The electron recoils at an angle φ with the X-axis with momentum P_e, energy E_e, and the photon scatters away making angle θ with its original direction and with momentum P_2, energy E_2, frequency f_2, and wavelength λ_2.

From Chapter 1, we know that the total energy E of a particle with momentum p and rest mass m_0 can be written as:

$$E = \sqrt{(m_0^2 c^4 + p^2 c^2)} \tag{2.10}$$

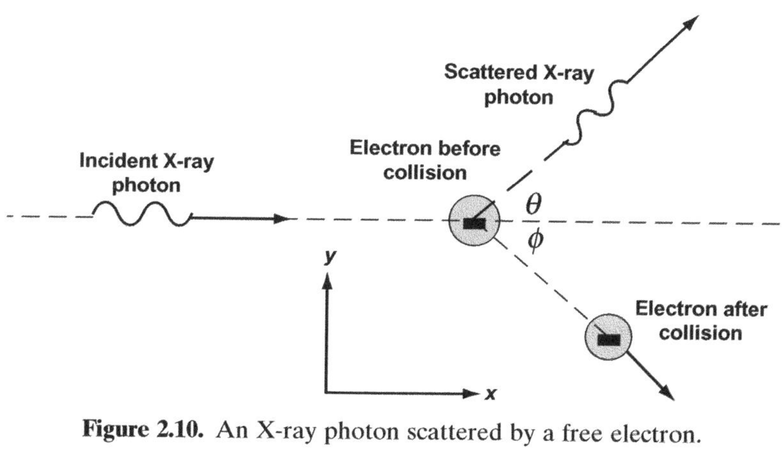

Figure 2.10. An X-ray photon scattered by a free electron.

Because the rest mass, m_0, of a photon is zero, the above equation leads to the photon energy E_p as:

$$E_p = pc$$

By applying energy conservation to scattering, shown in Figure 2.10

$$E_1 + m_0c^2 = E_2 + E_e$$

where m_0 is the mass of an electron at rest.
which implies

$$P_1c + m_0c^2 = P_2c + [m_0^2c^4 + P_e^2c^2]^{1/2}$$

By applying momentum conservation to scattering in Figure 2.10 in the x direction:

$$P_1 = P_2\cos\theta + P_e\cos\Phi$$

which implies

$$\begin{aligned} \frac{h}{\lambda_1} &= \frac{h}{\lambda_2}\cos\theta + P_e\cos\Phi \\ \frac{hc}{\lambda_1} &= \frac{hc}{\lambda_2}\cos\theta + P_ec\cos\Phi \\ hf_1 &= hf_2\cos\theta + P_ec\cos\Phi \end{aligned} \tag{2.11}$$

Similarly, applying momentum conservation in the y direction:

$$\begin{aligned} \frac{h}{\lambda_2}\sin\theta &= P_e\sin\Phi \\ hf_2\sin\theta &= P_ec\sin\Phi \end{aligned} \tag{2.12}$$

By eliminating P_e and Φ from Eq. 2.12 by using other equations, we obtain

$$\Delta\lambda = \lambda_2 - \lambda_1 = \frac{h}{m_0c}(1-\cos\theta) \tag{2.13}$$

The change in wavelength, $\Delta\lambda$, is called the Compton wave shift and depends on the scattering angle θ of the scattered particle, photon, and the quantity

$$\lambda_c = \frac{h}{m_0c}$$

is called the Compton wavelength and depends on the mass of the scattering particle: an electron in this case. Note from Eq. 2.13 that when $\theta = 0$, that is, no scattering, the change in wavelength is 0. On the other hand, when $\theta = 180°$, in the case of backscattering, the change in wavelength is $2h/m_0c$.

Equation 2.13, first derived by Compton during the early 1920s, accurately described the Compton effect, first observed by Compton. This effect and its explanation constituted a very strong case in favor of the quantum theory of light or radiation.

PROBLEM 2.6

Calculate the values for the Compton wavelength and the maximum wave shift for a photon striking an electron.

Solution:

$$\begin{aligned}
\text{Compton wavelength} &= \lambda_c \\
&= h/mc \\
&= (4.136\times10^{-15}\ \text{eV s})/(c\times0.511\ \text{MeV/c}^2) \\
&= 8.094\times10^{-21}\ s\times c \\
&= 8.094\times10^{-21}\ s\times2.998\times10^{8}\ \text{m/s} \\
&= 2.427\times10^{-12}\ \text{m} \\
&= 2.427\ \text{pm}
\end{aligned}$$

The maximum wave shift occurs when the scattering angle $\theta = 180°$:

$$\begin{aligned}
\Delta\lambda &= (h/mc)\ (1-\cos\theta) \\
&= 2.427\ \text{pm}\times2 \\
&= 4.854\ \text{pm}
\end{aligned}$$

As shown in Problem 2.6, the maximum wave shift of a photon scattered by an electron is ~4.85 pm. This is a very small number and is difficult to observe experimentally unless the wavelength of the incident photon, λ_1, is small enough to make the fractional change appreciable. As illustrated in Problem 2.16, you can achieve a much greater fractional (or percentage) wave shift by using X-rays as opposed to visible light.

Alert: Note the crucial difference between the end results of the photoelectric effect and the Compton effect: A photon is completely consumed in photoelectric effect, whereas only part of the energy of a photon is consumed by the electron in the Compton effect.

STUDY CHECKPOINT 2.6

A. What is the rest mass of a photon?
B. What is the effective mass of a photon?

Solution:

A. The rest mass of a photon is zero.
B. The effective mass can be calculated from the equation:

$$E = mc^2$$

which implies

$$m = E/c^2$$

So, now you know how the quantum picture of light is needed to explain phenomenon, such as black-body radiation, the photoelectric effect, and the Compton effect. You may ask What do these phenomena have to do with the real world? If with the help of quantum mechanics we can understand these phenomena and manipulate them, we may have an application of quantum mechanics in the real world.

2.7 LIVING IN THE QUANTUM WORLD

Because we live in the macroworld and quantum physics is the physics of the microworld, it does not fit well with our common sense. Yet there are clearly many things in the macroworld like the spectrum of the Sun, a macroobject, that fit nicely to the quantum mechanics prediction of black-body radiation. Nevertheless, quantum physics is still obscure to most of the world, absurd to many, and is generally considered an intellectual exercise for physicists. Recent developments in science and technology, such as the emergence of nanotechnology, are bound to change that picture of quantum physics, because to carry on those developments, scientists will need to understand and use quantum physics. This is not to say that quantum physics did not have any applications so far.

This section discusses some of the applications of quantum physics related to the topics covered in this chapter.

2.7.1 Using Black-Body Radiation

Thermal transformations among materials occur at atomic and molecular levels and therefore depend on the atomic and molecular structure of the material. The wavelengths of photons involved in such transformations fall in the nanoscale region (1–300 nm), the scale of nanotechnology. Therefore, by understanding and using quantum mechanics behind black-body radiation, a nanotechnologist can control black-body radiation. In other words, nanotechnology offers the promise of producing black-body radiations of specific wavelength (and hence frequency) rather than a spectrum. In the visible light spectrum, this means you can produce just a green or a blue black-body radiation. Here are some examples of the perceived applications of controlling and manipulating black-body radiation at the nanoscale level:

1. Fiber optical communication systems with better efficiency and effectiveness.
2. Better optical signs and displays.
3. Improved lighting techniques.
4. Windows with better radiation control in the visible and invisible (e.g., ultra) spectra.

So far, scientists have used an approximate black body in the lab because they did not have much control on the scale at which black-body radiation occurs. The understanding of black-body radiation in terms of quantum mechanics has helped nanotechnologists to control radiation and particles on the nano scale level. They are working on creating perfect black bodies that reflect no light at all. Such a material is expected to have a wide spectrum of applications including more efficient solar cells, brighter light emitting diodes (LEDs), and optical interconnects in computer applications. Such materials will also facilitate studying black-body radiations to finer details, and therefore testing and advancing quantum mechanics.

As already mentioned, a perfect black-body material (that reflects no light at all) is considered to be used to increase the efficiency of solar cells. Solar cells themselves are an application of the photoelectric effect explained by quantum mechanics.

2.7.2 Using the Photoelectric Effect

These days, there are lots of activities in the field of alternative energy, that is, alternative to using fossil fuels, such as oil. Solar power is a major sector in alternative energy and refers to conversion of sunlight into electricity by using devices, such as the photovoltaic cell.

> *Note*: The field of alternative energy is also known by other names, such as renewable energy, and clean technology, and includes solar power, wind power, and biofuels.

Before we dive any further into this topic, you need to understand the following three terms:

- *Photovoltaic Cell.* It is a device that is used to convert light (any light) energy into electrical energy by using the photoelectric effect. The term photovoltaic has a Greek origin: *phos* means light, and *voltaic* means electrical.
- *Solar Cell.* It is any cell, such as a photovoltaic cell, that converts energy from sunlight into electrical energy. Note the difference: photovoltaic is a more general term; photons can come from anywhere, not just from the sun.
- *Photovoltaic Panel.* It is a panel that consists of photovoltaic cells connected in series to increase the overall voltage, and hence power output.

The following is the physics behind the photovoltaic (photoelectric) process:

1. A photon hits the solar panel.
2. The photon is absorbed by the material, (a semiconductor material, e.g., Si) in the panel.
3. The energy of the photon knocks an electron off an atom to make it free to flow through the material; the atom is ionized.
4. The free electrons flow to produce electricity: direct current (dc).
5. The dc, if desired, can be converted into alternating current (ac) by using appropriate devices, such as inverters.
6. The ac power enters the utility panel from where it is distributed.

An illustration of the photovoltaic effect in a solar cell is presented in Figure 2.11. In this example, the solar cell consists of two types of material: p- and *n*-type Si. Light of suitable wavelengths ionizes the atom in the material (say Si), and as a result positive and negative charges are separated. The positive charges (called holes) are swept into the positive layer, the P-layer, and the negative charges are swept into the negative layer, the N-layer. Due to the potential barrier inside the material, these opposite charges, attracted to each other, can mostly only combine through an external circuit. In this circuit, the free electrons produce power as they pass through a load.

A quantity of interest related to the performance of the photovoltaic cell is quantum efficiency, ε_q defined as follows:

$$\varepsilon_q = \frac{n_e}{n_p}$$

where n_p is the number of photons used to produce n_e photoelectrons collected to produce the current.

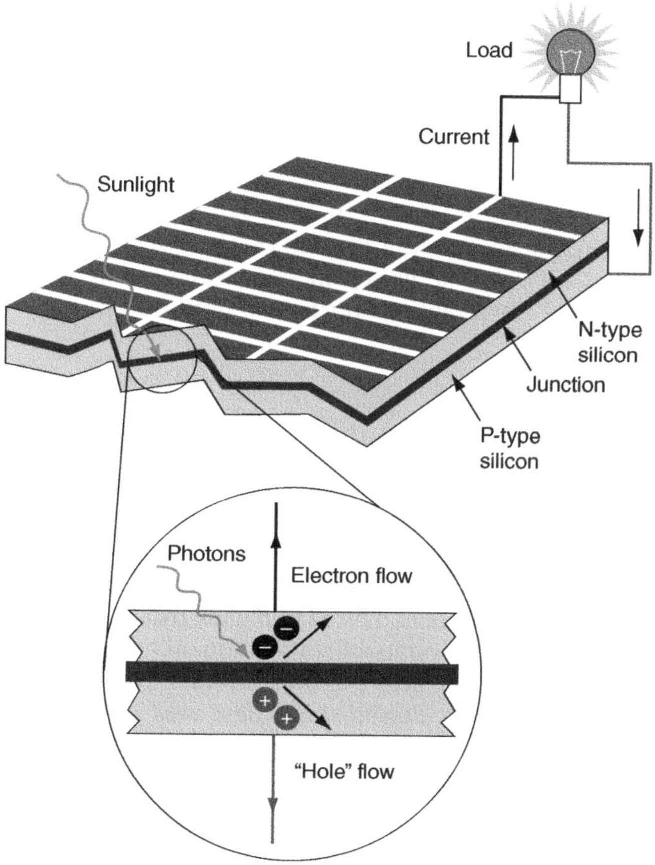

Figure 2.11. Illustration of how a photovoltaic cell processes sunlight. Courtesy of www.energyeducation.tx.gov.

As you already know from this chapter, the photoelectric effect is a quantum phenomenon and the photon itself is a quantum mechanical concept. So, the knowledge obtained by quantum physics about the photoelectric effect is at the heart of designing photovoltaic cells and improving their performance.

2.7.3 Using Compton Scattering

After Compton scattering (or Compton effect) was understood in terms of quantum mechanics, it became possible to make an effective use of it in several applications including the following:

- *Radiation Therapy.* Compton scattering is used as an interaction between high energy X-rays and atomic nuclei in living organisms. So it has enormous applications in radiobiology, for example, radiation therapy, commonly applied to cancerous tumors. In general, radiation therapy works by damaging the DNA of cancer cells by high-energy photons (X-ray); however, beams of electrons, protons, neutrons, or ions can also be used. The beam of these high-energy particles directly or indirectly ionizes the atoms that make up the DNA chain.

- *Gamma-Ray Spectroscopy.* Compton scattering, along with the photoelectric effect, are also applied in gamma-ray spectroscopy, used to measure the radiations emitted by radioactive substances.
- *Material Physics.* In material physics or chemistry, Compton scattering can be used to probe the wave function of the electrons in matter. You will learn about wave funtions further on in this book.

2.8 SUMMARY

Classical physics treats light (or EM radiation) as a wave whose energy is distributed over wavefronts and depends on the intensity of light. This classical picture was challenged by a number of experimental observations toward the end of the twentieth century. The theoretical predictions by classical physics could not reproduce the spectral density curves of radiation emitted by a black body. These curves were explained by Max Planck by assuming that the radiation is emitted in discrete amounts called light quanta and not in a continuous fashion, as predicted by classical physics. Another phenonmenon that was challenging classical physics during those days was called the photelectric effect, in which electrons are instantly emitted as a result of light incident on a metallic surface. The energy of the emitted electrons, called photoelectrons, was found to depend only on the frequency (or wavelength) of the incident light, and the number of the emitted photoelectrons depended on the intensity of the incident light. There was no explanation in the classical picture of light for these experimental observations. The explanation came from Albert Einstein, which further advanced the quantum picture of light introduced by Planck in explaining black-body radiation. According to Einstein's explanation, light consists of quanta, which carry discrete amounts of energy. In the photoelectric effect, a photon is completely absorbed by an electron bound to an atom, and as a result the electron has enough energy transferred by the photon to free itself from the surface on which the light is incident. This explains the instant emission of photoelectrons and the fact that the number of photoelectrons depended on the intensity of light and also that the energy of the photoelectrons depended on the frequency of the light. This is because increasing the intensity of light increases the number of photons, the energy of the photon depends on the frequency of light, and the interaction is between one photon and one electron.

Whereas a light quanta is completely absorbed by a bound electron, a free electron only absorbs part of the energy of the photon that hits it. As a result, the photon scatters with a different energy, and hence different frequency or wavelength. This phenomenon is called Compton scattering and was observed and explained by Arthur Compton in the early 1920s. This confirmed the physical reality of the light quanta introduced by Planck and Einstein, in terms of photons, that is, light consists of particles called photons. The existence of photons demonstrates the particle behavior of light that is considered a wave in classical physics.

In a nutshell, classical physics treats waves and particles as two separate types of entities reflecting the physical reality that we experience as macrobeings. However, the physical reality of the macroworld has its roots in the microworld of molecules and atoms and subatomic particles. We have seen in this chapter that the classical picture of the physical reality collapses in the microscopic world where waves can act as particles under certain conditions. A natural question to ask here would be Is the reverse true? That is, can particles exhibit wave behavior? This is the subject of Chapter 3.

2.9 ADDITIONAL PROBLEMS

Problem 2.7 Compare and contrast the contributions of black-body radiation and the photoelectric effect to the development of the quantum picture of light.

Problem 2.8 Argue the accuracy of the following statement:
The maximum KE of a photoelectron is proportional to the frequency of the absorbed photon.

Problem 2.9 Recall that the photoelectric effect is defined as the emission of an electron after a photon is completely absorbed as a result of one-on-one electron–photon interaction. Use this picture and the principles of momentum and energy conservation to prove that the photoelectric effect can only happen when a photon strikes a bound electron as opposed to a free electron.

Problem 2.10 Use the quantum picture to calculate the energy of a photon in blue light with a wavelength of 470 nm.

Problem 2.11 Calculate the threshold wavelength for Na with a work function of 2.75 eV. What is the stopping potential when an incident light of 400 nm is used.

Problem 2.12 The power of 2 mW from 450-nm light is incident on a photovoltaic cell. Calculate the number of photons incident on the cell per second.

Problem 2.13 The power of 2 mW from 450-nm light is incident on a photovoltaic cell. Calculate the number of photons incident on the cell per second. The quantum efficiency of the photovoltaic cell is 2%. Find the current produced by the cell.

Problem 2.14 Run an experiment to determine the value of Planck's constant and also to determine the work function of a metal surface. First, illuminate the surface with the light of frequency 1.40×10^{15} Hz, which produces photoelectrons with a maximum kinetic energy of 0.690 eV. Then, illuminate the surface with the light of 1.80×10^{15} Hz, and observe that the maximum kinetic energy of the emitted photoelectrons is 2.34 eV. Using the quantum picture of the photoelectric effect, calculate Planck's constant and the work function of the metal being used in this experiment.

Problem 2.15 Derive the formula for the Compton wave shift by using the following steps:

A. Use Eqs. 2.11 and 2.12 to derive the following:

$$P_e^2 c^2 = (hf_1)^2 + (hf_2)^2 - 2\,(hf_1)(hf_2)\cos\theta \tag{1}$$

B. The energy of the electron after scattering can be written as:

$$\begin{aligned} E &= \text{Rest mass energy} + \text{KE} \\ &= m_0 c^2 + (hf_1 - hf_2) \end{aligned}$$

where m_0 is the mass of electron at rest.
Use this equation and Eq. 2.10 to derive the following:

$$P_e^2 c^2 = (hf_1 - hf_2)^2 + 2mc^2(hf_1 - hf_2) \tag{2}$$

C. Compare Eqs. 1 and 2 derived in parts A and B to derive the following formula for the Compton shift:

$$\lambda_2 - \lambda_1 = (h/mc)\ (1 - \cos\theta)$$

Problem 2.16 Calculate the maximum percentage wave shifts for X-rays and visible light when it is scattered by an electron. Compare the two values to show that X-rays can have percentage wave shifts thousands of times greater than that of visible light.

Problem 2.17 Your friend had a sunburn because she exposed her skin to a UV light of wavelength in the vicinity of 330 nm.

A. What is the energy of the photon responsible for this sunburn?
B. What is the energy of 1 mol of these photons?
C. How many photons are in a 1.10-mJ burst of this radiation?
D. How many moles of photons are in the 1.10 mJ of radiation?

WAVE BEHAVIOR OF PARTICLES

3

A person with a new idea is a crank until the idea succeeds.

Mark Twain

Figure 3.0. Wave–particle duality of nature. Nature at the microscopic level challenges the classical nation of dividing the world into matter and waves. Courtesy of www.photoeverywhere.co.uk.

Wave–particle duality means that all energy and matter exhibit both wave- and particle-like behavior under different circumstances. The same entities exhibit particle-like behavior under one set of conditions and a wave-like behavior under another set of conditions. This concept, which is central to quantum physics, is a big deviation from classical physics, where a given entity is either a particle or a wave. Another way of looking at it is that the concepts of particles and waves as defined by classical physicists to describe the macroworld are inadequate. You can make a case for a missing concept

Quantum Physics for Scientists and Technologists, by Paul Sanghera

that is neither a particle nor a wave, or a unification of a particle and a wave. Otherwise, learn to live with the concept of wave-particle duality. For some of us, this is a good enough unification.

Physicists and chemists are already comfortable with this concept. For others, especially a major section of biologist, it may come as news: The wave–particle duality is making its way to your research field, get ready.

In This Chapter

3.1 INTRODUCTION

The explanation of experimental results from black-body radiation and the photoelectric effect, discussed in Chapter 2, pointed to the particle-like behavior of electromagnetic (EM) waves and perhaps to the fact that EM waves are composed of particles. Most scientists at the time of these discoveries took these explanations as mathematical tricks more than a physical reality. As you saw in Chapter 2, it took a couple of decades for the reality to set in, with the proposal from Compton that EM waves (light) actually consist of particles that he called photons. Around that time, believing in the generality, symmetry, and universality of the physical law, a graduate student named Louis de Broglie took a leap in the reverse direction by presenting the hypothesis that there is a wave (or a wave group) associated with each particle. This hypothesis, which withstood the experimental scrutiny, led to the particle–wave duality of matter, which in turn gave rise to the famous Heisenberg uncertainty principle; a cornerstone of modern science.

So, the three underlying streams of thought in this chapter are wave properties of particles, wave–particle duality of matter, and the uncertainty principle.

3.2 PARTICLES AND WAVES: THE BIG PICTURE

All matter is composed of atoms, which in turn are made up of protons, neutrons, and electrons: the subatomic particles. Protons and neutrons exist in the nucleus of the atoms, and electrons are configured around the nucleus. This configuration along with the number of protons and neutrons in the nucleus determine the atomic structure and the properties of the atoms and the corresponding elements, such as metals and nonmetals. In the classical picture, electrons, like protons and neutrons, are considered particles just like light is considered a wave.

In Chapter 2, you learned that light, which is classically a wave, could behave as particles. A natural question to ask is Could it be that electrons (and other particles for that matter) could behave as waves? If yes, what would it mean and what are the implications? This chapter is an answer to this question.

Table 3.1 is our calendar of some quantum events in chronological order.

In this chapter, you will see how the wave nature of particles leads to the so-called wave–particle duality, which in turn gives rise to equally surprising

TABLE 3.1. A Calendar of Quantum Events

Year	Nobel Scientific Event	Year of Nobel Prize Award
1927	Clinton Davisson and Lester Germer in the United States performed the famous Davisson–Germer experiment and confirmed the de Broglie hypothesis. G.P. Thomson did the same in England with a different experimental setup.	In 1937, Davisson and Thomson shared the Nobel Prize for their experimental work, which proved the de Broglie hypothesis.
1927	Werner Heisenberg developed his principle of uncertainty, which became the cornerstone of modern science.	Heisenberg received the Nobel Prize in 1932 for his contributions to developing quantum mechanics.
1924	Louis de Broglie, a graduate student in physics, in his Ph.D. thesis proposed the hypothesis	de Broglie received the Nobel Prize in 1929 for "his discovery of the wave nature of electrons".
1923	Arthur Holly Compton explained the results of his experiment called Compton scattering and coined the term "photon".	In 1927, Compton received the Nobel Prize for discovering and explaining the Compton effect, which demonstrated the particle nature of light waves.
1905	Albert Einstein explained the photoelectric effect in terms of energy quanta: light waves carry and deliver energy in discrete amounts called light quanta.	In 1921, Einstein receives the Nobel Prize for his explanation of the photoelectric effect in addition to his other contributions to theoretical physics.
1900	Max Planck developed a formula to explain experimental results from black-body radiation. It was based on the proposal that black body emits radiation only in discrete amounts called quanta.	In 1918, Planck received the Nobel Prize for his discovery of energy quanta.

effects, such as quantum confinement, and uncertainty in measuring certain pairs of physical variables simultaneously.

3.3 THE DE BROGLIE HYPOTHESIS

You learned in Chapter 2 that from 1900 to 1905, Planck and Einstein developed the quantum picture of light in order to explain black-body radiation and the photoelectric effect. The explanations for these experimental observations suggested that light, which was known to behave like a wave under certain circumstances, in these phenomena emitted, carried, and transferred energy in discrete amounts called light quanta. One obvious question one could have asked is If light carries energy in quanta, could it be that light is actually made of quanta: the particles? However, this emerging quantum picture was so new and strange at the time that nobody really believed in the physical reality of light quanta. In other words, nobody picked up the ball and ran with it. It took about two decades to realize that light quanta was real, when Arthur Compton observed Compton scattering and explained it in terms of photons in 1923. Slowly, the skeptics started to take it seriously that light may consist of particles that Compton called photons.

At about that time, a physics graduate student at the University of Paris, France, named Louis de Broglie, did pick up the ball and run with it. He asked a rather bold question: If light, which is considered to be a wave, can behave like particles, can particles under certain conditions behave like waves?

If a particle does behave like a wave, what would its wavelength be? You know from Section 1.21 that the energy E of a particle with rest mass m_0 and momentum p can be written as:

$$E = \sqrt{m_0^2 c^4 + p^2 c^2} \tag{3.1}$$

First, let us apply the particle as a wave idea on the photon. For a photon, which is massless, $m_0 = 0$, this equation becomes:

$$E = pc \tag{3.2}$$

For a photon with frequency f,

$$E = hf \tag{3.3}$$

From Eqs. 3.2 and 3.3:

$$pc = hf = hc/\lambda$$

which implies

$$f = \frac{E}{h} \tag{3.4}$$

$$\lambda = \frac{h}{p} \tag{3.5}$$

These equations were known from Plank's and Einstein's work before de Broglie came into the picture. In 1924, Louis de Broglie, in his Ph.D. thesis, made a bold proposal that Eqs. 3.4 and 3.5 were valid not only for photons, which are massless, but also for any material particles, such as electrons. This statement has a very serious implication: Equations 3.4 and 3.5 (frequency and wavelength), associated with particles suggest that any object or particle moving with momentum p has a wave of wavelength h/p associated with it. The wavelength of such a particle given by Eq. 3.5 is called the de Broglie wavelength and Eqs. 3.4 and 3.5 are called the de Broglie relations.

Alert: The relationship between the wavelength and momentum of all particles massless or massive, such as photons and electrons, is given by the same formula:

$$\lambda = h/p$$

Because momentum of a massive particle is given by its mass m multiplied by the velocity v, Eq. 3.5 for the de Broglie wavelength can also be written for a massive particle as:

$$\lambda = \frac{h}{mv} \tag{3.6}$$

Generally speaking, m in Eq. 3.6 is the relativistic mass related to the rest mass by

$$m = \frac{m_0}{\sqrt{1 - v^2/c^2}} \tag{3.7}$$

As argued in Chapter 1, we will assume in this book that $m = m_0$; that is, we are working with nonrelativistic quantum mechanics.

Note: A careful reader will note from Eq. 3.7 that when the velocity of a particle is small compared to the velocity of light, the relativistic mass of the particles is simply the rest mass. In many biotechnology and atomic physics applications, that is the case, which is the reason you get away with not grappling with relativity in those labs. In nuclear physics applications and in particle physics studies, one usually needs to include the relativistic correction.

Given the extremely small value of h compared to the mass and velocity of the macroobjects in Eq. 3.6, it is easy to imagine that the de Broglie wavelength for macroobjects is too small to be observed in the macroscopic experiments. However, given the small value of mass for a microobject, such as an electron, the value of the de Broglie wavelength may be significant and detectable in the microworld. The rule of thumb is that the wave properties of a particle are significant and detectable when the de Broglie wavelength is comparable with the dimensions of the particle itself or of the things it interacts with. Problem 3.1 illustrates this point. The wavelength associated with the soccerball in this problem, for example, is much smaller than any aperture or obstacle in the macroworld that wave effects, such as diffraction and interference, cannot be detected or observed.

PROBLEM 3.1

A. Find the de Broglie wavelength of a soccerball of mass 430 g that is moving with a speed of 25 m/s.
B. Find the de Broglie wavelength of an electron that is moving in a particle accelerator with a speed of 3.00×10^7 m/s.

Solution:

Because the speed in both cases is much smaller than the speed of light, we are going to treat these situations as nonrelativistic.

A.

$$\lambda = \frac{h}{mv} = \frac{6.63 \times 10^{-34}\ \text{Js}}{0.430\ \text{kg} \times 25.0\ \text{m/s}} = 6.17 \times 10^{-35}\ \text{m}$$

We used the following definition of the joule:

$$1\ \text{J} = 1\ \text{Nm} = 1\ \text{kg m}^2/\text{s}^2$$

As you can see, the wavelength of this object is so small compared to its material dimensions, such as size, that the wave aspects of its behavior will not be detectable.

B.

The mass of the electron, $m = 9.11 \times 10^{-31}$ kg.

$$\lambda = \frac{h}{mv} = \frac{6.63 \times 10^{-34}\ \text{Js}}{9.11 \times 10^{-31}\ \text{kg} \times 1.25 \times 10^7\ \text{m/s}} = 5.82 \times 10^{-11}\ \text{m}$$
$$= 0.582\ \text{Å}$$

This wavelength is not comparable to the dimensions of macroobjects, but is comparable to atoms of which electrons are the part. For example, the radius of the hydrogen atom is ~0.53 Å.

As an example in the microscopic world, consider an electron moving in a potential difference V with a momentum p and with mass m. Its kinetic energy and wavelength is given by the following equations:

$$E = \frac{1}{2}mv^2 = \frac{p^2}{2m} \tag{3.8}$$

Also, as mentioned in Chapter 1, this energy is provided by the potential difference as:

$$\mathrm{E} = 1\ \mathrm{eV} \tag{3.9}$$

where e is the electron charge.

By comparing Eqs. 3.8 and 3.9, we obtain

$$p = \sqrt{2m\,\mathrm{eV}}$$

Therefore, the de Broglie wavelength, h/p, can be written as:

$$\lambda = \frac{h}{\sqrt{2m\,\mathrm{eV}}} \tag{3.10}$$

As shown in Problem 3.2, this relationship can also be expressed as:

$$\lambda = \frac{1.226}{\sqrt{V}}\ \mathrm{nm} \tag{3.11}$$

where the potential V is in volts.

Just by looking at Eq. 3.11, you can imagine that this wavelength should be detectable because it is comparable to the voltage that can move the electron.

PROBLEM 3.2

Prove the relationship:

$$\lambda = \frac{1.226}{\sqrt{V}}\ \mathrm{nm}$$

Solution:

$$\frac{h}{\sqrt{2me}} = \frac{4.136 \times 10^{-15}\ \mathrm{eVs^{-1}}}{\sqrt{2 \times 0.511\ \mathrm{MeV}/c^2 \times 1.602 \times 10^{-19}\ \mathrm{C}}}$$

Use $c = 2.998 \times 10^8$ m/s
which implies

$$\frac{h}{\sqrt{2me}} = \frac{4.136\times10^{-15}\text{ eV s}^{-1}}{\sqrt{2\times0.511\text{ MeV}/c^2\times1.602\times10^{-19}\text{ C}}} = \frac{4.136\times10^{-15}\text{ eV s}^{-1}}{(1.822\times10^{-30}\text{ eV C s}^{-2}/\text{m}^2)^{1/2}}$$

$$= \frac{3.064m}{(\text{C}/\text{eV})^{1/2}} = 3.064\times(1.602\times10^{-19})^{1/2}\text{ mV}$$

$$= 1.226\text{ nm V}$$

In this calculation, remember that

$$1\text{ eV} = 1.602\times10^{-19}\text{ CV}$$

which implies

$$\lambda = \frac{h}{\sqrt{2\text{ meV}}} = \frac{1.226}{\sqrt{V}}\text{ nm}$$

Louis de Broglie (1892–1987)

Louis de Broglie was born in Dieppe, France on August 16, 1892. After studying history, he pursued his interest in mathematics and physics. In 1924, while studying at the University of Paris, de Broglie, in his Ph.D. thesis, proposed that each material object has a wave associated with it. In other words, each material object has a wave behavior that complements its particle behavior. His hypothesis was positively tested in 1927, and he received the Nobel Prize for this in 1929.

De Broglie spent most of his life in France, teaching and performing research in physics. He passed away on March 19, 1987 at the age of 94.

STUDY CHECKPOINT 3.1

According to the de Broglie hypothesis, which of the following exhibits wave characteristics?

A. Electrons only.
B. Photons only.
C. Tiny particles only.
D. Any matter.

Solution:

D

Whereas, Planck, Einstein, and Compton proposed their quantum hypotheses in order to explain experimental observations, de Broglie's hypothesis was speculative and no direct experimental evidence was present for its validity at the time. But to be accepted in science, a theoretical idea or hypothesis must face the test of experiments; that is where rubber meets the road. The first experiment confirming de Broglie's hypothesis by detecting and measuring the wavelength of an electron was conducted in 1927 by Davisson and Germer.

3.4 MEASURING THE WAVELENGTH OF ELECTRONS

You know from Chapter 1 that waves exhibit a phenomenon called diffraction, which has no analogue in the behavior of classical (or Newtonian) particles. One way of proving that a particle like an electron has a wave associated with it will be to demonstrate that electrons exhibit diffraction. That is exactly what Clinton Davisson and Lester Germer demonstrated in 1927 at Bell Labs in the United States. The original goal of their experiment, however, was not to test the electron waves of the de Broglie hypothesis. They were just studying the scattering of electrons from a metal surface, for example, measuring the energies of the scattered electrons, by using the experimental setup illustrated in Figure 3.1. In this setup, electrons from a heated filament are accelerated under an applied voltage before hitting a Ni block, the target specimen. The electron detector was mounted on an arc so that it could be rotated to detect the scattered electrons at different angles.

Davisson and Germer collected the data by varying the following parameters:

- The energy of the electron beam incident on the nickel (Ni) metal surface, called the primary beam.

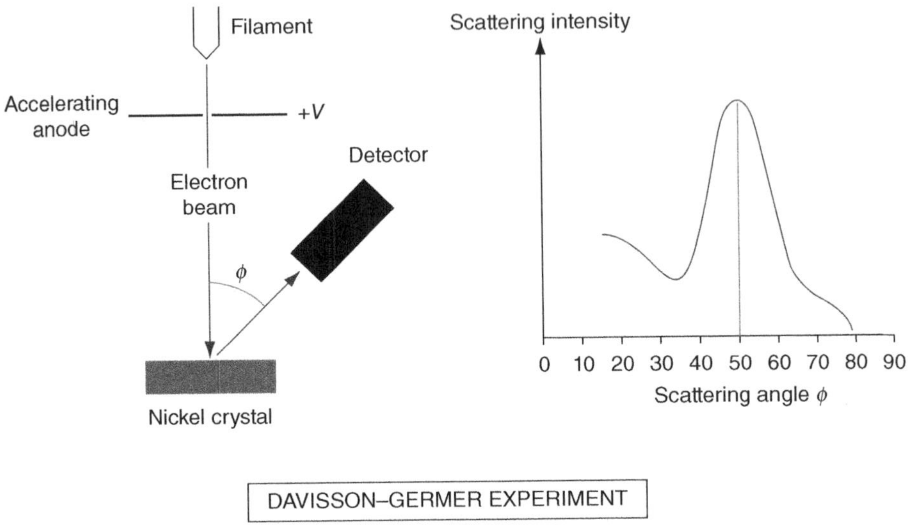

Figure 3.1. Illustration of the Davisson–Germer experiment. Courtesy of Greg Goebel vectorsite.net.

- The angle of the incident electrons with the target metal surface.
- The position of the electron detector.

The data was found to be in agreement with what was expected from classical physics. For example, they observed that the electrons were scattering in all directions and that the intensity of the scattered electrons only had a moderate dependence on the scattering angle; the intensity variation with angle was continuous, as predicted by classical mechanics. Furthermore, the energy of the primary electrons did not have much influence on the intensity spectrum of the scattered electrons. Everything was going as predicted by classical physics until an accident happened.

The experimenters noticed that the target Ni surface was oxidized by the air that accidentally entered into the apparatus. To reduce the target back to a pure metal, they simply baked it in an oven. After this operation, the target was moved back into the experimental setup and the experiment resumed to collect more data.

To their surprise, the results were dramatically different in this experimental run. Instead of continuous variation of electron intensity, they observed an intensity peak at certain scattering angles and intensity minima at others. There were clear distinct intensity maxima and minima whose positions were dependent on the energy of the incident electron beam. In order to explain these results, there were three dots that they needed to connect:

- These results only appeared after the target was heated. They thought it through and realized that upon heating the Ni target could behave like a large crystal whose atoms are arranged in a regular lattice.
- The diffraction of X-ray waves from crystals, discussed in Chapter 2, was known at that time and the diffraction pattern obtained from this experiment looked very similar to that obtained from Bragg's X-ray diffraction experiment.
- In light of the de Broglie hypothesis, particles, such as electrons, could exhibit wave behavior (e.g., diffraction), as the de Broglie wavelength had the same size scale as the separation of Ni atoms on the crystal lattice.

Davisson and Germer were able to connect these dots and conclude that what they were looking at was nothing less than the first evidence of electron waves, the confirmation of the de Broglie hypothesis. Let us work out a simple example to verify how the data collected by Davisson and Germer was in agreement with the prediction of the de Broglie hypothesis and Bragg's diffraction law. According to one data record, an intensity peak was obtained for an electron beam with energy $E = 54.0\,\text{eV}$ at an angle of 50° with the incident electron beam, which was equivalent to an angle of 65° made by the incident beam with the Bragg's plane of the Ni crystal. The spacing between two adjacent planes of the Ni crystal was measured by X-ray diffraction and was found to be 0.091 nm. If we assume that the incident electron was a wave, we can calculate the corresponding wavelength by using Bragg's law:

$$n\lambda = 2d\sin\theta$$

For $n = 1$, $d = 0.091\,\text{nm}$, and $\theta = 65°$:

$$\begin{aligned}\lambda &= 2d\sin\theta = 2\times 0.0910\times \sin(65°)\ \text{nm}\\ &= \theta = 2\times 0.091\times 0.906\ \text{nm}\\ &= 0.165\ \text{nm}\end{aligned}$$

On the other hand, we can calculate the de Broglie wavelength of an electron in this incident beam with energy 54.0 eV as follows:

$$\lambda = \frac{1.226}{\sqrt{54.0}}\ \text{nm} = 0.167\ \text{nm}$$

This value is in agreement with that calculated from Bragg's law, that is, the experimental data is consistent with predictions from Bragg's law of diffraction and the de Broglie hypothesis of the electron wave demonstrating that electrons are exhibiting wave behavior in this experiment in accordance with the de Broglie hypothesis.

Note: In England, G.P. Thomson ran an experiment with a different experimental setup and independently confirmed the de Broglie hypothesis. Davisson and Thomson received the Nobel Prize for their experimental work of electron diffraction in 1937.

Following the Davisson–Germer experiment, particle waves, also called material waves, have been observed for other material entities, such as neutrons, protons, and atoms, as well. Particle diffraction is now a very useful technique to study different structures. For example, Figure 3.2 presents the electron diffraction pattern obtained by striking gold (Au) atoms with the electron beam. Diffraction rings represent the constructive and destructive interference of electron waves from different planes of atoms. Such a pattern can be analyzed to study the structure of the specimen material.

PROBLEM 3.3

In the Davisson–Germer experiment, an intensity peak of scattered electrons was observed for an electron beam of energy 54.0 eV that was incident at an angle of 65° with Bragg's plane of the Ni target. Calculate the wavelength of the waves associated with these electrons by using the de Broglie relation, $\lambda = h/p$.

Solution:

$$E = \frac{1}{2}mv^2 = \frac{p^2}{2m}$$

which implies

$$p = \sqrt{2mE}$$

which implies

$$\begin{aligned}\lambda &= \frac{h}{p} = \frac{h}{\sqrt{2mE}} \\ &= \frac{4.136\times 10^{-15}\ \text{eV s}^{-1}}{(2\times 0.511\ \text{MeV}/c^2 \times 54.0\ \text{eV})^{1/2}} \\ &= \frac{4.136\times 10^{-15}\ \text{m}}{(6.14\times 10^{-10})^{1/2}} \\ &= 0.167\ \text{nm}\end{aligned}$$

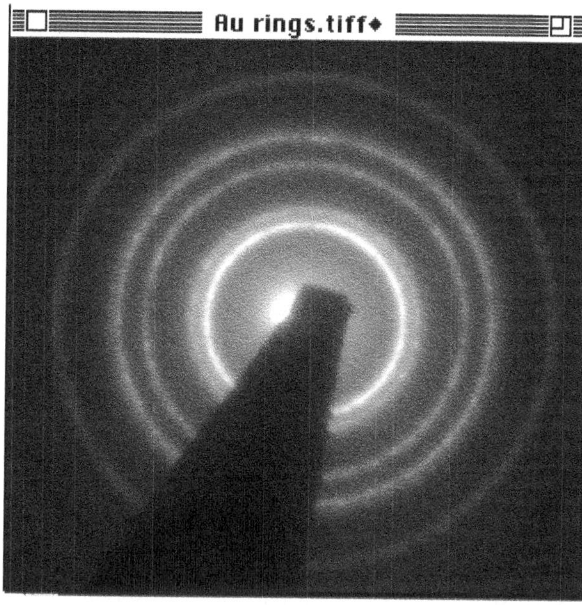

Figure 3.2. Electron diffraction pattern obtained by hitting Au atoms with a beam of electrons. Courtesy of Eric Steel, National Institute of Standards and Technology (NIST).

To summarize, the detection or the size of the effect of a particle wave depends on several factors, such as mass and speed of the particle. Nevertheless, each particle (or object) in the universe, large or small, has a wave associated with it. The wave nature of a moving particle gives rise to some very interesting characteristics when the particle is restricted to a certain region in space, a phenomenon called quantum confinement.

3.5 QUANTUM CONFINEMENT

Quantum confinement is the restriction on the motion of a particle in one or more dimensions within the scale in which quantum characteristics become important. It is an interesting effect that has significant, useful, and practical implications. Here, as an application of de Broglie hypothesis, we explore a simple situation of a confined particle by applying a rudimentary mathematical treatment. Although the spatial confinement can happen in all three dimensions, the simplest case is the confinement in one dimension (1D) that we explore here as an illustration. This effect will be treated in more detail in the forthcoming chapters.

As demonstrated in Figure 3.3, consider a point particle enclosed in a 1D box in which it experiences no force. One dimensional means that the particle exists on a line segment with the walls of the box as the two endpoints of the

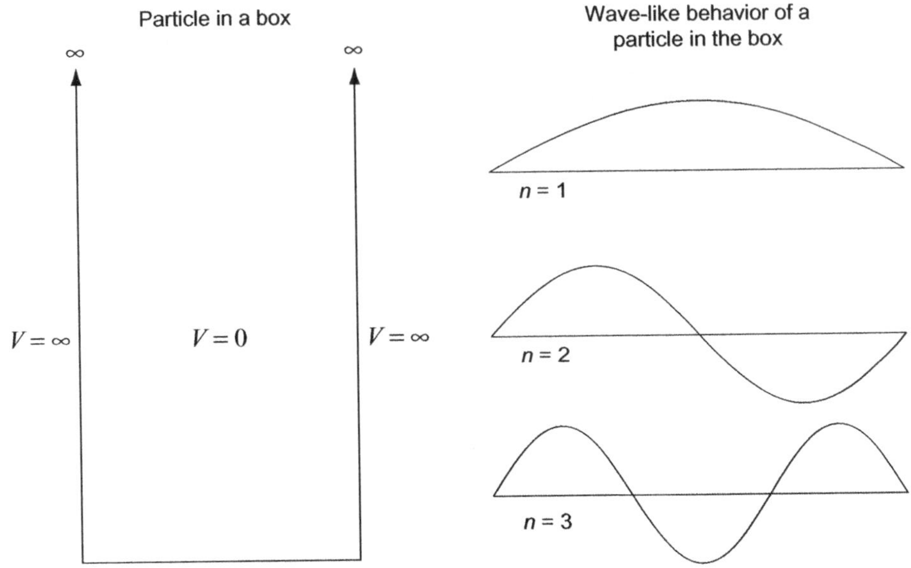

Figure 3.3. The wave shape of a confined particle corresponding to different values of n.

line. Each time it hits the wall, it reflects back without losing any energy. Assume the length of the line segment to be L.

> *Note*: To some biology and chemistry students, the term particle in a box may sound rather abstract. However, realize that this is just a fancy physics term to refer to the particle within a space beyond which the potential is infinite. As a result of the infinite outside potential, the particle will stay confined to the space in which it is at the present time. Think of atoms in a crystal lattice, electrons in an atom, and deoxyribonucleic acid (DNA) molecules in a cell. Confinement is everywhere in chemical and biological objects and phenomena.

In classical mechanics, this situation has a trivial explanation or solution: The confined particle behaves like a classical particle, and therefore moves according to Newton's laws of motion. For example, it moves in a straight line with a constant speed, until it scatters back from the wall that it hits, assuming elastic collisions. Also, it can assume any value of kinetic energy within the acceptable range. This may make sence for big balls in big boxes. Now, given the size of the particle, imagine shrinking the box so that the wavelength associated with the particle in the box is on the order of the same size as the length of the box. Now, we are talking about a quantum mechanical situation.

The particle wave must have a node at the boundary (wall) of the box because the probability of finding the particle there is approaching zero. A node is a point along the wave where it has minimal amplitude, and the amplitude represents the probability of finding the particle. Figure 3.3 illustrates the particle in a box and the wave associated with the particle for a different number of nodes. As mentioned earlier, the particle wave must have a node at the boundary (wall) of the box because the probability of finding the particle there is approaching zero. The implication of this statement is that at a minimum, the half-integer wavelengths must fit into the box. Mathematically, it can be expressed as:

$$L = n\frac{\lambda}{2} \tag{3.12}$$

The same equation can be derived from Bragg's law:

$$n\lambda = 2d\sin\theta$$

You can consider the two walls of the box as two Bragg's planes from which two waves are being reflected. Given that the particles are in the box, we can assume constructive interference between these waves. Furthermore, given $\theta = 90°$ and $d = L$, Bragg's law becomes:

$$n\lambda = 2L$$

which implies

$$L = n\frac{\lambda}{2}$$

The momentum and energy of the confined particle can be derived by applying the de Broglie hypothesis to Eq. 3.12. By using the de Broglie relation, $p = h/\lambda$, you can derive from Eq. 3.12:

$$p = n\frac{h}{2L}$$

which implies

$$E_n = \frac{p^2}{2m} = \frac{n^2h^2}{8mL^2} \tag{3.13}$$

where E_n is the energy of the confined particle. Each energy value corresponding to an integer n permitted by Eq. 3.13 is called an energy level, and the integer n is called the quantum number corresponding to the energy level.

As shown in Study Checkpoint 3.2, $n = 0$ is not an acceptable situation. For a very large value of n, you can realize from Figure 3.3 that the probability maxima of finding the particle will be so closely packed that the particle could be found anywhere in the box. Therefore, we are back to the classical world.

Alert: From our argument that the particle waves must have nodes at the walls of the box, you should realize that n is a positive integer.

STUDY CHECKPOINT 3.2

In Figure 3.3, what represents the probability of finding a particle at a specific point in the box?

Solution:

The amplitude of the particle wave at that point.

STUDY CHECKPOINT 3.3

Why can the value of n in Eq. 3.13 not be zero?

Solution:

To have $n = 0$ is not acceptable because it means from Eq. 3.12 that the length of the box is zero, that is, the box does not exist.

Equation 3.13 has the following three implications:

1. A confined particle can only have certain specific energies and no others, that can be calculated from Eq. 3.13 corresponding to $n = 1, 2, 3, \ldots$. This is called energy quantization.
2. Since n is a positive integer to keep the situation real, the confined particle cannot have a zero energy.
3. Given the extremely small value of Planck's constant, h, the energy quantization becomes conspicuous only when the mass of the particle, m, and the size of the confinement, L, is small enough, that is, in the same ballpark as h.

Note: Energy quantization and exclusion of zero energy for a confined particle have no counterparts or analogues in classical physics. This is an example of how new characteristics and properties of materials and systems are invoked at the nanoscale level, which is the scale of nanotechnology.

To summarize, a confined particle can only have discrete values of energy. In other words, the quantum confinement hypothesis forces the confined particles to exist in a certain set of discrete states. There is no such concept in classical physics, where a particle can have any energy or any state; that is, the behavior of the particle is continuous. As mentioned earlier, since n cannot be zero for a realistic situation, the zero energy for a trapped particle is not allowed. This zero-point energy issue can also be explained in terms of the uncertainty principle, the most famous and often misinterpreted aspect of quantum mechanics.

3.6 THE UNCERTAINTY PRINCIPLE

The physical observables used to study a particle or define its state, such as position, momentum, and energy, used in classical physics are also relevant in quantum physics. The principle of uncertainty states that there are a number of pairs of observables such that both observables in a pair cannot be simultaneously determined accurately. One such pair is the position x and momentum p of the particle.

The uncertainty principle, postulated by Heisenberg in 1927, can be understood as a natural consequence of the de Broglie hypothesis that each particle has a wave associated with it. First, let us ask a question As compared to waves known in classical physics, such as sound and EM waves, what really is this particle wave?

3.6.1 Understanding Particle Waves

You know from Chapter 1 that a wave is a disturbance that carries some kind of energy with it and travels through space and time. The disturbance is defined by something that changes as the wave travels. For example, the changing (or varying) parameters are the height of the water surface in a water wave, the air pressure in a sound wave, and EM field (electric and magnetic fields) in an EM wave. So, an obvious question that you can ask is What is the varying parameter in a matter (particle) wave, such as an electron wave?

As we know from the de Broglie hypothesis, a matter wave or the wavegroup represents a particle, and the particle exists somewhere within the wavegroup. That means each point in the wavegroup has a probability assigned to it for the existence of the particle. So, obviously the matter waves are the probability waves: The varying parameter is the probability of finding the particle represented by the wave function, the varying amplitude of the wave. To be precise, the square of the absolute value of the wave function, $|\psi|^2$, is called the probability density. The large value of probability density at a given point means the odds of finding the particle at the point are great, and a small value means the odds of finding the particle there are low.

Alert: The probability of finding an object at a point in space and time is proportional to $|\psi|^2$, where ψ is the wave function at that point of the wave associated with the object. Although you can look at ψ as the spectrum or spread of the probabilities of finding the particle at different places and times, it does not mean that the particle itself is spread out. Once an experiment is performed to detect the particle at a given point in space and time, either the whole particle is found or it is not.

So, a de Broglie wave associated with a particle is a wavegroup or wavepacket. The basic concept of a wavegroup is discussed in Chapter 1. Because the wavegroup represents the particle, it is trivial to realize that the group velocity, v_g, with which the wavegroup travels, is also the velocity of the particle that the wavegroup represents. Two other important parameters of a wavegroup are the length of the wavegroup, Δx, and its wavenumber, $\kappa \approx 2\pi/\lambda$.

STUDY CHECKPOINT 3.4

A. **True or False**: According to the uncertainty principle, you cannot make a precise measurement at the microscale level.

B. **True or False**: According to the uncertainty principle, it is not possible to measure the position or momentum of a particle precisely.

Solution:

A. False

B. False

As explained in Chapter 1, Δx and $\Delta \kappa$ are related to each other by the minimum value of their product:

$$\Delta x \Delta k = 2\pi$$

It turns out that the minimum value of their product goes even smaller, that is, ½ when the wavegroup has a Gaussian shape. Therefore, in general:

$$\Delta x \Delta \kappa \geq \frac{1}{2} \tag{3.14}$$

Gaussian wavepackets are often used to represent the position of particles as the integrals associated to them are rather easier to solve.

PROBLEM 3.4

The wavenumber of a wavegroup associated with a moving electron is $2.74\,\text{pm}^{-1}$. Calculate the following:

A. Wavelength of the electron.
B. Kinetic energy of the electron.
C. The speed with which the electron is moving.
D. The group velocity of the wavegroup.

Solution:

A.
By definition,

$$\kappa = \frac{2\pi}{\lambda}$$

$$\lambda = \frac{2\pi}{k} = \frac{2\pi}{2.74\ \text{pm}^{-1}} = 2.29\ \text{pm}$$

Rest energy of an electron = $E_0 = m_0c^2 = 0.511\,\text{MeV}$
By using the de Broglie relation, $\lambda = h/p$:

$$pc = \frac{hc}{\lambda} = \frac{4.136\times10^{-15}\ \text{eV s}\times2.998\times10^{8}\ \text{m/s}}{2.29\ \text{pm}} = 0.541\ \text{MeV}$$

B.
Total energy of the electron:

$$E = \sqrt{(m_0\,c^2)^2 + p^2c^2} = \sqrt{(0.511\ \text{MeV})^2 + (0.541\ \text{MeV})^2} = 0.744\ \text{MeV}$$

Kinetic energy of the electron:

$$E_k = E - E_0 = 0.744\ \text{MeV} - 0.511\ \text{MeV} = 0.233\ \text{MeV}$$

C.

$$E = mc^2 = \frac{m_0c^2}{\sqrt{1-\frac{v^2}{c^2}}} = \frac{E_0}{\sqrt{1-\frac{v^2}{c^2}}}$$

which implies

$$v = c\sqrt{1-\frac{E_0^2}{E^2}}$$

$$= 2.998\times10^8 \text{ m/s}\sqrt{1-\frac{(0.511)^2}{(0.744)^2}}$$

$$= 2.18\times10^8 \text{ m/s}$$

D.
Group velocity = velocity of the particle = 2.18×10^8 m/s

This equation is important in understanding the uncertainty principle.

3.6.2 Understanding the Uncertainty Principle

It is important to realize that uncertainty is the price you pay for assigning a wave to a particle. In other words, the uncertainty is built- in as we assign a wave to a moving particle because the wave (or wavegroup) represents the probability distribution of finding the particle anywhere within the wave: No more certainty of the classical approach. A wide wavegroup will have a less precisely specified particle position, but a more precisely defined wavelength, and therefore momentum. On an extreme, as illustrated in Figure 3.4*a*, a sine

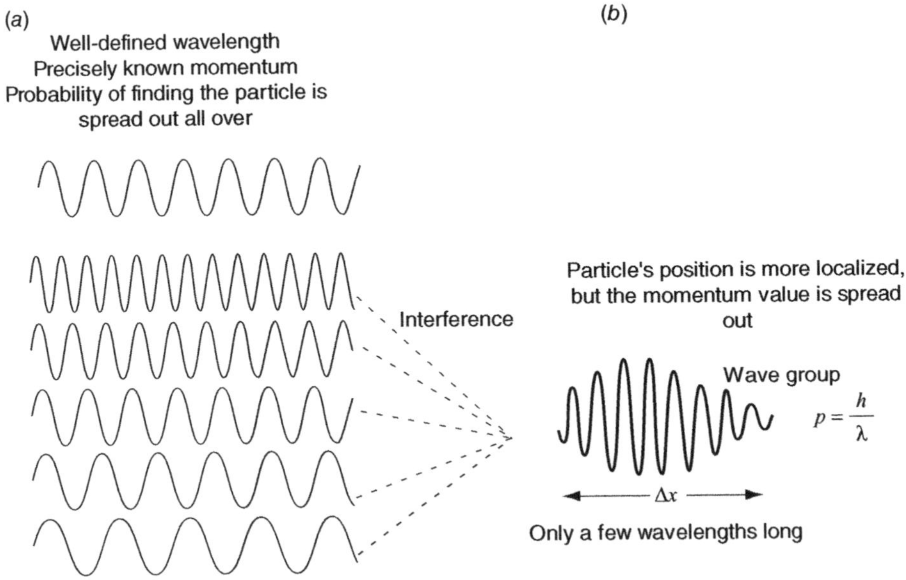

Figure 3.4. Momentum versus position of a particle as a wave.

wave of a well-defined wavelength λ has a precisely known momentum, $p = h/\lambda$. But where is the particle? The particle is somewhere in the wave, but we cannot say at a given time exactly where within the wave the particle can be found.

The reverse is also true. As illustrated in Figure 3.4*b*, interference of several waves of different wavelengths gives rise to a localized wave pattern, a wavegroup. The narrower the wavegroup, the more precisely the particle's position within the wavegroup can be specified. However, because now the wavegroup is composed of many different wavelengths and because it is only a few wavelengths long, it's wavelength cannot be known precisely. As a result, the probable momentum values $p = h/\lambda$, are spread out all over; that is, momentum is more uncertain.

So, you see, it is impossible to simultaneously know the exact position and exact momentum of a particle or object. Welcome to the uncertainty principle.

We can derive a more quantitative definition of the uncertainty principle. Recall that the de Broglie wavelength, λ, and the corresponding wavenumber, κ, can be written as:

$$\lambda = \frac{h}{p}$$

$$\kappa = \frac{2\pi}{\lambda}$$

which means

$$p = \frac{h\kappa}{2\pi}$$

which implies

$$\Delta p = \frac{h}{2\pi}\Delta\kappa$$

Combine this equation with the following equation from Section 3.6.1:

$$\Delta x \Delta\kappa \geq \frac{1}{2}$$

to obtain the uncertainty principle:

$$\Delta x \Delta p \geq \frac{h}{4\pi}$$

or

$$\Delta x \Delta p \geq \frac{\hbar}{2} \tag{3.15}$$

where $\hbar$, pronounced as *ech-bar*, is given by

$$\begin{aligned}\hbar &= \frac{h}{2\pi} \\ &= 1.054 \times 10^{-34} \text{ Js} \\ &= 6.58 \times 10^{-16} \text{ eV s}\end{aligned}$$

Werner Heisenberg (1901–1976)

Werner Heisenberg was born on December 5, 1901 in Wurzburg, Germany. He studied physics at the University of Munich, Germany. Along with others, he developed the theory of quantum mechanics based on the Matrix formulation. Heisenberg is best known for his uncertainty principle, popularly known as the Heisenberg's uncertainty principle, which he developed in 1927 while working at Bohr's Institute in Copenhagen.

Heisenberg received the Nobel Prize in physics in 1932 for his contributions to developing quantum mechanics. He spent his entire career in physics in Germany. Heisenberg died of cancer of the kidneys and gall bladder on February 1, 1976.

So, we have arrived at the principle of uncertainty from the wave picture of a particle. You can ask a question Can we realize the uncertainty principle from the particle behavior of a wave? The answer is yes. Here we go. Assume we want to measure the position and momentum of a particle. In order to measure any property of any object we interact with it. In this case, for example, we can shine light on the particle to "see" where it is and by knowing the incident and the scattered photons (of light), we can determine its momentum. However, you know from Compton scattering, the moment when the first photon hits the particle, the particle's momentum is going to change. This change in the particle momentum, Δp, cannot be precisely predicted, but it will be on the order of the momentum of the photon that hits the particle, which is given by the de Broglie relation, $p = h/\lambda$, where λ is the wavelength of the photon:

$$\Delta p \approx \frac{h}{\lambda}$$

This equation implies that the longer the wavelength of the probing photon, the smaller the uncertainty in the particle's momentum. However, the increase in wavelength also increases the uncertainty in the particle's position, Δx, which can be reasonably estimated to

$$\Delta x \geq \lambda / 2$$

which implies

$$\Delta x \Delta p \geq \frac{h}{2}$$
$$\geq \frac{\hbar}{2}$$

which is the uncertainty principle. However, this line of argument is sometimes stretched to jump to a misleading and wrong conclusion: It is possible for a particle to have precise position and momentum simultaneously; it is just the measurement process that introduces uncertainty. In other words, the wrong conclusion or claim is At any given instant, the precise position and momentum of a particle exists simultaneously; it is just that we cannot measure it. Yes, the measurement process does change the observables, however, the uncertainty in simultaneously measuring a pair of observables, such as position and momentum, is inherent in the nature of the particle; the wave–particle duality, an entity can act as both a wave and as a particle.

PROBLEM 3.5

A photoelectron is moving with a kinetic energy of 30 eV. At a given point in space and time, its momentum can be measured with 8.00% uncertainty. What is the minimum uncertainty with which its position could be measured?

Solution:

From Eq. 3.8:

$$p = \sqrt{2mE}$$

which implies

$$p = (2 \times 0.511\ \mathrm{MeV/c^2} \times 30\ \mathrm{eV})^{1/2}$$
$$= 5.54\ \mathrm{keV}$$

$$\Delta p = 0.0800 \times 5.54\ \mathrm{keV} = 0.443\ \mathrm{keV}$$

The uncertainty principle:

$$\Delta x \Delta p \geq \frac{\hbar}{2}$$

which implies

The minimum position of uncertainty is given by

$$\Delta x = \frac{\hbar}{2\Delta p} = \frac{6.58 \times 10^{-16}\ \mathrm{eV\ s}}{2 \times 0.443\ \mathrm{keV/c}}$$
$$= 0.223\ \mathrm{nm}$$

Position and momentum is one pair of observables whose members cannot be accurately measured simultaneously. There are other pairs of observable whose measurement is subject to the uncertainty principle.

3.6.3 Another Form of the Uncertainty Principle

Energy and time have the same relationship as momentum and space. Under this assumption, the uncertainty principle can be rewritten in terms of the uncertainty of the emitted energy, ΔE, during the time interval Δt:

$$\Delta E \times \Delta t \geq \frac{\hbar}{2}$$

The time interval Δt represents the uncertainty in time at which the energy measurement is made. Presented here is a crude derivation of this principle. Assume that the energy, E, that we are trying to measure in an interval Δt is in the form of waves of frequency f. It takes time to measure the frequency, which is the number of wave cycles divided by time. Assuming that the minimum uncertainty in counting the wave cycles is one wave cycle, we can write the corresponding frequency uncertainty as:

$$\Delta f \geq \frac{1}{\Delta t} \tag{3.15a}$$

From

$$E = hf$$

we can write

$$\Delta E = h\Delta f$$

which implies

$$\Delta f = \Delta E / h$$

By substituting this value of Δf in Eq. 3.15a, we obtain

$$\Delta E\Delta t \geq h$$

which implies

$$\Delta E\Delta t \geq \frac{\hbar}{2} \tag{3.16}$$

where ΔE is the uncertainty in measuring energy and Δt is the uncertainty in the time at which the measurement is made. It means it takes longer to accurately measure the frequency, and hence the energy.

There are two important things to note about these uncertainties. First, the uncertainties lie in the simultaneous measurements; that is, we can measure one quantity, say momentum, very precisely if we do not care about the position. Second, these uncertainties are not due to inadequate experimental apparatus, but are inherent due to the imprecise character of the physical quantities. Of course, the experimental errors will add to the overall uncertainties in the measurements. In a nutshell, the uncertainty principle has its roots in the wave–particle duality, discussed next.

3.7 WAVE-PARTICLE DUALITY OF NATURE

In Chapter 2, you learned how the quantum picture attributes particle properties to light, which is considered a wave in classical physics. In this chapter, you saw how the quantum picture attributes wave properties to material objects, such as electrons, which are considered particles in classical physics. So, what is going on here?

In other words, by now you might have begun struggling with two questions:

1. Is EM radiation really a wave or a particle? If it is made of photons, does it mean it is not a wave any longer?
2. Is an electron really a particle or a wave?

Here is the answer: EM radiations and electrons are both waves and particles or they are neither just waves nor just particles. In other words, the particle and wave pictures are complements of each other. Whether an entity, such as a radiation (light) or electron, will behave as a wave or a particle depends on what it is put through. For example, the particle picture explains Compton effect and the wave picture explains electron diffraction. In general, all matter and radiation, that is all entities, exhibits both particle- and wave-like behavior. This is called the wave–particle duality of nature.

STUDY CHECKPOINT 3.5

The waves of matter are

A. Pressure waves.
B. Electromagnetic waves.
C. Probability waves.
D. Mass waves.

Solution:

C Probability waves.

Scientists originally developed the concepts of particles and waves separately within the classical framework: Some things are particles and some other things are waves, and they stay that way under all methods of observing them. As a matter of fact, the classical framework does not recognize that the observation (measurement) can change the behavior of the observed and what you will measure is the changed behavior. Quantum mechanics exposes the inadequacy of each of these classical concepts in fully describing the behavior of physical entities or objects. Each particle and wave picture is therefore only part of the whole picture and can only explain certain effects. For example, in the classical picture, energy and momentum are the parameters that describe the dynamics of a particle, and frequency and wavelength are the defining parameters of a wave. Demonstrating the duality, quantum picture, as you have already seen, binds these particle parameters with the wave parameter through the following relationships:

$$E = hf$$

$$p = \frac{h}{\lambda}$$

So, physical entities are simultaneously waves and particles and they exhibit one or both of these behaviors depending on what we do to them. As you learned in Chapter 1, even in the classical picture, one thing common to both particles and waves is energy. Both matter and radiation carry energy. Wave–particle duality is an integral part of the quantum theory for all forms of energy.

The wave–particle duality is used in electron microscopy, where the small wavelengths associated with the electron are used to view objects of size much smaller than what is visible by using microscopes based on visible light.

3.8 LIVING IN THE QUANTUM WORLD

Quantum mechanical effects are prominent at the nanoscale level and are essentially used in studying and building nanostructures. For example, wave–particle duality is used in microscopy and diffraction in order to study the nanostructures.

> *Note*: Nanostructure is the structure that reveals details at the nanoscale, which is the scale of size typically in the range of 0.1–100 nm.

3.8.1 Seeing the Nanoworld with Electron Waves

We, human, are macrobeings and can only see the macroworld with our naked eyes. To see the micro (nano included)-world, we need the help of the science of the microworld: quantum physics. This is how we see things: Light waves (rays) reflected from an object brings the information about the object to our eyes. To see the object clearly, your eye must be able to distinguish the object from its surroundings, a characteristic called resolution. The approximate resolution of the human eye is ~0.05 mm; that is, an object of 0.1 mm size is the smallest object the naked eye can see from a distance of ~25 cm. This also means that the naked eye can see two objects separately (distinguish them from each other) if they are at a minimum of 0.1-mm apart, which is 100 μm), and 100,000 nm, far from the nanoscale. This minimum size that we can see can carry about 1 million atoms in it that we cannot see apart from each other.

We can see things the size of a few microns with the help of a light microscope, which uses light just like our eyes do, but magnifies the image of the object for us to see. That is how biologists see bacteria and cells in the lab. For example, you can see some details of the cells of an organism by using a light microscope, which are a few microns in size. The resolving power, P_R, of an optical (light) microscope is given by

$$P_R = \frac{\lambda}{2N_A}$$

where λ is the wavelength of light used in the microscope, and N_A is the numerical aperture of the microscope's objective lens. The numerical aperture depends on the refractive index n of the medium in which the lens is working and the angle θ of the maximum cone of light that can enter or exit the lens:

$$N_A = n \sin\theta$$

As a typical example, suppose the numerical aperture of the objective lens of your light microscope is 1.25. The average wavelength of white light is ~500 nm. What is the minimum size of a particle that you can see with this microscope as a clear and distinct particle?

$$\begin{aligned} P_R &= \frac{\lambda}{2N_A} \\ &= \frac{500\ \text{nm}}{2 \times 1.25} \\ &= 200\ \text{nm} \\ &= 0.200\ \mu\text{m} \end{aligned}$$

So, the average resolving power of this light microscope is ~0.200 μm, which is 200 nm. Therefore, on average, the smallest size of a particle that can be clearly and distinctively seen with this microscope is ~0.200 μm.

> *Alert*: Distinguish between two related but different concepts: magnification and resolution. Magnification is the factor by which the image of an object is enlarged as compared to the object size, whereas resolution is the smallest size of the object that can be distinguished from its surrounding material. Resolution is related to how much detail of the structure can be seen.

This means that the optical microscopes using visible light are not powerful enough to help you see the molecules, which are of nanosize. Therefore, to observe nanostructures, you need to use electron microscopes, which use a beam of electrons instead of a beam of photons (light), to generate an image of the target specimen. Because electrons typically have more momentum than photons of visible light, their de Broglie wavelength ($\lambda = E/p$) will be smaller, resulting in a greater spatial resolution.

In this chapter, you learned that an electron can behave like a wave with a wavelength given by the following formula:

$$\lambda = h / mv$$

The wavelength of an electron going through the electric potential V is given by

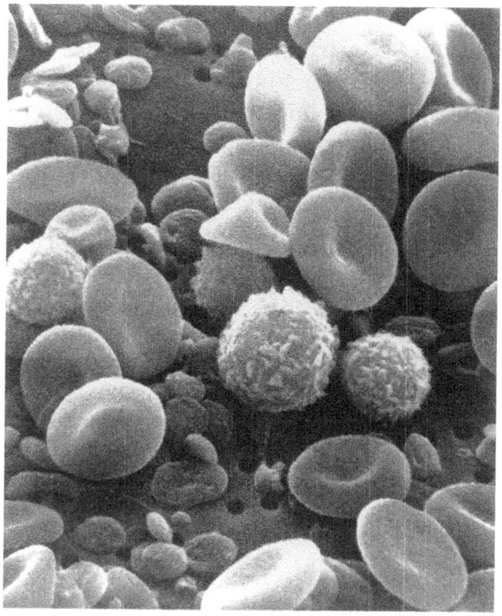

Figure 3.5. An SEM image of blood cells from normal circulating human blood, showing details. Courtesy of Bruce Wetzel and Harry Schaefer, National Cancer Institute.

$$\lambda = 1.23/(V)^{1/2} \text{ nm}$$

For $V = 60\,\text{kV}$

$$\begin{aligned}\lambda &= 1.23/(V)^{1/2} \text{ nm} \\ &= 0.005 \text{ nm}\end{aligned}$$

Compare this wavelength of 0.005 nm to 200 nm, the maximum theoretical resolution of an optical microscope, and you will realize that electrons can probe way deeper into the structure of matter to probe finer details. Figure 3.5 presents an image of normal circulating human blood taken by a scanning electron microscope (SEM). It shows the details, such as red blood cells, white blood cells, and small disc-shaped platelets. Platelets are tiny cells formed in bone marrow and are necessary for blood clotting.

The invention of SEM in 1981, based on the wave–particle duality, was a milestone in the development of the field called nanotechnology. In this chapter, you also explored how particles can exhibit diffraction, a wave property. The diffraction of particle waves is also used in studying nanostructures.

3.8.2 Seeing Nanostructures with the Diffraction of Particle Waves

Owing to particle–wave duality discussed in the previous sections, particles can exhibit wave properties, such as diffraction. Neutron and electron diffraction has been frequently used to study ordered structures, such as crystals, where you can get constructive and destructive interference. Electron diffraction is a technique used to study the structure of matter by hitting a sample with electrons and observing the resulting interference pattern. From this chapter, you already know that it was an electron diffraction experiment that confirmed the validity of the de Broglie hypothesis.

Currently, more precise instruments, such as transmission electron microscope (TEM) are used to exploit electron diffraction for studying structures at the nanoscale level. The basic idea is that the electrons transmitting through the specimen will be scattered by the constituent particles of the specimen, and the scattering will expose the internal pattern (or nanostructure) of the specimen, which will be captured in an image. The imaging process used in a TEM is described in the following:

1. A source, such as an electron gun, generates a beam of electrons.
2. By using EM field, the electron beam is accelerated and focused to the target specimen.
3. The incident electrons are scattered by the internal structure of the specimen as they transmit through the specimen.
4. The scattered electrons, after they transmit through the specimen, are focused by objective lens and amplified by a magnifying lens to generate an enlarged image of the specimen's structure.
5. The image strikes a phosphor screen, which generates light, which in turn enables a user to see the image.

As an example, Figure 3.6 presents an image taken with TEM.

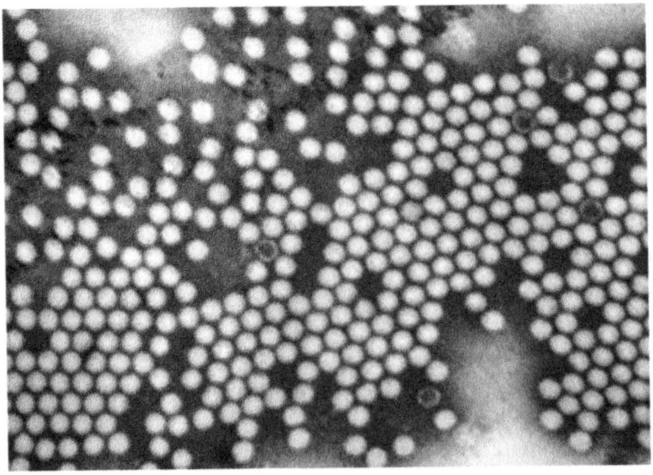

Figure 3.6. A TEM image of the polio virus. The polio virus is 30 nm in size.[1] Courtesy of Dr. Fred Murphy and Sylvia Whitefield, Centers for Disease Control and Prevention.

3.8.3 Using Atomic Waves to Navigate Your Way

In Chapter 1, you learned about the interference of light waves that produce a pattern of black and white strips: White strips correspond to the areas where the interfering waves line up and add to each other, whereas black strips correspond to the areas where the interfering waves cancel out. Interference pattern can be used to calculate the difference in the distance traveled by each of the interference waves. This difference of distance can be measured more precisely by using the waves of an atom because their wavelength is much smaller than the typical wavelength of light. The navigation systems based on atomic interferometry are being designed, built, and tested. Such systems facilitate navigation by measuring the rotation of the Earth and are intended to be used both for land navigation and in airplanes, and may improve or replace global positioning systems (GPS).

Atomic interferometry is also intended to be used for making accurate measurements of some physical constants, for example, those constants that represent the strength of natural forces, such as gravity and EM forces.

3.9 SUMMARY

The three big takeaway lessons from this chapter are the following:

1. All material objects have a wave associated with them, that is, they can exhibit wave-like properties, such as diffraction. This leads to the wave–particle duality of matter.
2. One implication of the wave particle duality is quantum confinement: A particle whose motion has been restricted in one or more spatial dimensions only has discrete values of energy.
3. According to Heisenberg's uncertainty principle, there is an unavoidable uncertainty in simultaneously measuring both variables of certain pairs of variables, such as position and momentum of a moving particle.

Classical physics treats waves and particles separate from each other; two different kinds of entities. A given entity is either a particle or a wave. For example, light and sound are waves, and a marble and a ball are particles (objects). This classical picture of physical reality perceived in the macroworld breaks down in the microscopic world in which the physical reality of the macroworld has its roots. Black-body radiation and the photoelectric effect helped to realize the particle nature of light that was considered a wave in classical physics. Arthur de Broglie realized that the reverse could be true as well: Particles, such as electrons, could behave as waves just like waves, such as light, could behave like particles. The de Broglie hypothesis led to the wave–particle duality of nature.

Light or EM radiation are waves because under certain circumstances it exhibits phenomena, such as diffraction and interference. Electromagnetic radiations are also particles because under other circumstances it exhibits the characteristics that we associate with particles. In the same fashion, particles, such as electrons, also behave as both particles and waves under suitable circumstances. This wave–particle duality erases the line between particles and waves drawn by classical physics.

Quantum confinement and the uncertainty principle, two salient features of quantum mechanics, have their roots in wave–particle duality. Quantum confinement means that a particle whose motion has been restricted in one or more spatial dimensions only has discrete values of energy. The uncertainty principle states that there is an unavoidable uncertainty in simultaneously measuring both variables of certain pairs of variables, such as position and momentum of a moving particle.

Electrons in an atom are an example of a confined particles. This confinement defines the structure of atoms explored next in Chapter 4.

3.10 ADDITIONAL PROBLEMS

Problem 3.6 In a photoelectric effect experiment, the kinetic energy of an electron is found to be 1.30 eV. What is the de Broglie wavelength of this electron?

Problem 3.7 The incident electron beam in an electron microscope has been accelerated by applying a voltage of 100 kV. What is the de Broglie wavelength of an electron in the beam?

Problem 3.8 Use de Broglie relations to prove that a confined particle cannot have zero energy.

Problem 3.9 Consider an electron in an atom of diameter 0.500 Å. Assume that the uncertainty in the position of an electron within the atom is on the order of its diameter.

A. Calculate the minimum uncertainty in the momentum of the electron.
B. Assuming the momentum of the electron is equal to the momentum uncertainty found in part A, calculate the kinetic energy of the electron.

Problem 3.10 The principle of uncertainty has its roots in the wave–particle duality built into the de Broglie hypothesis. Use the following de Broglie relations to present a crude derivation of the uncertainty principle:

$$f = \frac{E}{h} \tag{1}$$

$$\lambda = \frac{h}{p} \tag{2}$$

Problem 3.11 Consider a virus 10 nm in diameter confined to a shell of the same size. Assume that the average density of the virus is $1\,g/cm^{-3}$. What is the minimum speed of restlessness of the virus if the shape of the virus is spherical?

Problem 3.12 Use the uncertainty principle to calculate the maximum wavelength of the wave associated with the virus in Problem 3.11.

Problem 3.13 A bacterium called *Escherichia coli* is commonly found in the lower intestine of warm-blooded animals. Assume that a linear-shaped *E. coli* $2\,\mu m$ in size is confined in a hole of its size. Calculate the maximum wavelength associated with this bacterium.

ANATOMY OF AN ATOM 4

Wings are a constraint that makes it possible to fly.

Robert Bringhurst

Figure 4.0. Neon lights. The quantum reality lightening the classical world. Courtesy of www.photoeverywhwere.co.uk.

From Las Vegas to Tokyo, neon (Ne) lights and signs make cities, big and small, lively all over the world. The electrons in the Ne or other atoms in the Ne tube are excited to a higher energy state by electricity, and they emit light of a characteristic color when they drop back to the lower energy state. You will see in this chapter how quantum mechanics explains this electronic dance inside atoms.

Quantum Physics for Scientists and Technologists, by Paul Sanghera

In This Chapter

4.1 INTRODUCTION

Ancient Greek philosophers, as early as 400 BC, proposed that all matter was composed of some fundamental constituents, which they called atoms. Based on these ideas, an English chemist John Dalton presented his atomic theory of the chemical elements in 1808. The suggestions that atoms themselves were not fundamental (indivisible) came as early as 1820 from experiments revealing that atoms consist of smaller particles: the subatomic particles. By the late nineteenth century, scientists had discovered that all matter was made of (chemical) elements, which in turn were composed of atoms, and atoms might have structure too. But at that time they did not know much about the atom itself, such as its internal structure and dynamics. The study of the structure of an atom has been crucial for the development of chemistry because atomic structure plays a significant role in determining the properties of the chemical elements. Later, the subatomic particles that compose atoms were discovered to be electrons, protons, and neutrons. Various experiments also revealed the structure of the atom: An atom is a stable self-contained entity comprised of protons and neutrons in its nucleus, which is surrounded by a cloud of electrons that exactly match the number of protons in the nucleus.

Whereas classical physics failed to explain experimental observations, quantum theory, which was in its infancy at the time, thrived on these experiments by successfully explaining their findings.

So, the three underlying streams of thought in this chapter are the atomic theory of elements, the classical theory of atomic structure, and the quantum theory of atomic structure.

> *Note*: By now, the scientists know that protons and neutrons that make up the nucleus of an atom also have structures, and they are made up of even smaller particles called quarks. This level of the study of matter is important for the advancement of physics.

4.2 QUANTUM MECHANICS OF AN ATOM: THE BIG PICTURE

A physicist's probe of matter goes as deep as possible in the search for the fundamental constituents of matter: from the universe to an atom, to protons and neutrons, to quarks, and beyond. A biologist is interested in studing living matter to the details of a molecular level because that is where life begins its activities in terms of molecules of life, such as deoxyribonucleic acid (DNA), proteins, carbohydrates, and lipids. A chemist is interested in studying matter at the molecular and atomic levels.

All matter is made up of a finite number of elements. Each element consists of unique atoms that are different from any other element. These unique atoms exhibit (or reveal) their structure in the spectrum of light (electromagnetic, EM, radiation) that they emit and absorb. You learned about black-body radiation in Chapter 2. In addition to measuring the spectra of black-body radiation, experimental physicists at that time were also familiar with the spectra emitted by gases. Toward the end of the nineteenth century, it was discovered that when an atomic gas was heated by passing an electric current through it, it emited a spectrum of radiation corresponding to only certain specific wavelengths. Such a spectrum is called a line spectrum. Figure 4.1 presents such spectra for a few elements: hydrogen (H_2), helium (H_e), mercury (H_g), and neon (N_e).

It was also discovered that each element produced its own unique line spectrum. Classical physics had no explanation for this behavior. These line spectra were trying to tell us something about the structure of atoms, which was eventually understood through the quantum mechanical model of atomic structure. The journey to get there begins from Dalton's atomic theory and goes through the Thomson, the Rutherford, and the Bohr models.

Figure 4.1. Visible line spectra of some elements. Courtesy of Jan Homann, University of Pennsylvania.

4.3 DALTON'S ATOMIC THEORY

In a chemical reaction, one set of chemical substances is converted into another set of chemical substances. It turns out that these reactions occur at the atomic level, and Dalton's atomic theory accounts well for chemical reactions. Before Dalton proposed his atomic theory, chemists knew the following two principles:

1. *Law of Conservation of Mass.* Total mass involved in a chemical reaction is conserved; it is neither created nor destroyed. This means

   $$\text{Total mass of products} = \text{Total mass of reactants}$$

   In other words, it means that each atom and its type, which element it belongs to, remain intact. The atoms only change their bonds with other atoms: Old bonds are broken and new bonds are made.
2. *Law of Definite Proportion.* The percentage by mass of each element in a compound is always the same regardless of its source: It is part of the identity of the compound. For example, pure water always contains 11.1 g of H_2 and 88.9 g of O_2. It makes no difference if water comes from Furwala, a remote village in the Punjab state of India, or from Lake Tahoe located along the border between California and Nevada in the United States, as long as the water is pure. If the water is not pure, then chemically it is not water, in the strict sense of chemistry.

In 1808, John Dalton fused these two principles with the concept of atoms to propose the atomic theory of matter with the following main features:

1. Each element is made up of tiny particles called atoms.
2. All atoms that compose a given element are identical. For example, each atom in an element has the same mass as every other atom of the same element.
3. Atoms of a given element are different from atoms of all other elements. For example, atoms of two different elements have different masses.
4. To form chemical compounds, atoms of one element can combine with atoms of other elements in a definite ratio of whole numbers, so that the proportion of atoms from different elements in a compound is always the same.
5. Atoms are fundamental particles, that is, they are indivisible: They cannot be created or destroyed. For example, atoms of one element cannot be changed into atoms of a different element in a chemical reaction.

Dalton formulated his theory based on chemical reactions performed in the macroworld of the laboratory. Neither Dalton nor those scientists who followed him for a century after his work was published had direct experimental evidence for the existence of atoms. In light of Dalton's theory, atoms are neither created not destroyed in a chemical reaction; they are simply rearranged to break the existing bonds and build new bonds to create products. In order to understand the creation or destruction of arrangements or bonds, you need to know more about atoms. In the pursuit of knowing more, scientists discovered that atoms are not fundamental after all. That means they are composed of smaller particles, which is equivalent to saying that atoms have a structure.

4.4 THE STRUCTURE OF AN ATOM

During the 1820s, one feature of Dalton's atomic theory that atoms are fundamental particles was challenged by experiments suggesting that the atom consists of even smaller particles, today collectively referred to as the subatomic, elementary, or fundamental particles. The work of several scientists, such as Michael Faraday and William Crookes, led to the discovery of a negatively charged subatomic particle called an electron. The first step toward speculating on the structure of an atom came from the experimental discovery that atoms contain electrons. Furthermore, because atoms overall were neutral, it was obvious to speculate that if there were electrons (negatively charged particles) in the atom, there must be some positively charged matter in it as well to balance the negative charge and make the atom neutral.

Note: According to Dalton's atomic theory, all atoms of an element are identical. This was challenged more than 100 years after the theory was first suggested. It was realized that some atoms of an element have more mass than other atoms of the same element. Now, we know that all atoms of a particular element have the same number of protons, called the atomic number of the element. The mass difference between different atoms of a specific element is caused by a different number of neutrons. The total number of protons plus neutrons in an atom is called the mass number. Atoms with identical atomic numbers, but different mass numbers, are called isotopes of the same element. To complete the story, the number of electrons in an atom is equal to the number of protons.

In 1897, a British physicist J. J. Thomson, who had played a great role in the discovery of the electron, put together the properties of electrons, and by combining these properties with the known properties of atoms proposed the first model of atomic structure. This structure is illustrated in Figure 4.2. According to the Thomson model, an atom is a lump of positively charged matter with electrons embedded in it.

Thomson's model was challenged by the results of a series of experiments conducted under the supervision of Ernest Rutherford, a former student of

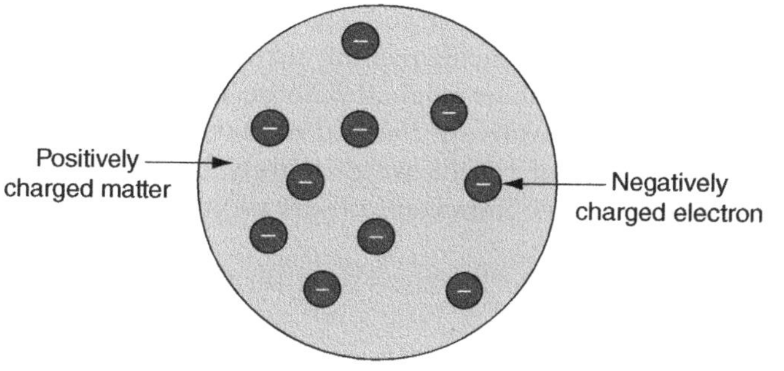

Figure 4.2. The Thomson model of atomic structure.

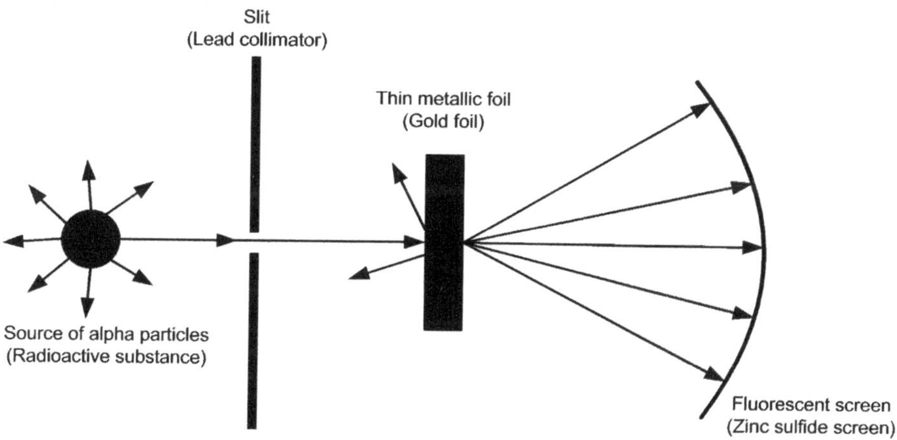

Figure 4.3. Illustration of the Rutherford scattering experiment.

none other than Thomson himself. The schematic diagram presented in Figure 4.3 illustrates the experimental setup for those experiments, which were conducted in 1911. A narrow beam of alpha particles is produced by using a source of alpha particles, such as a radioactive substance, and letting these particles pass through a hole, such as a slit in a lead collimator. This narrow beam of alpha particles is then directed at a thin metallic foil, such as a gold (Au) foil. The alpha particles are expected to be scattered by the foil, and a screen made of fluorescent material, such as zinc sulfide (ZnS) is set to receive the scattered particles. The screen would give off a visible flash of light when an alpha particle hits it.

> *Note*: An alpha particle is the nucleus of the He atom, that is, it is essentially the He atom that has lost both of its electrons. So, it carries a positive charge of $+2e$, that is, the positive charge that will balance the negative charge carried by two electrons.

Alpha particles are positively charged. Therefore, according to the Thomson model, most of the alpha particles passing through the metallic foil will be moderately deflected from their original path because they will experience a repulsive force from the positively charged matter everywhere in the atoms. However, the experimental results were entirely in disagreement with this expectation. There were two important observations:

1. Most of the alpha particles passed right through the foil with no change in their direction.
2. A few alpha particles were actually deflected, but with large angles; some of them even bounced straight back: a deflection of 180°.

These results challenged the Thomson model and led to the nuclear model of atomic structure with the following main features:

1. An atom contains an extremely small and dense portion called the nucleus.
2. All of the positive charge of the atom (and almost all of the mass) is concentrated in the nucleus.
3. The nucleus is surrounded by a much larger volume of empty space.
4. The empty space around the nucleus is thinly populated by a number of negatively charged electrons to balance out the positive charge of the nucleus.

Ernest Rutherford (1871–1937)

Ernest Rutherford, a student of J. J. Thomson, was born in Brightwater, New Zealand on August 30, 1871. In 1898, he was appointed the Chair of department of Physics at McGill University in Canada. Niels Bohr was one of his students. In physics, he is known for the Rutherford model of the atom, Rutherford scattering, and discovery of the proton. He won the 1908 Nobel Prize in chemistry for his investigations into the disintegration of the elements, and the chemistry of radioactive substances.

Rutherford died on October 19, 1937 at the age of 66 in Cambridge, England following an operation for an umbilical hernia.

Caution: The results from the Rutherford experiment demonstrated that an atom has a structure and thereby is not a fundamental particle. This means that atoms can be created and destroyed as opposed to Dalton's atomic theory. However, note that the processes that create or destroy atoms occur during nuclear reactions, which are a much higher energy processes than the atomic processes or chemical reactions. In chemical reactions, Dalton's theory still works pretty well.

The calculations based on the results of these experiments estimated that the diameter of empty space around the nucleus was 10,000–100,000 times greater than the diameter of the nucleus. When relatively massive particles, protons and neutrons, were discovered, scientists realized that these particles make up the nucleus of the atom. It was natural to wonder how the electrons might be behaving in the vast empty space around the nucleus. The most popularly held speculation was that they traveled in circular orbits around the nucleus, much like planets move in orbits around the sun.

Note: The proton was discovered by Rutherford in 1919 and the neutron by Chadwick in 1932.

Obviously, the roots for this kind of speculation were in classical physics.

4.5 THE CLASSICAL COLLAPSE OF AN ATOM

The Rutherford atomic model emerged from the Rutherford experiment, which revealed that a tiny but massive and positively charged nucleus is surrounded by electrons at relatively great distances. How would you explain this in classical physics? One thing was sure: Electrons had to move to escape protons, otherwise according to Coulomb's law of electrostatic force of attraction between opposite charges, they will crash into the nucleus. Scientists knew how planets revolving around the sun escape crashing into the sun due to gravitational force. So it was natural to think electrons revolved around the nucleus in circular or elliptical orbits.

In the spirit of classical physics, let us assume that an electron (negatively charged) in an atom under the influence of Coulomb (electrostatic) attraction toward the positively charged nucleus revolves in a circular orbit around the nucleus. Let us calculate some predictions of classical mechanics for this planetary model of the hydrogen atom, the simplest atom.

Note: It may be more reasonable to assume that the electron would revolve in an elliptical orbit, but it is simpler to work with the circular orbit. The basic conclusions from this classical approach will stay the same regardless of whether the orbits are assumed to be circular or elliptical.

Recall from Chapter 1 that the electrostatic force, F_e, applied on the electron with mass m_e and charge e by the nucleus with one proton is

$$F_e = \frac{1}{4\pi\varepsilon_0}\frac{e^2}{r^2}$$

where r is the radius of the electron orbit, ε_0 is a constant called the permittivity, and its value is given by

$$\varepsilon_0 = 8.854 \times 10^{-12}\,\frac{\text{C}^2}{\text{N m}^2}$$

or

$$\frac{1}{4\pi\varepsilon_0} = 8.988 \times 10^9\,\frac{\text{N m}^2}{\text{C}^2}$$

Assuming the electron is moving in the orbit with velocity v, according to Newton's laws of motion (see Chapter 1), the centripetal force F_c experienced by the electron is

$$F_c = \frac{m_e v^2}{r}$$

In order for the electron not to collapse into the nucleus, the centripetal and electrostatic forces must be equal to each other:

$$F_c = F_e$$

which implies

$$\frac{m_e v^2}{r} = \frac{1}{4\pi\varepsilon_0}\frac{e^2}{r^2}$$

From this equation, we can derive some important properties of the electron in the orbit, such as its velocity and energy:

$$v = \sqrt{\frac{1}{4\pi\varepsilon_0}}\frac{e}{\sqrt{m_e r}} \tag{4.1}$$

which leads to the kinetic energy:

$$E_k = \frac{1}{2}m_e v^2 = \frac{e^2}{8\pi\varepsilon_0 r}$$

As shown in Problem 4.1, the potential energy of the electron is

$$E_p = -\frac{e^2}{4\pi\varepsilon_0 r} \tag{4.1a}$$

Therefore, the total energy of this electron is

$$E_e = E_k + E_p = \frac{e^2}{8\pi\varepsilon_0 r} - \frac{e^2}{4\pi\varepsilon_0 r} = -\frac{e^2}{8\pi\varepsilon_0 r} \tag{4.2}$$

Note that the total energy of the electron is negative, which means the electron is bound to the nucleus.

PROBLEM 4.1

Calculate the potential energy of the electron revolving around a proton with radius r in a hydrogen atom.

Solution:

Potential energy exists due to the position of the electron at a distance r from the proton and is provided by the force of electrostatic attraction between the electron and the proton, which is given by Coulomb's law to be

$$F = k\frac{q_1 q_2}{r^2} = -k\frac{e^2}{r^2} = -\frac{1}{4\pi\varepsilon_0}\frac{e^2}{r^2}$$

Here the proton and electron contain an electric charge of $+e$ and $-e$, respectively. Potential energy can be looked upon here as the work needed to be done to bring an electron from the center of the atom to a distance r away from the center against this force and that will be

$$E_p = W = F \times r = -\frac{1}{4\pi\varepsilon_0}\frac{e^2}{r^2} \times r = -\frac{1}{4\pi\varepsilon_0}\frac{e^2}{r}$$

STUDY CHECKPOINT 4.1

What would happen to the electron if the total energy of the electron turned out to be positive?

Answer:

In this case, the electron would not follow a closed orbit around the nucleus. In other words, it would not be bound to the nucleus.

In these calculations, we have applied two major laws of classical physics, Newton's laws of motion and Coulomb's law of electrostatic attraction, we have come to the conclusion that atoms are stable, which of course is in agreement with experimental observations. However, the problem starts when we realize that there is yet another major law of classical physics that applies to this situation. The electron revolving in an orbit is in constant acceleration to stay in the orbit, and according to classical EM theory (by Maxwell), any charged particle in acceleration radiates EM energy, called *synchrotron radiation*. Therefore, the electron revolving around the nucleus must lose its energy to radiation and ultimately collapse into the nucleus. If and when it happens, it will eventually combine with the proton and all sorts of atomic and nuclear hell would break loose resulting in the collapse of the universe as we know it.

PROBLEM 4.2

It is experimentally known that an energy of 13.6 eV is required to knock out the electron of the hydrogen atom. Using the classical calculations performed in this section, answer the following questions:

A. What is the radius of the orbit in which the electron moves around the nucleus of the hydrogen atom in angstrons?
B. What is the velocity of the electron moving around the nucleus of the hydrogen atom?
C. Is the nonrelativistic treatment of the electron mass in the classical model justified?

Solution:

A.

$$E_e = 13.6\ \text{eV} = 13.6\ \text{eV} \times \frac{1.60 \times 10^{-19}\ \text{J}}{1\ \text{eV}} = 2.18 \times 10^{-18}\ \text{J}$$

From Eq. 4.2:

$$r = -\frac{e^2}{8\pi\varepsilon_0 E_e}$$

Therefore,

$$r = -\frac{(1.60 \times 10^{-19}\ \text{C})^2}{8\pi \times 8.85 \times 10^{-12}\ \text{N.}\frac{\text{m}^2}{\text{C}^2} \times 2.18 \times 10^{-18}\ \text{J}} = 5.28 \times 10^{-11}\ \text{m} = 0.528\ \text{Å}$$

B.
From Eq. 4.1

$$v = \frac{e}{\sqrt{4\pi\varepsilon_0 mr}} = \frac{1.60\times10^{-19}\ \text{C}}{\sqrt{4\pi\times8.85\times10^{-12}\ \text{N.}\frac{\text{m}^2}{\text{C}^2}\times9.11\times10^{-31}\ \text{kg}\times5.28\times10^{-11}\ \text{m}}}$$
$$= 2.19\times10^{6}\ \text{m/s}$$

C.
Speed of light:

$$c = 3.00\times10^{8}\ \text{m/s}$$

$$\frac{v}{c} = \frac{2.19\times10^{6}\ \text{m/s}}{3.00\times10^{8}\ \text{m/s}} = 7.30\times10^{-3}$$

The speed of the electron is only 0.73% of the speed of light. Therefore the non-relativistic treatment is justified.

An obvious theoretical way out of this classical disaster is to assume that the electron revolving around the nucleus in a fixed orbit somehow cannot lose any energy. This is exactly where the quantum atomic model starts.

4.6 THE QUANTUM RESCUE

In 1913, a Danish physicist Niels Bohr, fused the work of Planck and Einstein, discussed in Chapters 2 and 3, with the results from the Rutherford experiment to propose the model of the hydrogen atom (see Fig. 4.4), which is considered the first quantum model of atomic structure.

4.6.1 Bohr's Model

Rutherford's experiment suggested that an atom can be thought of as a microscopic solar system in which electrons orbit the nucleus. With this in the background, Bohr suggested the model with the following main features:

1. An atom consists of an extremely dense nucleus that contains all the positive charge of the atom and nearly all the mass.
2. Negatively charged electrons travel in circular orbits around the nucleus.
3. There are certain orbits around the nucleus, called stationary orbits, in which the revolving electrons do not radiate energy. Electrons moving in these orbits are in stationary states; they possess a constant amount of energy as long as they stay in their orbit, which means they will stay

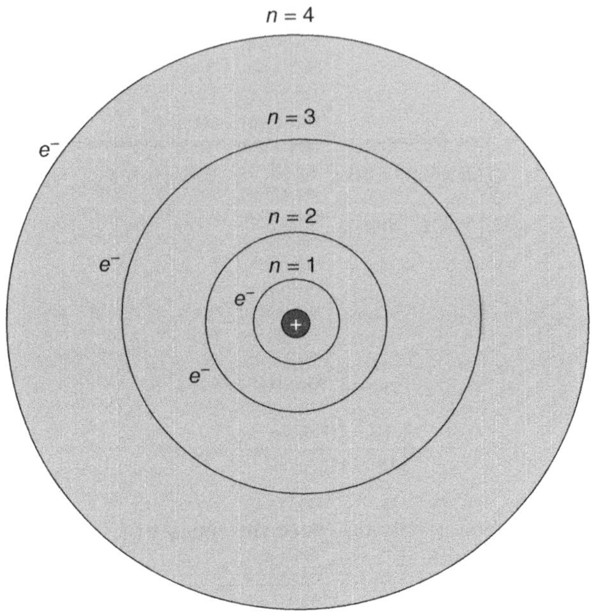

Figure 4.4. The schematic diagram of the Bohr model of an atomic structure.

in that state unless they are disturbed by an outside source, such as a photon.

4. An electron radiates energy when it jumps from one stationary state (orbit) at higher energy to another stationary state (orbit) at lower energy. They absorb energy in order to rise from a lower to a higher energy state.

You can see that features 1 and 2 are directly from the results of the Rutherford experiment, and features 3 and 4 are there as a defense against classical EM theory wrongfully predicting the collapse of the atom as discussed in Section 4.6. An electron in a stationary orbit does not radiate, that means, it has a constant energy. But what is that energy? To answer this question, Bohr leaned on the following two giant physicists of his time:

1. Planck, who had successfully explained black-body radiation, discussed in Chapter 2, by proposing that energy is proportional to frequency and is emitted in discrete amounts.
2. Einstein, who advanced Planck's idea to successfully explain the photoelectric effect, also discussed in Chapter 2.

You can see from Chapter 2 that the work of Einstein and Planck can be summarized in one mathematical formula:

$$E = nhf$$

Although Bohr did not believe at the time in the existence of a photon, he did use quantization in order to explain the spectrum of light emitted by hydrogen atoms. He proposed that the angular momentum of the electron in a stationary orbit around the nucleus is given by

$$mvr = n\hbar$$

where h is Planck's constant, used in previous chapters, and n is an integer.

This formula fixes radii of stationary orbits as:

$$r_n = \frac{n\hbar}{mv}$$

By substituting the value of v from Eq. 4.1:

$$r_n = \frac{n\hbar}{mv} = \frac{\sqrt{4\pi\varepsilon_0 r_n}\, n\hbar}{e\sqrt{m}}$$

which implies

$$r_n = \frac{4\pi\varepsilon_0 n^2\hbar^2}{me^2} \qquad (4.3)$$

The radius of the innermost orbit ($n = 1$), called the Bohr radius, is derived from Eq. 4.3 as:

$$r_1 = \frac{4\pi\varepsilon_0\hbar^2}{me^2} \qquad (4.4)$$

PROBLEM 4.3

Calculate the value of the Bohr radius.

Solution:

The Bohr radius is given by

$$r_1 = \frac{4\pi\varepsilon_0\hbar^2}{me^2}$$

Values of the constants involved in this equation are

$$4\pi\varepsilon_0 = 1/(8.988 \times 10^9 \text{ N.m}^2/\text{C}^2)$$

$$e = 1.602 \times 10^{-19} \text{ C}$$

$$m = 0.511 \text{ MeV/c}^2$$

Substituting these constants into the equation, we obtain

$$\text{Bohr radius} = 5.292 \times 10^{-11} \text{ m} = 0.53 \text{ Å}$$

STUDY CHECKPOINT 4.2

Rewrite the classical equation for the total energy of an electron moving in an orbit around the nucleus (Eq. 4.2) in light of Bohr's atomic model.

Solution:

The equation is

$$E_e = -\frac{e^2}{8\pi\varepsilon_0 r_n}$$

where

$$r_n = \frac{4\pi\varepsilon_0 n^2 \hbar^2}{me^2}$$

Niels Bohr (1885–1962)

Niels Bohr, a student of Ernest Rutherford, was born in Copenhagen, Denmark on October 7, 1885. In 1916, Niels Bohr became a professor at the University of Copenhagen. Werner Heisenberg, the inventor of uncertainty principle was one of his students. Bohr also worked at the top-secret Los Alamos laboratory in New Mexico, United States, on the Manhattan Project.

In 1922, Bohr received the Nobel Prize in physics for his investigation of the structure of atom. During the last five years of his life, Bohr took a keen interest in the developments in molecular biology. His latest thoughts on the problem of life appeared in his final (unfinished) article published after his death, titled "Light and life revisited." Bohr died on November 18, 1962 in Copenhagen of heart failure.

So, the Bohr model explains the results of the Rutherford experiment by proposing that electrons move around the nucleus of an atom in *stationary* orbits and they do not radiate or absorb any energy until they jump from one orbit to another. The Bohr model was also successful in explaining another problem that existed toward the end of nineteenth century, the line spectra of light emitted by different elements.

4.6.2 The Bohr Model Meets the Spectral Series

These days we know that light absorbed or transmitted by an element reveals significant information about the atomic structure of the element. When a gas of atoms is heated up, for example, by passing an electric current through it, it emits light. It was discovered that if you view this light through a narrow-slit aperture, for example, by using a spectrometer, the light would appear as a discrete set of lines of different color; that is, of different wavelengths or frequencies. As described earlier in this chapter, every element, when a sample of it in vapor phase is excited, displays a unique discrete set of lines, called the line spectrum.

It was discovered that the lines in the emission spectrum of an element falls into sets called the spectral series. The first of such a series was discovered by a Swiss school teacher J. J. Balmer in 1885, when he was studying the light spectrum emitted by hydrogen atoms. This spectrum is now known as the Balmer series. As shown in Figure 4.5, the Balmer series represent wavelengths of visible light emitted by hydrogen. Subsequently, other series were discovered that represent emission of light of specific wavelengths by hydrogen in infrared (IR) and ultraviolet (UV) regions. The electron transitions giving rise to these spectral lines according to the Bohr model are illustrated in Figure 4.6, which also presents a series that fall outside the visible light spectrum.

> *Note*: The spectral series discovered in the UV region is called the Lyman series and those discovered in the IR regions are called the Paschen, Brackett, and Pfund series.

By analyzing the data from experiments on different elements, J.R. Rydberg and Walter Ritz wrote down the following empirical formula to explain the data:

$$\frac{1}{\lambda} = \Re\left(\frac{1}{l^2} - \frac{1}{n^2}\right) \quad n > l \tag{4.5}$$

Figure 4.5. Balmer series observed for the hydrogen atom. Courtesy of Jan Homann, University of Pennsylvania.

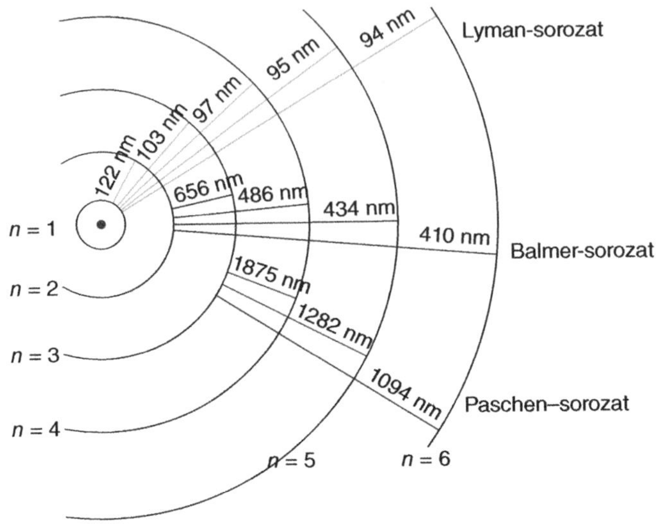

Figure 4.6. Electron transitions and their resulting wavelengths for hydrogen. Courtesy of S. Z. Dori.

where l and n are integers and $\mathfrak{R}$ is a constant called the Rydberg constant whose value stays the same for all series of an element, and varies slightly from element to element.

Note: The value of the Rydberg constant for hydrogen is $10.96776 \mu m^{-1}$, and approaches $10.97373 \mu m^{-1}$ for massive elements.

In formula 4.5, values of l represent different series, and values of n for the same value of l represent different lines in the same series. This general formula represented a different series, as shown in Table 4.1.

STUDY CHECKPOINT 4.3

You have discovered a spectral series that can be described by the Rydberg formula with $l = 7$. What will be the possible value of n to represent the lines in the series and why?

Answer:

$n = 8, 9, 10 \ldots$
Because in Rydberg formula, $n > l$

TABLE 4.1. Characteristics of the Spectral Series

Region	Series	l	Formula	n
Ultraviolet	Lyman	1	$\frac{1}{\lambda}=\Re\left(\frac{1}{1^2}-\frac{1}{n^2}\right)$	2, 3, 4, ...
Visible	Balmer	2	$\frac{1}{\lambda}=\Re\left(\frac{1}{2^2}-\frac{1}{n^2}\right)$	3, 4, 5, ...
Infrared	Paschen	3	$\frac{1}{\lambda}=\Re\left(\frac{1}{3^2}-\frac{1}{n^2}\right)$	4, 5, 6, ...
	Brackett	4	$\frac{1}{\lambda}=\Re\left(\frac{1}{4^2}-\frac{1}{n^2}\right)$	5, 6, 7, ...
	Pfund	5	$\frac{1}{\lambda}=\Re\left(\frac{1}{5^2}-\frac{1}{n^2}\right)$	6, 7, 8, ...

PROBLEM 4.4

Calculate the wavelength of the highest frequency line in the Balmer series.

Solution:

Consider the Rydberg formula:

$$\frac{1}{\lambda}=\Re\left(\frac{1}{l^2}-\frac{1}{n^2}\right)$$

For the Balmer series, $l = 2$

The highest frequency corresponds to the lowest wavelength, which means the highest value of l/λ, which in turn will be achieved for $n = \infty$.

Substituting these values of l and n in the Rydberg formula, we obtain

$$1/\lambda = \Re/4$$

which implies

$$\begin{aligned}\lambda &= 4/\Re = 4/(10.97\ \mu\text{m}^{-1})\\ &= 0.365\ \mu\text{m}\\ &= 365\ \text{nm}\end{aligned}$$

which is near the highest frequency end of the visible light spectrum.

The greatest success of the Bohr model has been to explain these spectral series. In other words, we can derive the Rydberg formula from the Bohr model, as we will demonstrate here.

Equation 4.2 can be rewritten to represent the energy of an electron in the nth orbit with radius r_n:

$$E_n = -\frac{e^2}{8\pi\varepsilon_0 r_n}$$

Substituting the value of r_n from Eq. 4.3 into this equation, we obtain

$$E_n = -\frac{m_e e^4}{8\varepsilon_0^2 h^2}\left(\frac{1}{n^2}\right) = -\frac{E_0}{n^2} \tag{4.6}$$

where E_0 is a constant and can be calculated to give the value of 13.6 eV. From Eq. 4.6:

$$E_0 = \frac{-m_e e^4}{8\varepsilon_0^2 h^2} = \frac{-m_e e^4}{32\pi^2\varepsilon_0^2\hbar^2} = -13.6\ \text{eV} \tag{4.6a}$$

STUDY CHECKPOINT 4.4

What is the energy of the electron in the ground state of a hydrogen atom?

Answer:

The ground state corresponds to $n = 1$

Substitute this value of n into Eq. 4.6:

$$E_1 = -E_0/1^2 = -13.6\ \text{eV}$$

Now assume that the electron in the nth orbit jumps to a lower orbit l. According to the Bohr model, the energy released by this electron will be the difference of the energy in the n and l orbit, which, by using Eq. 4.6, is

$$\begin{aligned} E &= E_n - E_l \\ &= E_0\left(\frac{1}{l^2} - \frac{1}{n^2}\right) \end{aligned} \tag{4.6b}$$

The emitted energy can be written in terms of its frequency and wavelength as:

$$E = hf = hc/\lambda$$

which implies

$$\frac{1}{\lambda}=\frac{E_0}{hc}\left(\frac{1}{l^2}-\frac{1}{n^2}\right) \tag{4.7}$$

If you compare Eq. 4.7 with Eq. 4.5, you realize that these are identical equations with:

$$\Re = E_0/hc = 13.6\ \mathrm{eV/hc} \tag{4.8}$$

Hence, the Rydberg formula is predicted by Bohr's model and it explains the spectral lines well. It turns out that this equation holds well within experimental uncertainty of measurements, and therefore the Bohr model explains the spectral lines successfully. So, according to the Bohr model, a spectral series results from the energies emitted by the electrons at higher energy levels (stationary orbits) jumping to a specific lower energy level. As illustrated in Figure 4.7, the Lyman series corresponds to the electrons jumping from higher orbits to the $n = 1$ orbit. The Balmer series results from electrons jumping from higher energy orbits to the $n = 2$ orbit, and so on.

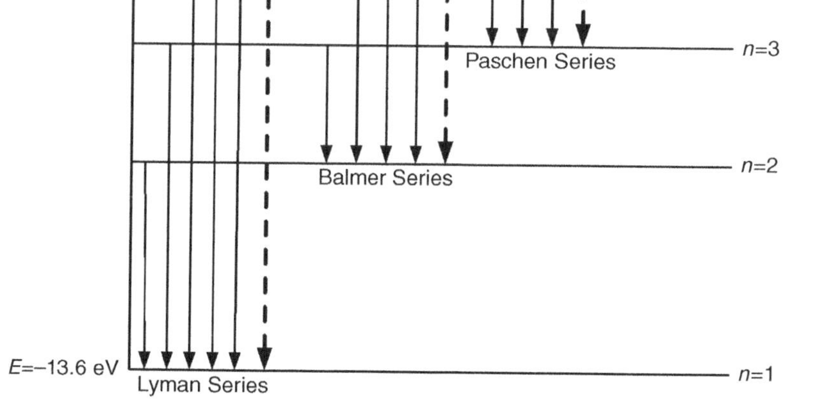

Figure 4.7. Spectral series originating from the jumping of electrons from higher stationary orbits to a specific stationary orbit.

PROBLEM 4.5

Calculate the value of the Rydberg constant $\mathfrak{R}$ predicted by the Bohr model and compare it with the measured value $10.96776\,\mu\text{m}^{-1}$.

Solution:

From Eq. 4.8:

$$\begin{aligned}\mathfrak{R} &= E_0/hc = 13.6\text{ eV}/hc\\ &= 13.6\text{ eV}/(4.136\ 10^{-15}\text{ eV s}\times 2.998\times 10^{8}\text{ m/s})\\ &= 1.096/(10^{-7}\text{ m})\\ &= 10.96/(10^{-6}\text{ m})\\ &= 10.96\,\mu\text{m}^{-1}\end{aligned}$$

which agrees with the given measured value.

As demonstrated by Eq. 4.3 and 4.6, both the radius of the electron in an orbit in the atom and its energy are quantized in the Bohr model due to the presence of the integer n. It means an electron in an atom can orbit around the nucleus only in certain specified distances from it, and therefore can have only certain specified energies called energy levels or states. Yes, the electron can never be found between these specified orbits, called stationary orbits or stationary states, but it can jump from one stationary state to another. This means an electron can absorb or emit a photon only if the value of the photon energy corresponds to the energy difference between these states.

Here is an example. Consider an electron that is in energy level n_i and drops to energy level n_f. You can calculate the frequency of the energy emitted by this electron during this jump by using Eq. 4.6b:

$$\begin{aligned}E &= hf = hc/\lambda\\ &= E_0\left(\frac{1}{n_f^2}-\frac{1}{n_i^2}\right)\end{aligned} \tag{4.9}$$

In a nutshell, the Bohr model had two great contributions to the development of the understanding of atomic structure:

1. It successfully explained the spectral series of the hydrogen atom.
2. It introduced the idea of quantized energy levels of an electron in an atom; that is, electrons in an atom exist only in certain discrete energy levels. Obviously, energy is involved in moving an electron from one orbit to another.

These quantized energy levels are represented by the principal quantum number n in the modern quantum mechanics of atomic structure.

STUDY CHECKPOINT 4.5

Discuss qualitatively how the Bohr atomic model is consistent with the principle of quantum confinement introduced in Chapter 3.

Solution:

According to the principle of quantum confinement, a confined particle can only have certain specific energies, that is, its energy is quantized. In the Bohr model, the energy of the electron in an atom is quantized. This is consistent with quantum confinement because the electron is confined within an atom, that is, it is bound to the nucleus of the atom.

However, there were problems with the Bohr model.

4.6.3 Limitations of the Bohr Model

Even though the Bohr model was an important first step in the way toward developing quantum theory, it had limitations, such as the following:

- Although it explained the spectra of the hydrogen atom well, it failed to explain much of the spectra of larger atoms.
- Even in the hydrogen atom, it wrongfully predicts the angular momentum of the electron in the ground state to be $\hbar$, whereas it should truly be zero.
- It also avoids addressing the problem of why the negatively charged electron will not collapse into the positively charged nucleus. It simply assumes it will not happen by postulating the stationary states, but does not explain why.

Nevertheless, by introducing the concept of discrete stationary states, the Bohr model played an important role in developing the more comprehensive models and quantum theory of atomic structure and behavior.

4.7 QUANTUM MECHANICS OF AN ATOMIC STRUCTURE

You learned in Chapter 3 how Louis de Broglie introduced the wave nature of matter, such as the wave nature of electrons. Between 1925 and 1928, Erwin Schrödinger developed wave mechanics for subatomic particles based on this idea. Out of these efforts emerged the quantum mechanical model of the atom that has been withstanding the test of the experiments to date. In this section, we briefly explore important elements of this model: principle energy levels, sublevels, and orbitals of electrons in an atom.

4.7.1 Principle Energy Levels

The Bohr model has made its way into the quantum mechanical model in the following two ways:

1. The stationary orbits in the Bohr model are identified as principal energy levels defined by the principal quantum number n with the same values as in Bohr model: 1, 2, 3. …
2. Just like the Bohr model, the energy of an electron in the atom depends on the principal energy level, that is, the principal quantum number n.

There is no theoretical (or mathematical) limit on the maximum number of principal levels, but $n = 7$ is the highest that can be occupied by the ground-state electrons in an atom of any known element. The maximum number of electrons allowed in a principal energy level with a principal quantum number n is given by $2n^2$. For example, the lowest energy level ($n = 1$), can hold up to two electrons, the next higher energy level ($n = 2$), can hold up to eight electrons, and so on.

PROBLEM 4.6

What is the maximum number of electrons that can be held in first seven energy levels?

The maximum number of electrons in first-seven energy levels

$$
\begin{aligned}
&= 2\times1^2 + 2\times2^2 + 2\times3^2 + 2\times4^2 + 2\times5^2 + 2\times6^2 + 2\times7^2 \\
&= 2 + 8 + 18 + 32 + 50 + 72 + 98 \\
&= 280
\end{aligned}
$$

You can ask two questions here

1. Where are the electrons in the principal energy level (orbit)?
2. What determines the maximum number of electrons in a given principal energy level?

The answer to the first question is in a given principal level, the electrons live at sublevels whose names proceed as follows: s, p, d, f, g, h. … A sublevel can only have so many electrons, and a principal energy level can have only so many sublevels, therefore there is a limit on the maximum number of electrons in a principal energy level.

4.7.2 Sublevels

As mentioned earlier, corresponding to each principal energy level, there are one or more sublevels with types: s, p, d, f, g, h… .

> *Note*: The names of the sublevels s, p, d, and f come from the formal terms used in spectroscopy, the study of an interaction between matter and EM radiation. Theoretically, there are sublevels beyond f, but the first-four sublevels (s, p, d, and f) are enough to describe the configuration of an electron in atoms of the known elements.

A sublevel is identified by its name prefixed with the value of the principal quantum number *n*. For example, sublevel s in the $n = 1$ energy level is called 1s, and sublevel s in the $n = 2$ level is called 2s. The maximum numbers of sublevels that can exist in a principal energy level *n* are equal to *n*. For example, for principal energy level $n = 1$, there is only one sublevel s called 1s, and for energy level n = 2, there are two sublevels s and p called 2s and 2p. For energy level $n = 3$, there are three sublevels s, p, and d called 3s, 3p, and 3d, whereas the $n = 4$ energy level contains four energy sublevels s, p, d, and f called 4s, 4p, 4d, and 4f. Table 4.2 presents more examples.

What do the sublevels look like and how many electrons can a sublevel hold? This is where orbitals come into the picture.

4.7.3 Electron Orbitals

As you learned in Chapter 3, it is not possible to precisely determine the velocity and position of an electron simultaneously. Because we cannot simultaneously determine the position and velocity of an electron in an atom, unlike the Bohr model, there are no clearly defined orbits. However, we can mathematically describe regions in an orbit where electrons can be found with high probability. These regions are called orbitals. There are two important things you need to remember about orbitals:

TABLE 4.2. Sublevels and Electrons in Principal Energy Levels

Principle Quantum Number *n*	Number of Sublevels	Names of Sublevels	Maximum Number of Electrons ($=2n^2$)
1	1	1s	2
2	2	2s, 2p	8
3	3	3s, 3p, 3d	18
4	4	4s, 4p, 4d, 4f	32
5	5	5s, 5p, 5d, 5f, 5g	50
6	6	6s, 6p, 6d, 6f, 6g, 6h	72
7	7	7s, 7p, 7d, 7f, 7g, 7h, 7i	98

1. Each kind of sublevel contains a certain number of orbitals: sublevel s has 1 orbital, p has three, d has five, and f has seven. This sequence of odd numbers continues through higher sublevels.
2. The maximum number of electron that can occupy an orbital is two.

Table 4.3. presents the number of orbitals and maximum number of electrons at different sublevels.

TABLE 4.3. Orbitals and Electrons in Sublevels

Sublevel	Number of Orbitals	Maximum Number of Electrons
s	1	$2 \times 1 = 2$
p	3	$2 \times 3 = 6$
d	5	$2 \times 5 = 10$
f	7	$2 \times 7 = 14$

STUDY CHECKPOINT 4.6

Question 1: What are the maximum numbers of electrons in sublevel h?

Answer 1:

Number of orbitals in g = 9
Number of orbitals in h = 11
Maximum number of electrons in h = $2 \times 11 = 22$

Question 2: How many orbitals are there in an orbit n = 3 in an atom?

Answer 2:

The $n = 3$ orbit contains s, p, and d sublevels.
Number of orbitals in sublevel s = 1
Number of orbitals in sublevel p = 3
Number of orbitals in sublevel d = 5
Total number of orbitals in $n = 3$:

$$1+3+5=9$$

Why are there different numbers of orbitals in different sublevels? For example, why is there only one s and 3 p orbitals? The reason is because different kinds of orbitals have different shapes, as shown in Figure 4.8. An s orbital is spherical with nucleus at the center. This shape can only have one

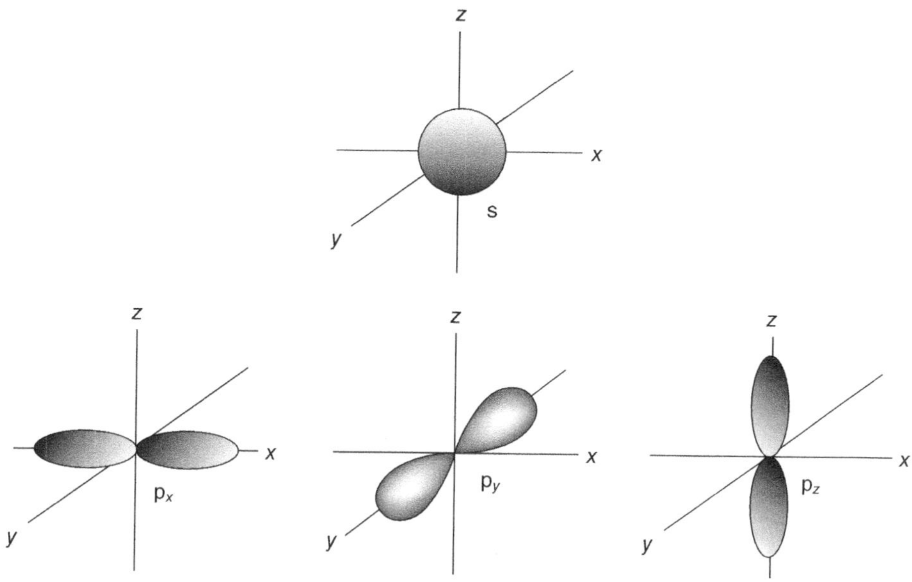

Figure 4.8. Illustrations of s and p orbitals.

orientation. A p orbital has two lobes meeting at the nucleus. This shape can have three independent orientations: along the x, y, and z dimensions. Therefore, there are three p orbitals. The d, f, and higher orbitals have a more complex geometry resulting in more degrees of freedom, and hence a higher number of orbitals.

Why does each orbital hold only two electrons at maximum? This can be understood in terms of the Pauli exclusion principle according to which no two electrons can occupy the same quantum state simultaneously. An orbital has only two quantum states, and therefore it can have only two electrons at maximum. We will revisit orbitals further on in this book, for example, in Chapter 7.

Although the quantization idea of the Bohr model made it to the quantum mechanical model of the atomic structure, the Bohr model as a whole is at the interface of the classical and quantum approach.

4.8 CLASSICAL PHYSICS OR QUANTUM PHYSICS: WHICH ONE IS THE TRUE PHYSICS?

As you have seen so far, classical physics breaks down at the atomic scale, and quantum physics explains well the phenomenon at this scale. You may be asking the following kinds of questions:

1. So, which physics is the true physics that explains physical phenomenon at all scales?
2. Is it true that classical physics is the physics of the macroscopic world and quantum physics is the physics of the microscopic world? If so, does it not mean that the correctness of physics is scale dependent, or that nature follows different laws at macroscopic and microscopic levels?

Starting with the second question, here is the short answer: A true theory of science should not be scale dependent, its approximations may. We already know that classical physics breaks down at the atomic scale, so it cannot be an accurate theory. If quantum physics has to be an accurate theory, it must work at the macroscopic level as well. We already know that classical physics works well at the macroscopic level, so if quantum physics is an accurate theory, it must yield the results of classical physics when applied to the macroscopic level. This requirement is called the correspondence principle.

Let us work out an example of an electron in a hydrogen atom, with both classical and quantum physics. According to classical physics, an electron revolving around the nucleus of this atom in an orbit of radius r and velocity v, emits radiation of frequency equal to its rotational frequency, f_{rot}:

$$\begin{aligned} f_{\text{rot}} &= 1/\text{time taken to complete one cycle} \\ &= \text{electron speed / orbit circumference} \end{aligned}$$

which implies

$$f_{\text{rot}} = \frac{v}{2\pi r}$$

Condition for the stability of the orbit assumed by Bohr is

$$mvr_n = n\hbar$$

which implies

$$v = \frac{n\hbar}{mr_n}$$

which implies

$$f_{\text{rot}} = \frac{v}{2\pi r_n} = \frac{n\hbar}{2\pi m r_n^2}$$

By substituting the value of r_n from Eq. 4.3 into this equation, we obtain

$$f_{\text{rot}} = \frac{n\hbar}{2\pi m}\frac{m^2e^4}{16\pi^2\varepsilon_0^2 n^4\hbar^4} = \frac{me^4}{32\pi^2\varepsilon_0^2\hbar^2}\frac{1}{\hbar\pi n^3} = \frac{E_0}{\pi\hbar}\frac{1}{n^3} = \frac{E_0}{h}\left(\frac{2}{n^3}\right) \tag{4.10}$$

We have used Eq. 4.6 to introduce E_0 into this equation.

According to quantum theory, which is based on the Bohr model, the electron then jumps from energy level n_f to energy level n_i, and emits a radiation of frequency f_n given by

$$f_n = \frac{E_0}{h}\left(\frac{1}{n_f^2} - \frac{1}{n_i^2}\right) \tag{4.11}$$

We have used Eq. 4.9 to write this equation.

Now, at first glance, the results represented by Eq. 4.10 and 4.11 are obviously inconsistent with each other. But consider the limit when n_f and n_i are very large numbers; this is true in the macroscopic world. In this limit, consider, for example,

$$n_f = n$$

$$n_i = n_f + 1 = n + 1$$

This means the electron jumps to the next lower level.

$$\frac{1}{n_f^2} - \frac{1}{n_i^2} = \frac{1}{n^2} - \frac{1}{(n+1)^2} = \frac{(n+1)^2 - n^2}{n^2(n+1)^2} = \frac{2n+1}{n^4(1+1/n)^2} = \frac{2+1/n}{n^3(1+1/n)^2} \to \frac{2}{n^3}$$

By substituting this result in Eq. 4.11, we obtain

$$f_n = \frac{E_0}{h}\left(\frac{2}{n^3}\right) \tag{4.12}$$

This equation is consistent with Eq. 4.10, which means in the classical limit of $n \to \infty$, the classical and quantum laws agree. This example explains that when the electron emits radiation, it is actually making a transition between adjacent energy levels of very large quantum numbers. Due to the large quantum number, the energy differences between the consecutive energy levels are infinitesimally small and go unmeasured. Therefore, the energy quantization is not visible.

Note: The requirement that quantum physics yields the same results as classical physics in the limit of the quantum number approaching infinity is called the correspondence principle.

In a nutshell, as this example demonstrates, quantum mechanics is an accurate theory and classical mechanics is an approximation of quantum mechanics. This approximation works well at the macroscopic level because the quantum effects are negligible at this scale. This bending makes sense because as we mentioned earlier in this book, the roots of physical reality perceived in the macroworld are in the microworld. This occurs because the macroworld is composed of the objects of the microworld.

Applications of a scientific theory in the real world are always the ultimate evidence of the correctness of the theory. Classical physics allows for easier computation and is an excellent approximation for the macroworld. For example, today when trajectories are computed for spacecrafts, the equations used are from Newtonian mechanics, and they work quite well.

4.9 LIVING IN THE QUANTUM WORLD

Butadiene (C_4H_6) is an important industrial chemical, which is used as a monomer in the production of useful polymer materials, such as synthetic rubber. It is used in the mixture with other materials, such as styrene or acrylonitrile, to produce polymers that are both strong and elastic. Styrene–butadiene rubber is the material most commonly used for the production of automobile tires.

We can put such chemicals to more and efficient use if we understand them at the molecular level. Here is an example of how the butadiene molecule can be understood in terms of quantum mechanics.

4.9.1 Free Electron Model for Pi Bonding

A pi (π) bond is a covalent bond in which two lobes of one electron orbital involved in the bond overlap with two lobes of the other electron orbital. However, only one of the orbital's nodal planes pass through both the nuclei of the involved electrons. Electrons in p orbitals usually participate in this kind of bonding. These bonds are usually very weak. As an example, consider a molecule of C_4H_6 with an alternate single and double bonds, as illustrated in Figure 4.9. In this molecule, there are a total of 22 valence electrons: 16 from four C atoms and 6 from 6 H atoms.

Recall that the electron configuration in a C atom is $1s^2$, $2s^2$, $2p^2$. The 18 electrons in the butadiene molecule occupy the lower nine orbitals involved

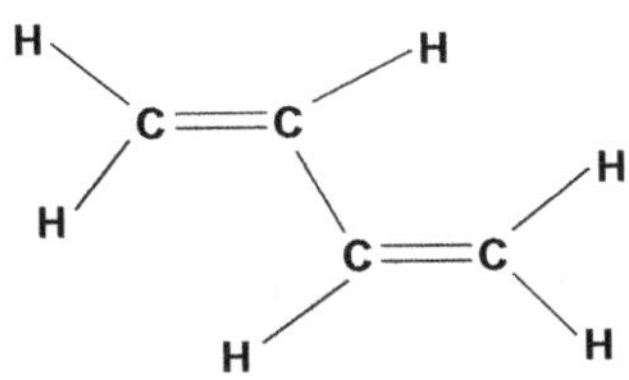

Figure 4.9. The butadiene molecule.

in bonds called σ-bonds: six C–H bonds and three C–C bonds. The four remaining valence electrons participate in π–bonds and can be considered essentially free due the weakness of π–bonds. We can model this situation as free electrons in a one-dimensional (1 D) box with an infinite potential outside the box. In other words, these electrons move freely over the length of the C bonds, as illustrated in Figure 4.10. Because the electron in the terminal C atoms will go beyond location of the terminal C, we add one-half of the bond length at each terminal C. So the box length will be equal to four times the average bond length: $4 \times 0.140\,\text{nm} = 5.60\,\text{nm}$.

The valence electrons in the C atom are in $n = 2$. Therefore, the lowest unoccupied energy state corresponds to $n = 3$. This means the lowest energy absorption band corresponds to the transition from $n = 2$ to $n = 3$. Under the particle in a box model, the energy of a π–electron is given by

$$E_n = n^2 \frac{h^2}{8\,ma^2} \tag{4.13}$$

Therefore, the first electron absorption band is predicted to be

$$\Delta E = E_3 - E_2 = (3^2 - 2^2)\frac{h^2}{8\,ma^2} \tag{4.14}$$

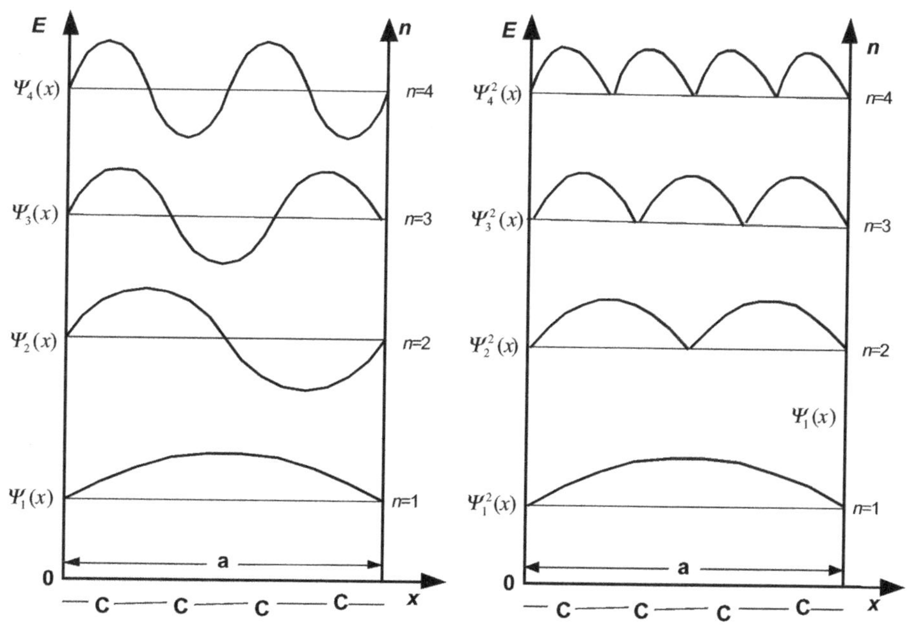

Figure 4.10. Wave functions and probability densities for butadiene in the particle in a box model. Read $\psi^2(x)$ as $|\psi(x)|^2$.

where m is the mass of the electron and $a = 5.60\,\text{nm}$, as estimated above. Substituting $\Delta E = hc/\lambda$ in Eq. 4.14, we obtain $\lambda = 207\,\text{nm}$, which compares well with the experimental value of $\lambda = 210\,\text{nm}$ for the first absorption band of the butadiene molecule.

4.10 SUMMARY

Elements absorb and emit light (EM radiation) in characteristic patterns called spectra, which reflects their atomic structure. Toward the end of the nineteenth century, emission of line spectra by atoms, such as hydrogen, was one of several problems scientists were struggling to explain. It was successfully explained by the Bohr model by proposing that negatively charged electrons in an atom revolve around the positively charged nucleus only in certain specific orbits called stationary orbits. The reason is because as long as the electron is in its stationary orbit it does not radiate any energy. An electron in an atom radiates energy only when it jumps from a higher stationary orbit to a lower stationary orbit, and absorbs energy only to jump from a lower stationary orbit to a higher stationary orbit. The stationary orbits of the Bohr model made their way into the modern quantum mechanical model of atomic structure, where they are called principal energy levels. These principal energy levels contain sublevels, such as s, p, d, and f, which contain orbitals. Orbitals are the regions in an atom where there is a high probability of finding electrons. An orbital can contain at most two electrons.

Quantum physics is the true physics, whereas classical physics is its approximation at the macroscopic level. It works because the quantum effects are relatively too small at the macroscopic level to be observed or to make a noticable difference to the measurements.

So far in this book, you have developed a good sense of how the quantum approach differs from the classical approach and how historically it was introduced by the need of explaining phenomena at the atomic scale that could not be explained with the classical approach. It is about time that we put all the cards of quantum physics on the table; that is, put together the main principles and formalism of quantum mechanics in one place to get an integrated picture. That is exactly what we do in Chapter 5.

4.11 ADDITIONAL PROBLEMS

Problem 4.7 Discuss how isotopes of an atom challenge Dalton's atomic theory.

Problem 4.8 Dalton proposed the atomic theory to explain the Law of Conservation of Mass and the Law of Definite Composition. Discuss how you can derive these two laws from Dalton's atomic theory.

Problem 4.9 A photon is absorbed by a hydrogen atom in its ground state. As a result of this absorption, the hydrogen atom goes into its excited state $n = 2$. What was the frequency of the photon?

Problem 4.10 Calculate the longest wavelength of a line in the Lyman series.

Problem 4.11 Find the shortest wavelength of a photon emitted by an electron in the hydrogen atom.

Problem 4.12 How many sublevels are there in an atom with $n = 4$? List all of them.

Problem 4.13 How many orbitals are there in an atom with $n = 4$? List all of them.

Problem 4.14 What is the largest number of electrons that can occupy the following orbitals:

A. 5s

B. 4p

C. 3d

Problem 4.15 The highest frequency spectral line in the Balmer series in the visible spectrum is observed when an electron from $n = 6$ jumps to $n = 2$. Think of an organism whose eyes can see a slightly wider region of the EM spectrum than the human eye can see, so that the organism could see a fifth line in the Balmer series visible emission spectrum. Calculate the wavelength corresponding to this fifth line.

PRINCIPLES AND FORMALISM OF QUANTUM MECHANICS 5

Mathematical science is in my opinion an indivisible whole, an organism whose vitality is conditioned upon the connection of its parts.

David Hilbert

Figure 5.0. Portrait of a Philosopher, 1915. Courtesy of Lyubov Popova. From Petrova, bassner, Burliuk-Holt, *Russian Futurism*.

Quantum Physics for Scientists and Technologists, by Paul Sanghera

Formalism is a process to capture the truth in a given field in terms of a set of rules without dealing with any truth that may exist outside the framework of these rules. Therefore, formalism lends itself well to disciplines that are based upon axiomatic systems, where an axiom, also called a postulate, is a proposition that is not proved or demonstrated, but is generally considered to be either self-evident or true. The word "axiom" comes from the Greek word, which means "to deem worthy". For example, from the viewpoint of a formalist, mathematics is no more than symbols with certain meanings, connected according to some logic and a few elementary rules. The truth of the postulate is taken for granted, and serves as a starting point for deducing and inferring other truths, which in science we call theoretical predictions. This is where the postulates are put to the test by testing the predictions with experiments. However, be warned that outside the fields of logic and mathematics, the term "axiom" is sometimes used loosely for any established principle of a field.

In this chapter, you will venture through the formalism of quantum mechanics.

In This Chapter

5.1 INTRODUCTION

In the previous chapters, you have learned how experimental observations toward the end of the nineteenth century and the beginning of the twentieth century suggested that some physical observables, such as energy and momentum, can only have certain permissible values. Those variables, also called canonical variables, are said to be quantized. To postulate this quantization was necessary to explain the fundamental behavior of matter and radiation exposed by these experiments. Obviously, postulates and models based on empirical studies, such as these, provide useful insight into the problem under study. However, a fundamental theory will go farther to explain the origin of these postulates and will have a greater predictive power. The origin of quantization was not really understood until the formal theory of quantum mechanics was developed during the 1920s. This theory would predict more than just the quantized values of certain observables. For example, you will see in the forthcoming chapters that other postulates and principles in addition to quantization, which was proposed in the previous chapters to explain experimental results, will also naturally arise from this theory.

To some, this chapter may come across as somewhat abstract. However, in the forthcoming chapters, we will be applying these principles to real-world physical examples and show how useful tools emerge from these principles that are applied to explain a wide spectrum of physical phenomena.

5.2 HERE COMES QUANTUM MECHANICS

The history of the development of quantum mechanics to explain the behavior of the microscopic world is a very good example of how we typically make progress in science. When we see some experimental observations that cannot be explained by existing scientific knowledge, we hypothesize. A sound hypothesis usually makes predictions. We test the predictions of the hypothesis and modify it to accommodate any new information learned from our experiment. If we gather more (and enough) related experimental information outside the realm of the existing hypothesis, we make another hypothesis. Eventually, over time and with enough experimental information, the big picture starts emerging. Subsequently, we try to fit all this information and interrelated hypotheses into what is called a theory and base it on a small number of assumptions. The beauty of such a theory lies in the fact that all the previously scattered hypotheses and principles can now be derived from the theory based on a very few assumptions. No, we do not stop even at this point. We keep on challenging the theory with new facts to see if they fit into the theory in order to keep building our confidence in it. One of the strengths of science is that theories are always under scrutiny and are up for revision or even a one-way trip to the junkyard. This underlines the difference between science and a belief system, such as a religion or astrology. Scientists thrive on discovering contradictions and do not shy away from saying goodbye to a theory if the facts can be better explained by a new theory. As a matter of fact, scientists on one hand develop theories and on the other hand spend their lives proving them wrong. This is the story of the theory of evolution in biology, of earth-centric versus heliocentric theories in cosmology, and of classical and quantum mechanics. The history of the development of quantum mechanics is a very good case study to explain how theories are developed. As you have seen in the previous chapters, many elements of quantum mechanics, such as Planck's quantization and the de Broglie particle wave, were first presented as hypotheses. After thorough testing and experimental verifications, these hypotheses became an integral part of quantum theory.

The theory of quantum mechanics, once put together, should not only explain the experimental results discussed earlier, it should also have a more predictive power. In fact, it should be able to derive all the possible information from a mechanical (or physical) system just like classical, Newtonian physics. This is because classical physics works well for macroscopic objects with speed substantially less than the speed of light. Under those conditions, predictions of quantum mechanics must be consistent with those of classical physics.

Note: In this book, as physicists often do, we divide the world into two parts: macroscopic, which generally means the world of objects that are large enough that we can see with our naked eyes; and microscopic, which means the world of objects that are so small we cannot see them with our naked eyes. For example, we need a microscope to see a cell of an organism; so the cells belong to the microworld. The scale of the microworld, starting from a few hundred micrometers, goes all the way down to nanometers and even smaller; as small as we can theoretically or experimentally go to examine matter.

The original formulation of such a quantum theory is generally credited to Werner Heisenberg and Erwin Schrödinger, who presented essentially the same theory in two different mathematical formulations. Heisenberg's formulation, developed in 1925 and 1926 by Heisenberg along with Max Born and Pascual Jordan, is in terms of matrices, whereas Schrödinger's formulation developed in 1926 by Erwin Schrödinger is in the form of wave mechanics. Paul Dirac, Schrödinger, and others later demonstrated that both formulations yield identical results even though they use a different mathematical structure. Since its mathematical framework is similar to the classical wave description in introductory physics that the students might already be familiar with, we use Schrödinger's approach to present quantum mechanics in this book. Also, this approach allows us to explain quantum mechanics by using simple calculus that most of the physical science students study as part of their lower division science courses.

So, in this book we consistently use the wave mechanics approach to quantum mechanics. The heart of wave mechanics is the wave function.

5.3 WAVE FUNCTION: THE BASIC BUILDING BLOCK OF QUANTUM MECHANICS

As you saw in Chapter 1, the wave function in an introductory physics course is introduced as a solution to a wave equation. In quantum mechanics, that is a good starting point. Also, you learned in the previous chapters about the de Broglie hypothesis, which states that particles can behave as waves. If there is a wave, there has to be a wave function to represent that wave. To accommodate the de Broglie hypothesis, the theory of quantum mechanics starts with the wave function.

5.3.1 It Is All about Information

The importance of information is not limited to information technology or computer science. Science, in general, is largely defined by the three-prong information activity: collect factual information, such as experimental observations and results on a topic, let an existing theory or a new hypothesis predict the information about the same topic, and compare the information from these two sources: (theory and experiment).

I just happened to learn quantum mechanics before I got into biology. Very early on, I heard in a biology class that deoxyribonucleic acid (DNA) is an information molecule: It contains coded within it the information on what proteins are to be made and how to make them. When I heard it, I immediately thought of the wave function in quantum mechanics as an analogy. In much the same way that DNA is an information molecule in a cell, a wave function is an information entity in quantum mechanics that contains all the information about the physical system it represents. In a nutshell, the wave function is the know-it-all information entity. However, it comes with a price tag: probability or uncertainty, as opposed to the certainty of classical mechanics.

5.3.2 Introducing Probability in Science

In the previous chapters, we introduced the wave–particle duality of nature. If a particle (or a physical system) can be represented by a wave, it must have a wave function associated with it. In quantum mechanics, such a wave function, Ψ, represents the probability of finding the particle, such as an electron, at a given point in space and time. To be precise, you will learn later that because the value of Ψ in general turns out to be a complex number (the number that includes the square root of −1), Ψ itself does not have any direct physical interpretation. However, the square of its absolute magnitude $|\Psi|^2$ evaluated at a specific point in space and time represents the probability density of experimentally finding the particle at that point in space and time. Of course, because the state of the particle can evolve over time due to changing conditions that the particle is subject to, the wave function will accordingly change over time as well.

> *Note*: Do not be confused between the terms particle and system. A system refers to one or many particles plus the conditions they are in, such as potential energy or confinement. Even a single free particle with zero potential energy can be referred to as a system. So, the specific meaning of system depends on the context.

Because the particle must be somewhere, the integration of $|\Psi|^2$ over total space and time must be equal to 1. This can be represented mathematically as

$$\int_{-\infty}^{\infty} |\Psi|^2 \, dV = 1 \tag{5.1}$$

where dV represents infinitesimal volume at time t in which the particle can be found with the probability of $|\Psi|^2$. This is also called the *normalization condition* for the wave function. The wave function that meets this condition is called a normalized wave function.

Note: It is not hard to imagine that the people who believe in the deterministic nature of scientific measurements should freak out at the probabilistic nature postulated by quantum mechanics. And they, including the scientists who developed quantum mechanics, did freak out. People still do. This is the aspect of quantum mechanics that disturbed Albert Einstein, a pioneer of quantum mechanics. Throughout his life he expressed his disbelief in the probabilistic nature of quantum mechanics, which lives on in his famous statement: God does not play dice with the universe. That said, there are all kinds of misinterpretations of the probabilistic nature of quantum mechanics out there against which you need to guard your understanding.

QUANTUM MECHANICS POSTULATE 1

The state of a physical system (or particle) is completely specified by an entity associated with it called a wave function, Ψ, that in general depends on the spatial coordinates of the system and time. The square modulus of this wave function is the probability density for finding the system with a specified set of values for the spatial and temporal coordinates.

The fact that the wave function contains all the information about the particle it represents should not sound like magic. Recall from classical mechanics reviewed in Chapter 1 that all the mechanical (or physical) properties of an object, such as velocity, momentum, and energy, are either a function of its location (position) in space and time or derivatives of it. Because the wave function is a function of space and time, all the dynamical information about the particle can be derived from it. Some properties only depend on spatial coordinates, others also depend on time. Accordingly, we have time-independent wave functions and time-dependent wave functions.

Equation 5.2 below is an example of a very simple time-independent wave function:

$$\Psi = A \sin \sqrt{\frac{2mE}{\hbar^2}} x \tag{5.2}$$

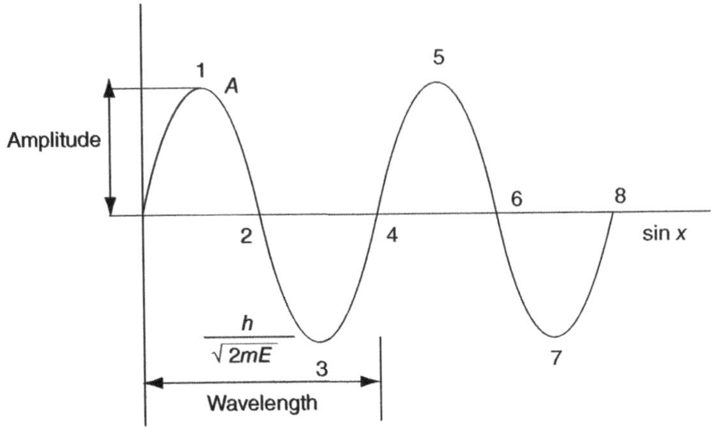

Figure 5.1. Sketch of the wave function $\Psi = A\sin\sqrt{\frac{2mE}{\hbar^2}}x$.

It has some visible information in it about the system in terms of energy E and position x. As sketched in Figure 5.1, this wave function represents a wave with a maximum amplitude A and the wavelength is given by

$$\lambda = \frac{h}{\sqrt{2mE}} \tag{5.3}$$

as shown in Problem 5.5.

Certain points on the plot of the wave function in Figure 5.1 are numbered for the purpose of Study Checkpoint 5.1.

STUDY CHECKPOINT 5.1

Consider the points numbered on the plot of the wave function presented in Figure 5.1. Suppose this wave function represents a particle.

A. At what points is the probability of finding the particle maximum?
B. At what points is the probability of finding the particle minimum?

Solution:

A. **1, 3, 5, 7.** Remember, the probability of finding the particle is represented by the positive magnitude of the amplitude of the wave function. At these points, the amplitude is maximum.
B. 2, 4, 6, 8.

This is just a toy function to elaborate the concept; a realistic function will look more like a wavegroup.

In a nutshell, a wave function has all the information about the system that it represents. How do we extract information from it? That is where operators come into the picture.

5.4 OPERATORS: THE INFORMATION EXTRACTORS

Continuing with the DNA analogy, enzymes operate on the DNA molecule to extract information from it and express the information in terms of messenger RNA (mRNA). Analogously, in quantum mechanics there are quantum mechanical operators that operate on the wave function to extract information from it and express it in terms of values sometimes called eigenvalues. There is nothing mystic about operators. You have all taken the derivative of a function, say df/dx. Think of d/dx as an operator operating on the function f.

QUANTUM MECHANICS POSTULATE 2

A measurable physical observable is represented by a quantum mechanical operator that operates on a wave function to predict the value of the measurements of the observable.

As an example, in quantum mechanics consider an operator:

$$\hat{A} = i\hbar \frac{\partial}{\partial x}$$

This operator operating on a function Ψ means:

$$\hat{A}\Psi = i\hbar \frac{\partial}{\partial x}\Psi$$

where $\hat{A}$ is just a shortcut notation for this operator and $i = \sqrt{-1}$. It is important to know that the operators mathematically extract information from the wave functions. But what does quantum mechanics says about how to relate this information with experimental information, that is, measurement?

5.5 PREDICTING THE MEASUREMENTS

A theory must predict what will be the possible results of a measurement. In the framework of quantum mechanics, the only possible results from the measurements of an observable A are the eigenvalues of the corresponding operator $\hat{A}$ according to the following equation:

$$\hat{A}\Psi_n = a_n\Psi_n$$

Any measurement of A can yield only one of the eigenvalues a_n. A system may have more than one possible eigenvalue with some probability attached to each of them. The wave function Ψ_n is called an eigenfunction. These *eigen* terms come from the German word *eigenwert,* which means proper or characteristic value. If there are a number of identical systems represented by a particular eigenfunction Ψ_a, the measurement of A made on all of these systems will yield the same values: a_k.

QUANTUM MECHANICS POSTULATE 3

The only possible results in the measurement of a physical observable A are the eigenvalues a_n of the corresponding operator $\hat{A}$ operating on the wave function Ψ_n in accordance with the following eignevalue equation:

$$A\Psi_n = a_n\Psi_n$$

where a_n is a real number; that is, it does not include $i \equiv \sqrt{-1}$, and n is an integer.

Postulate 3 captures one of the most salient features of quantum mechanics, quantization, discussed in previous chapters.

Caution: Quantum mechanics predicts quantization unlike classical mechanics, but it does not exclude a continuum of eigenvalues. You will see an example of continuum values in the case of a free particle, that is unbounded.

We will get back to eigenvalues in Section 5.7. For now, consider the following issue. When we perform an experiment, we get a definite value and not a probability. For example, the particle will either be found here or there, but not a little bit here and a little bit there. But the predictions of quantum mechanics are in terms of probabilities. So, then, how do we associate these predictions with experimental measurements? In other words, we need the formalism to express the quantum mechanical predictions so that they can be tested.

This is where the concept of expectation values comes into the picture. We will derive the expectation value from the wave function and compare it to experimental results. First, what is the expectation value?

5.5.1 Expectation Values

Expectation value of an observable is the expected value of an observable derived from the wave function and compared with the average of several experimental measurements of that observable. To explain the concept, let us consider a particle that is constrained to move along the x-axis. Assume that we are trying to determine the position of a particle along the x-axis. Let us say we made several experimental measurements, which found the particle N_1 times at x_1, N_2 times at x_2, N_3 times at x_3, and in general N_i times at x_i. The average of these measurements is given by

$$\bar{x} = \frac{N_1x_1 + N_2x_2 + N_3x_3 + \cdots}{N_1 + N_2 + N_3 + \cdots} = \frac{\sum_i N_i x_i}{\sum_i N_i} \tag{5.4}$$

PROBLEM 5.1

Calculate the average of the following set of measurements:

$x = 5, 7, 8, 7, 6, 6, 5, 9, 8, 6, 8, 5$

Solution:

$5 \times (3 \text{ times}) = 15$
$6 \times (3 \text{ times}) = 18$
$7 \times (2 \text{ times}) = 14$
$8 \times (3 \text{ times}) = 24$
$9 \times (1 \text{ time}) = 9$
$N = 3 + 3 + 2 + 3 + 1 = 12$

$$\bar{x} = \frac{\sum_i N_i x_i}{\sum_i N_i} = \frac{15+18+14+24+9}{12} = 6.67$$

Now, assume that the results used in Eq. 5.4 were not measurements, but given probabilities predicted by a theory. For example, it is predicted that if an experiment is performed, x_1 will appear N_1 times, x_2 will appear N_2 times, and so on. Theoretically, you can predict infinite values; that is, a whole continuous curve of values. In other words, we can move from this discrete approach to a continuous approach represented by the probability $P(x)$ that the particle will be found at a specific point x. Replace N_i in Eq. 5.4 with P, and the average value denoted by $\bar{x}$ with the expected value of x denoted by $<x>$ to obtain

$$<x> = \frac{\int_{-\infty}^{\infty} xP(x)dx}{\int_{-\infty}^{\infty} P(x)dx} \tag{5.5}$$

As mentioned earlier in this chapter, $|\Psi|^2$ is the probability density of finding the particle in the interval dx. This means

$$P(x)dx = |\Psi|^2\, dx \tag{5.6}$$

By using this information, Eq. 5.5 becomes:

$$\langle x \rangle = \frac{\int_{-\infty}^{\infty} x|\Psi|^2\, dx}{\int_{-\infty}^{\infty} |\Psi|^2\, dx} \tag{5.7}$$

If the wave function is normalized, the denominator on the right-hand side (rhs) of Eq. 5.7 will be equal to 1, and therefore the equation will reduce to

$$\langle x \rangle = \int_{-\infty}^{\infty} x|\Psi|^2\, dx \tag{5.8}$$

If Ψ is a complex function, then

$$|\Psi|^2 = \Psi\Psi^*$$

where Ψ^* is called the complex conjugate of Ψ and is obtained by replacing i in Ψ with $-i$.

STUDY CHECKPOINT 5.2

Calculate $|\Psi|^2$ if

$$\Psi = a + ib$$

where a and b are real numbers.

Solution:

$$\Psi = a + ib$$

$$\Psi^* = a - ib$$

$$|\Psi|^2 = \Psi\Psi^* = a^2 - (ib)^2 = a^2 + b^2$$

which is a real number.

So for a complex wave function, Eq. 5.8 can be rewritten as:

$$\langle x \rangle = \int_{-\infty}^{\infty} \Psi^* x \Psi dx \tag{5.9}$$

We have placed x in the middle of the two wave functions for reasons that will become apparent later. Equation 5.9 can be generalized to find the expectation value of any observable that is a function of x, say $f(x)$:

$$\langle f(x)\rangle = \int_{-\infty}^{\infty} \Psi^* f(x)\Psi dx \tag{5.10}$$

This equation is true even if f, and hence Ψ, varies with time because it will be determined at a specific time.

STUDY CHECKPOINT 5.3

Write down the equation for the expectation value of potential $V(X)$ if the corresponding wave function $\Psi(x)$ is normalized.

Solution:

$$\langle V(x)\rangle = \int_{-\infty}^{+\infty} \Psi^* V(x)\Psi dx$$

In a nutshell, the average value of an observable is the weighted average of several measurements, whereas the expectation value is the predicted value of this average by quantum theory. When we say that the wave function $\Psi(x)$ contains all the information about the system it represents, we mean that we can calculate the expectation values of various observables from the wave function. In addition to the location of the particle, there are two more important observables: momentum and energy. How exactly do we calculate expectation values for these variables? To answer this question, we need to use the concept of operators introduced earlier in this chapter.

5.5.2 Operators

We know from Sections 5.4 and 5.5 that operators act on a wave function to extract information for the system they represent. We also know that the predicted value of an observable is represented in terms of the expectation value. Combine these two pieces to realize that the expectation value can be calculated by using operators.

To be more specific, let us ask the question How do we determine the expectation values of momentum $p(x, t)$, and energy $E(x, t)$? The answer is simple: Rewrite Eq. 5.10 by replacing $f(x)$ with p and E to obtain

$$\langle p\rangle = \int_{-\infty}^{\infty} \Psi^*(x,t)\, p\, \Psi(x,t)dx \tag{5.11}$$

$$\langle E \rangle = \int_{-\infty}^{\infty} \Psi^*(x,t) E \Psi(x,t) dx \tag{5.12}$$

But, what really are the expressions for p and E that we can use in these equations? Let us try to derive them from the wave function itself. You know from wave mechanics that a wave moving in the $+x$ direction can be described by a wave function Ψ as follows:

$$\Psi(x,t) = A[\cos(kx - \omega t) + i\sin(kx - \omega t)] = Ae^{i(kx-\omega t)} \tag{5.13}$$

where A is the wave amplitude and is constant, and

$k = \frac{p}{\hbar}$ is the wavenumber

$\omega = \frac{E}{\hbar}$ is called the angular frequency.

Therefore,

$$\frac{\partial \Psi}{\partial x} = ikAe^{i(kx-\omega t)} = ik\Psi = \frac{ip}{\hbar}\Psi$$

which implies

$$p\Psi = -i\hbar \frac{\partial \Psi}{\partial x} \tag{5.14}$$

Consider the rhs of Eq. 5.14. In this expression, the quantity $-i\hbar(\partial/\partial x)$ is operating on the wave function in the sense that it is telling us to take the partial derivative of the function that follows it and multiply the result by $-i\hbar$. Such a quantity is called an operator. So, an operator is a theoretical entity that makes a mathematical operation upon a function and transforms it into another function. So, Eq. 5.14 defines the momentum operator, which can be written as:

$$\hat{p} = -i\hbar \frac{\partial}{\partial x} \tag{5.15}$$

As Problem 5.2 demonstrates, the energy operator can be derived as:

$$\hat{E} = i\hbar \frac{\partial}{\partial t} \tag{5.16}$$

PROBLEM 5.2

From the solution of the time-dependent equation, prove that the energy operator can be written as:

$$\hat{E} = i\hbar \frac{\partial}{\partial t}$$

Solution:

The time-dependent Schrödinger equation has the following solution, as given in Eq. 5.13:

$$\Psi = Ae^{i(kx-\omega t)}$$

where A is a constant, and

$$k = \frac{p}{\hbar}$$

$$\omega = \frac{E}{\hbar}$$

Therefore,

$$\frac{\partial \Psi}{\partial t} = -i\omega\Psi = \frac{-iE}{\hbar}\Psi$$

which implies

$$E\Psi = i\hbar \frac{\partial \Psi}{\partial t}$$

which implies

$$\hat{E} = i\hbar \frac{\partial}{\partial t}$$

An operator is a very general concept and is not limited to momentum and energy. Each physical observable f has an operator $\hat{f}$ associated with it. The expected value of this observable can be found very much like Eq. 5.10, but by replacing the observable's function with its operator:

$$\langle f \rangle = \int_{-\infty}^{\infty} \Psi^* \hat{f} \Psi dx \tag{5.17}$$

The generality of an operator allows quantum mechanics to handle any observable. This fact is helpful because there may be some observables at the microscopic scale that did not exist in classical mechanics.

QUANTUM MECHANICS POSTULATE 4

If the state of a system is described by the wave function Ψ, the expectation value of an observable f for this system can be computed by using the following equation:

$$\langle f \rangle = \int_{-\infty}^{\infty} \Psi^* \hat{f} \Psi dx$$

where $\hat{f}$ is the operator corresponding to the observable f, and Ψ is normalized.

Now, you understand why an operator should be in the middle of the expression (Eq. 5.17). The reason is it needs to operate on Ψ. Now that we have derived the definitions of momentum and energy operators, we can replace Eqs. 5.11 and 5.12 for the momentum and energy expectation values with the following equations:

$$\langle p \rangle = \int_{-\infty}^{\infty} \Psi^*(x,t)\, \hat{p}\, \Psi(x,t) dx = -i\hbar \int_{-\infty}^{\infty} \Psi^*(x,t) \frac{\partial \Psi(x,t)}{\partial x} dx \tag{5.18}$$

$$\langle E \rangle = \int_{-\infty}^{\infty} \Psi^*(x,t)\, \hat{E}\, \Psi(x,t) dx = i\hbar \int_{-\infty}^{\infty} \Psi^*(x,t) \frac{\partial \Psi(x,t)}{\partial t} dx \tag{5.19}$$

To summarize, a general formalism here is that an observable in quantum mechanics is represented by an operator that operates on the wave function to predict values of the observable in terms of expectation value. This generalization is needed to deal with the microscopic world because it may have observables (in addition to the observables realized in the macroworld, such as position, momentum, and energy) that do not have an analogue in the macroscopic world. Electron spin is an example of such an observable that we will discuss later.

So, in developing quantum mechanics, we have so far introduced two tools: the wave function and the operator. These tools can be used to build an equation; the equation that defines quantum mechanics.

5.6 PUT IT ALL INTO AN EQUATION

In this section, we will use the definitions of momentum and energy operators defined in this chapter, to reach an equation that defines quantum mechanics.

We begin with a fundamentally true physical equation about the total energy E of a particle both in classical and quantum mechanics:

$$E = T + V$$

where V is the potential energy of the particle and T is the kinetic energy that can be written in terms of momentum:

$$T = \frac{p^2}{2m}$$

where p is the momentum and m is the mass of the particle.

This leads to

$$E = \frac{p^2}{2m} + V$$

In this equation, replace energy E and momentum p with the corresponding operators to obtain

$$\hat{E} = \frac{\hat{p}^2}{2m} + V$$

Let the operators on the rhs and left-hand side (lhs) of the equation operate on the same wave function Ψ.

$$\hat{E}\Psi = \frac{\hat{p}^2}{2m}\Psi + V\Psi$$

Substituting for $\hat{p}$ *and* $\hat{E}$ from Eqs. 5.15 and 5.16, we obtain

$$i\hbar\frac{\partial\Psi}{\partial t} = -\frac{\hbar^2}{2m}\left(\frac{\partial^2\Psi}{\partial x^2}\right) + V\Psi \tag{5.20}$$

Voila, you have obtained the famous Schrödinger equation in its most general form: the time-dependent Schrödinger equation. This is the fundamental equation of quantum mechanics the same way as the equation of motion representing Newton's second law of motion is the fundamental equation of Newtonian mechanics, which you met in Chapter 1. However, do not consider our effort here as a derivation of the Schrödinger equation. We have simply shown that postulating momentum and energy operators in Eqs. 5.15 and 5.16 is equivalent to postulating the Schrödinger equation.

QUANTUM MECHANICS POSTULATE 5

The time development of the wave function, also called the state function, is governed by the following equation, called the Schrödinger equation:

$$i\hbar\frac{\partial\Psi}{\partial t}+\frac{\hbar^2}{2m}\left(\frac{\partial^2\Psi}{\partial x^2}\right)-V(x,t)\Psi(x)=0$$

The time-independent form of this equation in one dimension is

$$\frac{\hbar^2}{2m}\left(\frac{d^2\Psi}{dx^2}\right)+(E-V(x))\Psi(x)=0$$

The Schrödinger equation cannot be derived from fundamental principles because the equation itself is a fundamental principle. We hold this principle to be fundamentally true. Although like any other scientific theory or principle, it needs to be continuously tested against experimental results and observations. Since its invention, the Schrödinger equation has withstood experimental tests; more on this in the forthcoming chapters.

As mentioned earlier in this chapter, the power of a scientific theory is twofold: It explains the existing experimental observations and it makes falsifiable predictions. In this context, the Schrödinger equation will only be useful if it provides a mechanism to determine the values of physical observables, that is, measurable quantities, of a physical system, such as position, momentum, and energy. For that to happen, first you set up the Schrödinger equation for a given problem, and then you solve that equation to obtain the wave function. The wave function, the solution to the Schrödinger equation, contains all the information about the system that the equation represents.

As mentioned earlier, once we found an operator corresponding to an observable, we operate it on the wave function that contains the information about the system. This operation yields the predicted possible values of the measurements of that observable.

You may notice that Eq. 5.20, the Schrödinger equation, resembles many other differential equations in physics. The trick is to use this equation to formulate a physical problem in such a way that yields a physically acceptable solution, that is, the wave function. This is where eigenfunctions and eigenvalues come into the picture.

5.7 EIGENFUNCTIONS AND EIGENVALUES

Have you noticed that we are dealing with complex wave functions (involving i) and complex operators corresponding to physical observables. However, the measurements of physical observables are always real. For example, the measured energy value could be 2 MeV and never (2 + i3) MeV. Therefore we are interested in those wave functions that when operated upon by the operators yield real values. These wave functions and values are called eigenfunctions and eigenvalues. Generally speaking, for an eigenfunction Ψ and operator $\hat{A}$:

$$\hat{A}\Psi = a\Psi$$

where a is an eigenvalue if it is real.

Now, we establish a practical example of eigenvalues. By substituting $E\Psi$ for $i\hbar(\partial\Psi/\partial t)$ in Eq. 5.20 and rearranging the terms, we obtain the following:

$$\left(-\frac{\hbar^2}{2m}\frac{\partial^2}{\partial x^2} + V(x)\right)\Psi(x) = E\Psi(x) \tag{5.21}$$

Note: Although, for convenience, we have written the Schrödinger equation only in one spatial dimension, its form is very general and can be easily extended to three spatial dimensions as:

$$\left(-\frac{\hbar^2}{2m}\left(\frac{\partial^2}{\partial x^2} + \frac{\partial^2}{\partial y^2} + \frac{\partial^2}{\partial z^2}\right) + V(x, y, z)\right)\Psi(x, y, z) = E\Psi(x, y, z)$$

This is called the time-independent (or steady-state) Schrödinger equation. Equation 5.21 can be interpreted this way: An operator on the lhs of the equation operates on the wave function $\Psi(x)$ and yields the real value E (rhs). It turns out that the steady-state Schrödinger equation 5.21 can only be solved for certain values of E for real physical problems, as you will see in the upcoming chapters. These discrete values of E are called eigenvalues and the corresponding wave functions are called eigenfunctions.

Recognize that the operator in Eq. 5.21 represents the total energy ($E = T + V$) of the system. Therefore, Eq. 5.21 can be rewritten as:

$$\hat{H}\Psi_n(x) = E_n\Psi_n(x) \tag{5.22}$$

We say that E_n are the eigenvalues of the Hamiltonian operator, when the operator operates on the eigenfunctions Ψ_n of a system. We will not be making use of the Hamiltonian in this book; just use it to impress your friends.

STUDY CHECKPOINT 5.4

Show that the operator $-\frac{\hbar^2}{2m}\frac{\partial^2}{\partial x^2}+V(x)$ represents the total energy of the system.

Solution:

$$E=T+V=\frac{p^2}{2m}+V=\frac{1}{2m}(-i\hbar\frac{\partial}{\partial x})^2+V(x)=-\frac{\hbar^2}{2m}\frac{\partial^2}{\partial x^2}+V(x)$$

If you are getting overwhelmed with the new terminology, here is some good news. The terminology of operators, eigenfunctions, and eigenvalues is part of the framework and formalism that physicists use to study more complex systems. Because this book is more aimed on understanding the fundamentals of quantum mechanics with the aim of applying them to real-world applications with minimal jargon and complexity of formalisms, we will not be using this terminology much in this book. It is presented here mostly for the sake of completeness, and also to help the advanced reader to reach a more theoretical depth of quantum mechanics.

5.8 DOUBLE SLIT EXPERIMENT REVISITED

Look at the Schrödinger equation and convince yourself that it is linear in the wave function Ψ, which means that it has terms that contain the first power of Ψ and its derivative, and does not contain the higher powers of Ψ or its derivative. For this reason, a linear combination of solutions of this equation is also a solution. For example, if Ψ_1 and Ψ_2 are two solutions of the Schrödinger equation, so is Ψ, given by the following equation:

$$\Psi=\Psi_1+\Psi_2$$

This means that the wave represented by the Schrödinger equation obeys the principle of superposition just like any other waves, such as light and sound waves; and that the solutions to the Schrödinger equation are subject to interference effects. In other words, de Broglie waves associated with particles are subject to superposition and interference effects discussed in Sections 1.7 and 1.10.

With this in mind, let us revisit the double slit experiment covered in Section 1.10 for particles.

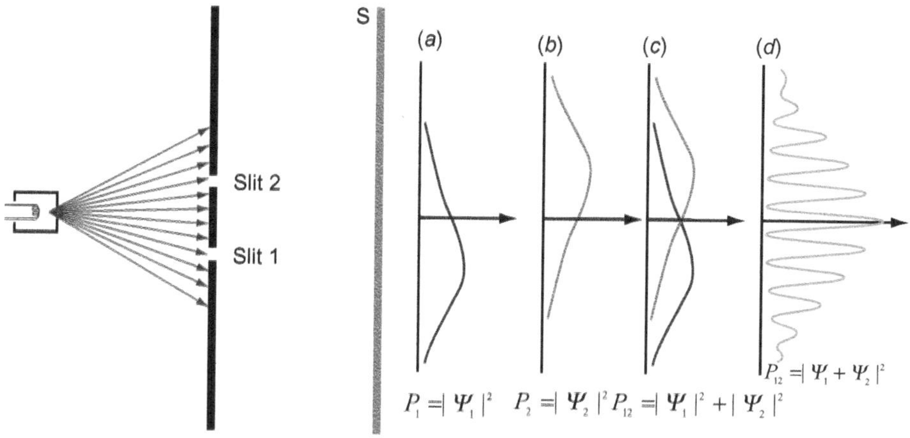

Figure 5.2. Double slit experiment for particles.

5.8.1 Double Slit Experiment for Particles

Recall from Section 1.10 that the double slit experiment performed by Thomas Young for light waves contributed to establishing that light is a wave. If particles behave as waves as well, then the double slit experiment performed on particles, such as electrons, should also produce the interference pattern, and the experiments show that it does.

A simple illustration of such an experiment is presented in Figure 5.2, which shows a pair of slits (also sometimes referred to as holes) through which beams of electrons of the same energy pass. The electrons are to be detected on the screen *S*. Consider the following three cases:

1. If slit 1 is open and slit 2 is closed, the intensity distribution of the electrons measured by the electron detector screen is shown in graph (*a*) and can be understood in terms of the probability density P_1:

$$P_1 = |\Psi_1|^2$$

   where P_1 is the probability that an electron will be detected at the screen and Ψ_1 is the wave function of the electron to pass through slit 1.
2. If slit 2 is open and slit 1 is closed, the intensity distribution of the electrons measured by the electron detector screen is shown in graph (*b*) and can be understood in terms of the probability density P_2:

$$P_2 = |\Psi_2|^2$$

   where P_2 is the probability that an electron will be detected at the screen and Ψ_2 is the wave function of the electron to pass through slit 2.

3. When both slits are opened simultaneously, classically thinking, we might expect the intensity distribution shown in graph (*c*) represented by the probability density *P*:

$$P = P_1 + P_2 = |\Psi_1|^2 + |\Psi_2|^2 \tag{5.23}$$

However, experiments show that the intensity distribution looks more like the one shown in graph (*d*), which supports the quantum mechanics superposition principle. Here the wave functions, and not the probabilities, add. In other words, when both slits are open, the intensity at the screen will be represented by the probability distribution *P*:

$$P = |\Psi|^2 = |\Psi_1 + \Psi_2|^2 = |\Psi_1|^2 + |\Psi_2|^2 + 2\Psi_1\Psi_2 \tag{5.24}$$

The cross-term $2\Psi_1\Psi_2$ is responsible for constructive and destructive interference.

> *Caution*: In quantum mechanics, it is the wave function and not the probability that adds for superposition.

Some of us with mathematical minds will focus on the wave equation that governs the motion of the electron and will readily accept Eq. 5.24. After all, in waves it is the wave functions (amplitudes) that add, not the probabilities; this is true even in classical physics. Others with more physical minds will argue for Eq. 5.23. There argument will sound like this: There are only two slits. A given electron can either pass through slit 1 or 2 (there is no third option) to reach the screen. We already have the information about the probability of an electron arriving at the screen, and therefore on how many electrons will arrive at the screen, when only slit 1 is open. We also have this information when only slit 2 is open. It just makes sense to add up the numbers when both slits are open.

So, we have a conflict here. It turns out that both sides are correct in their argument in one or another sense; all we need here is a conflict resolution presented by particle–wave duality.

5.8.2 Chasing the Electron

The argument in support of Eq. 5.23 is based on focusing on electrons passing through the slits. So, let us go to the bottom of this approach and ask a question Which slit does a given electron passes through? The supporters of Eq. 5.23 will say either slit 1 or 2. However, that is the classical way of looking at the situation. The answer from quantum mechanics is The electron has a $|\Psi_1|^2$ probability of passing through slit 1 and $|\Psi_2|^2$ probability of passing through slit 2. There is a third probability that does not exist in classical physics; the probability represented by the cross term in Eq. 5.24. The interpretation of this third probability may vary; one being that an electron passes through both slits. This interpretation will drive those people nuts who still have a classical picture of an electron in mind. Another response to this situation will be The very question of which hole the electron will pass through is a classical question, you cannot ask it in quantum mechanics.

To further appease the supporters of Eq. 5.23, let us get back to an experiment. Put a light source behind both slits to detect which slit the electrons come from. Also keep track at what positions the electrons arrive at on the screen. In such an experiment, we find that graph (c) is observed. The interference is lost. Now we have a real conflict here, an experimental conflict. Electrons follow Eq. 5.23 when we are watching them and they follow Eq. 5.24 when we leave them alone.

The resolution of this apparently conflicting situation lies in the wave particle duality. Electrons here travel as waves, but behave as particles when we try to observe them. We are in the midst of the situation that is much misunderstood, misinterpreted, and highly misused in the non-physics world. This is the situation from which people draw incorrect and silly conclusions, such as the following:

- An object can exist at two places simultaneously (an electron passing through two holes simultaneously as particles).
- An object only exists when you watch it (an electron behaves as a particle only when you try to observe it). This is stretched as far as to claim that for the universe to exist, we need a conscious observer, the so-called theory of a biocentric universe.

To dispel these myths right here where they are born, you need to understand this situation with due diligence. First, an electron still existed when we were not observing it; it just existed as a wave. But then why did it turn into a particle when we tried to observe it? To appreciate the answer to this question, first recall from previous chapters that an electron just like any other entity can exist both as a particle or as a wave. It was traveling as a wave, we disturbed it with our photons from the detecting light source to force it into the particle behavior. That changed behavior of electrons gave rise to the intensity pattern on the screen represented by graph (c). In other words, we changed the paths of electrons in the process of observing (measuring) them. We just happened to be conscious being; if the photons were there to hit the electron by some other means even in our absence, the result will be the same. So in such a situation, when you think of an observer, think of the entity, such as a photon, that disturbs the entity being observed, think of the environment that disturbs the entity. There are particles in our body, such as protons, electrons, and ions, which are constantly being observed and measured by their environment and they deal with their environment through wave–particle duality to keep us alive. This is the quantum reality of our lives.

To take advantage of the situation we are exploring, let me introduce two more concepts that you may run into out there: collapse of the wave function and quantum decoherence. When both slits are open, our electron is traveling as a wave and can be represented by the superposition of two wave functions:

$$\Psi = a\Psi_1 + b\Psi_2 \tag{5.25}$$

When we tried to observe the electron to find out which hole it came from, we reduced the superposition of states to a single state. This is called collapse of the wave function from a superposition of several states (eigenstates) to a single one of these states. This happened due to the interaction of the electron with the observer, the photon in this case. Quantum decoherence is any process in which systems interact with their environment to give the appearance of the wave function collapse and thereby justify the classical intuition.

5.9 THE QUANTUM REALITY

At the heart of quantum mechanics are three concepts: superposition of multiple states, wave particle duality, and the uncertainty principle. Sometimes these three concepts together are referred to as quantum reality.

Superposition of states arises from the fact that we cannot precisely predict the state of a particle, for example, the particle will be found at location x with momentum p at time t. As discussed earlier, we can only make statements like this one: There is a probability a_1 that the particle will be found in state Ψ_1, a probability a_2 that it will be found in state Ψ_2, and so on. This can be expressed in terms of the following equation:

$$\Psi = c_1\Psi_1 + c_2\Psi_2 + c_3\Psi_3 + \ldots = \sum_n c_n\Psi_n \tag{5.26}$$

where

$$a_n = |c_n|^2$$

represents the probability that the system will be found in state Ψ_n. The coefficient c_n, in general, is a complex number and the probability a_n, of course, is a real number. Here is what Eq. 5.26 means: At any given instant in time, the wave function Ψ of a particle (or an isolated system) can be expressed as a linear superposition of a complete orthonormal set of wave functions Ψ_n. Orthonormal means that the total probability of finding the particles in various states is one, which makes sense. Quantum superposition is a very fundamental concept of quantum mechanics, which may refer to the superposition of position (location) states, momentum states, energy states, or basically any property of a quantum system.

QUANTUM MECHANICS POSTULATE 6

At any given instant in time, the wave function Ψ of a particle (or an isolated system) can be expressed as a linear superposition of a complete orthonormal set of wave functions $\{\Psi_n\}$:

$$\Psi = c_1\Psi_1 + c_2\Psi_2 + c_3\Psi_3 + \ldots = \sum_n c_n\Psi_n \tag{5.26}$$

where

$$a_n = |c_n|^2$$

represents the probability that the system will be found in state Ψ_n.

The linear superposition of states is the prediction of quantum mechanics. However, when a specific measurement is made on a property of the system, it will have a definite value, that is, it will be found in one state and not in many states. This is called the collapse of the wave function. For example, quantum mechanics predicts that at any given instant of time the momentum of a system is the superposition of various momentum states. However, if you make a momentum measurement, the result will be a specific value for the momentum. Because you have precisely measured the momentum of the system, the uncertainty principle kicks in to spread its position in space like a wave, and hence wave–particle duality. Similarly, if we measure the position of a system, the uncertainty principle forces the momentum to spread out like a wave in terms of the superposition of alternative momentum states.

In a nutshell, all matter and radiation exhibit a mixture of two characters: wave and particle; and these two complementary characters are linked by the uncertainty principle.

STUDY CHECKPOINT 5.5

Chemistry and some biology students should be familiar with the two possible (equivalent) Lewis structures, called resonance structures, shown in Figure 5.3 1*a* and 2*a*. Each structure shows three C–C single bonds and three C=C double bonds at different alternate places. Classical wisdom will predict that both structures 1*a* and 2*a* will exist, and that the three single bonds in the molecule will be longer than the three double bonds. However, experiments show that all six C–C bonds in a benzene (C_6H_6) molecule are of equal length, 1.40 Å, which is inbetween the typical length of a C–C single bond (1.54 Å) and the C=C double bond (1.34 Å). In other words, structures 1*a* and 2*a* (or 1*b* and 2*b*) in Figure 5.3 do not exist. What exists is structure 3*b*. Can you explain it, qualitatively, by using quantum mechanics?

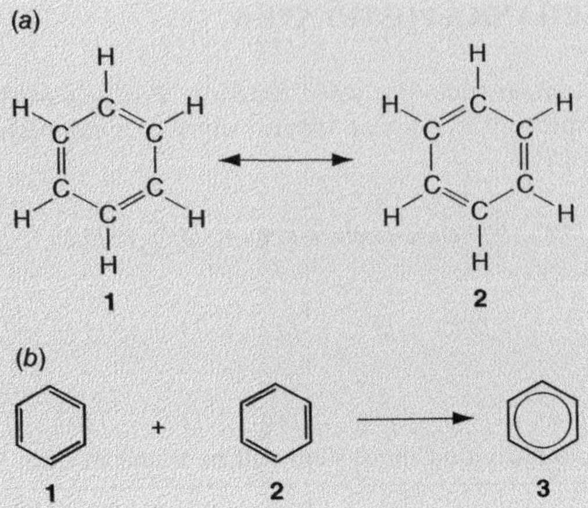

Figure 5.3. Resonance structures of the benzene molecule.

Solution:

This experimental observation can be interpreted as evidence of quantum superposition of states. The benzene molecule exists in the form of 3*b* as a quantum superposition of two states 1*b* and 2*b*.

5.10 LIVING IN THE QUANTUM WORLD

In this chapter, you studied the basic postulates of the theory of quantum mechanics. So what?, you say, What has it to do with the real world, for example, with biology, the science of life? The answer is In principle, everything.

Living organisms are made up of exactly the same fundamental material as inanimate objects. The basic unit of life is a cell, which is made of four types of molecules [carbohydrates, proteins, lipids, nucleic acid (DNA and ribonucleic acid RNA)] which in turn are made from the same set of elements and fundamental particles as any nonliving material. In addition to this, there are fundamental particles directly involved in helping these macromolecules to perform their job. We will elaborate on this point by exploring the actions of a special type of proteins called enzymes.

Enzymes, which catalyze biological reactions inside cells, are present in every living organism, and may continue to function long after the organism is medically (technically) dead. The reactions that enzymes catalyze and will not occur properly without their help are essential to the life process. As you are enjoying reading this book, right at this moment, hundreds of thousands of reactions are taking place in the cells of your body to keep you alive and well. While your central nervous system is busy processing the information

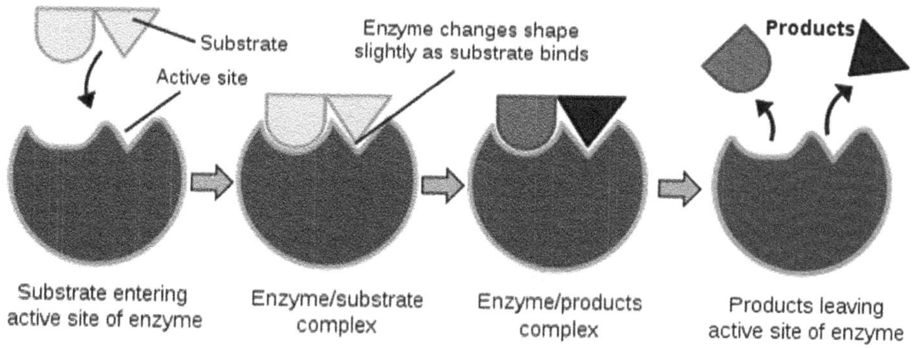

Figure 5.4. An illustration of the role of an enzyme in a biochemical reaction. Courtesy of Dr. Tim Vickeis.

gathered from this page, your digestive system is processing your last meal. Your immune system is busy dealing with many particles, such as germs, which have made their way into your body; your heart is of course pumping; and many new cells are being made, and many old cells are being replaced as other internal systems carry out a multitude of intricate processes. All of these functions are made possible and performed effectively with help from a large number of enzymes.

As illustrated in Figure 5.4, enzymes perform their job by binding their spot called an active site to the reactant molecule called a substrate. Internally, all enzymes perform their actions by directing the movement of protons and electrons (fundamental particles) along the paths that lead to the product of the reaction.

One can argue about the following: At what level below the cell is the border where classical physics fails and quantum mechanics must be applied to understand life? However, there should be no doubt that to understand the role of fundamental particles, such as protons and electrons, directly working right inside the living cell, we need quantum mechanics. You can imagine that the cell (or the cell environment) on one hand provides choices (multiple states) to these particles, which exist in general as a superposition of multiple states. On the other hand, cells also acts as an observer to force these particles at a certain moment to collapse their wave function into one of the possible states. Here is an example to illustrate this point in the context of enzymes. As mentioned earlier, all enzymatic reactions involve the motion of single electrons and protons; the enzymes perform their function by directing these particles along specific paths in the cell. For instance, the enzyme lactate dehydrogenase (LDH) is an enzyme that catalyzes the conversion of a molecule called lactate to another molecule called pyruvate; which is an important step in energy production in cells. Many different types of cells in the body contain this enzyme; some of the organs relatively rich in LDH are the heart, kidney, liver, and muscle. The LDH catalyzes a reaction by directing a proton along a path between a histidine amino acid on the protein (enzyme) and the lactate

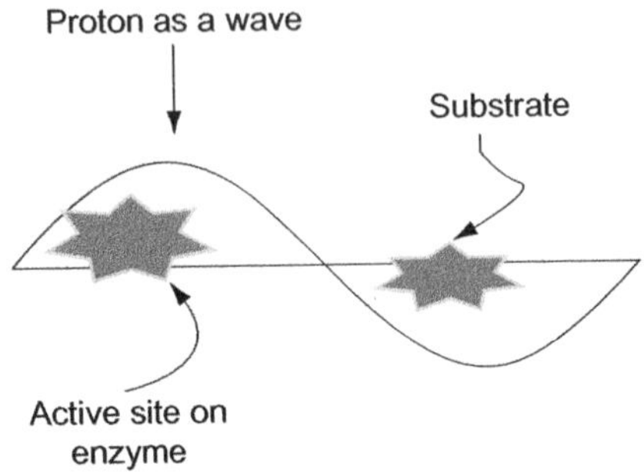

Figure 5.5. A proton directed by the enzyme to the substrate in order to bind with the substrate. The sizes of the substrate, enzyme, and proton wave are not in the right proportion.

substrate molecule. Classical thinking would go like this: At a given moment, the proton is either attached to the histidine, made its way to the lactate, or on its way to the lactate.

From a quantum mechanics viewpoint, the proton is capable of existing as a superposition of both states: attached to histidine and attached to lactate; being at both places at the same time. As we learned in this chapter, the wave character of protons allows the proton to be able to exist at both places at the same time. As illustrated in Figure 5.5, roughly speaking, you can visualize it by considering a proton (or the wave function of the proton) as a two-hump wave, with one hump on histidine and the other hump on lactate. Initially, the hump on the histidine is much larger than the hump on the lactate. As the reaction starts, the hump on the histidine starts getting smaller and the hump on the lactate gets progressively larger. This is the quantum travel of a proton in this case. Of course, during this process, if a measurement is made by the cell through some event, the proton function will collapse to a single state: either attached to the histidine or to the lactate. Now this approach is quite different than the classical approach and will undoubtedly yield different results. Thereby we will be able to understand the process at a much better and realistic level.

In a nutshell, all actions performed by living cells are based on the motion of fundamental particles, which can only be understood by using quantum mechanics. In other words, quantum mechanics is at work at the very origin of every living organism.

5.11 SUMMARY

The theory of quantum mechanics presented in this chapter answers the basic question What framework and formalism does quantum mechanics offer to predict the possible values of observables of a physical system? First, the theory does it by proposing that the wave function associated with a physical system, such as a particle, contains all the information about the system. It also introduces the mathematical machinery of quantum mechanics that will be used in the forthcoming chapters to solve real-world problems. An observable of the system is represented by an operator and the operator acts on the wave function to extract the information about the observable. Because the wave function represents the probability, the theory must answer the question of how to present the predictions in a form so that they can be directly compared with experimental results. Theory does it in the form of expectation value.

Information from the wave function is extracted in terms of an expectation value that corresponds to the average value of several experimental measurements.

This formulation of quantum mechanics can be put into an equation called the Schrödinger wave equation or simply the Schrödinger equation. Solving a physical problem in quantum mechanics then reduces to setting up the Schrödinger equation for the problem and finding its physical solutions. These physical solutions are called eigenfunctions and an operator corresponding to a physical observable operates on the eigenfunction to predict the value of the observable called an eigenvalue. There will be a set of eigenvalues corresponding to a set of possible eigenfunctions.

You might wonder by now: How can a whole theory be put into just one simple looking equation? In response to this question, you will be given a guided tour of the Schrödinger equation in Chapter 6.

5.12 ADDITIONAL PROBLEMS

Problem 5.3 Which of the following functions are the eigenfunctions of the momentum operator?

A. e^{ax}

B. e^{iax} where a is a real number; a constant.

C. $ax^2 + bx$, where a and b are real numbers.

Problem 5.4 Is the following function an eigenfunction of $\partial^2/\partial x^2$?

$$\Psi = \cos\theta + i\sin\theta$$

Problem 5.5 Determine the wavelength of the wave represented by the following function:

$$\Psi = A\sin\sqrt{\frac{2mE}{\hbar^2}}x$$

THE ANATOMY AND PHYSIOLOGY OF AN EQUATION

6

Politics is for the present, but an equation is for eternity.

Albert Einstein

Figure 6.0. One of six sculptures of *Walk of Ideas* in Berlin displaying the equivalence of mass and energy, unveiled on 19 May 2006. Photo courtesy of Lienhard Schulz.

Einstein's equation unifying the concept of mass and energy is the most famous equation in and outside the field of science. This chapter will introduce a not that famous equation, called Schrödinger equation, which captures the core principles of quantum mechanics. It may not be as popular as Einstein's $E = mc^2$ equation, but it is as central to quantum mechanics as Newton's laws are to classical mechanics.

In This Chapter

Quantum Physics for Scientists and Technologists, by Paul Sanghera

6.1 INTRODUCTION

As described in Chapter 5, a theory bases itself on a minimum number of assumptions from which other empirical principles and laws can be derived that may have previously been proposed to explain the experimental results and observations. Equations, the means for efficiently encapsulating and describing the relationships among physical quantities and variables, are very powerful tools in many fields, especially in scientific disciplines, such as mathematics, physics, and chemistry. A good scientific theory is often expressed in terms of a few equations. For example, as you saw in Sections 1.15 and 1.16, the whole classical electromagnetic (EM) theory can be put into four equations, called Maxwell's equations, and the whole of classical mechanics can be expressed in one equation, Newton's second law of motion. Such is the case with quantum mechanics, which is often expressed through the Schrödinger equation introduced in Chapter 5.

Just as seen in classical mechanics, Newton's laws, especially the second law, governs the motion of objects in the macroscopic world; the Schrödinger equation governs the dynamics of particles and is used to derive the particle's physical properties, such as location (position), momentum, and energy. This chapter will demonstrate the power of this equation by applying it to two general situations: a particle in a box and harmonic oscillators, which represent countless real-world applications. We will show how solving the Schrödinger equation for these situations naturally yields some of the principles that were previously postulated to explain experimental results, as discussed in Chapters 2–4. So, the main goal of this chapter is to understand and appreciate how the Schrödinger equation represents quantum mechanics. To that end, we will explore three avenues: introduction to the Schrödinger equation, solving this equation, and applying it to some simple systems, such as a particle in a box and a harmonic oscillator.

6.2 THE SCHRÖDINGER WAVE EQUATION

The Schrödinger wave equation holds the quantum mechanical description of the dynamics of a physical system. The story of the invention of this equation goes back to that graduate student Louis de Broglie, who proposed that there is a wave characteristic associated with each particle; just as there is a particle nature to light, which is discussed in Chapter 3. In November 1925, an Austrian physicist Erwin Schrödinger was presenting a seminar at the University of

Zurich on the de Broglie's model of particles behaving as waves. From among the audience, Peter Debye suggested that if there were a wave there ought to be a wave equation describing that wave's behavior. Within a few weeks after the seminar, Schrödinger fused his knowledge of the quantum mechanical ideas and models existing at that time with what he knew about wave optics, resulting in his wave equation. Without going into the history of its evolution, here is the most general form of the Schrödinger wave equation expressed in Cartesian coordinates:

$$i\hbar\frac{\partial \Psi}{\partial t}+\frac{\hbar^2}{2m}\left(\frac{\partial^2\Psi}{\partial x^2}+\frac{\partial^2\Psi}{\partial y^2}+\frac{\partial^2\Psi}{\partial z^2}\right)-V(x,y,z,t)\Psi(x,y,z,t)=0 \qquad (6.1)$$

where $i=\sqrt{-1}$ is an imaginary number whose square will be a real number. Note that this equation, also known as the time-dependent Schrödinger wave equation, describes the behavior of a physical system, such as a particle of mass m as it moves in a potential V over time represented by t in three spatial dimensions: x, y, and z. The potential V is also a function of space and time.

Erwin Schrödinger (1887–1961)

Erwin Schrödinger was born in Vienna, Austria on August 12, 1887. He received his doctorate in physics in 1910 at the University of Vienna. He developed the wave formulation of quantum mechanics including his famous equation during his stay at the University of Zurich from 1920 to 1926. In 1933, he shared the Nobel Prize with Paul Dirac "for the discovery of new productive forms of the atomic theory".

Schrödinger worked in several fields including physics, mathematics, biology, history, philosophy, language, and literature. He was at Dublin's Institute in Advanced Study from 1939 until 1956, when he returned to Vienna. It was at Dublin that he became interested in Biology. His book *What Is Life*, published in 1944, introduced biologists to the way of looking at life from the perspective of physics. It was this book that started James Watson and Francis Crick (a physicist himself) on their journey to search for the secrets of genes. This led to the discovery of the deoxyribonucleic acids (DNAs) double helix structure in 1953, for which Francis and Crick were awarded 1962 Nobel Prize in physiology or medicine.

Schrödinger died on January 4, 1961 in Vienna at the age of 73 of tuberculosis.

The wave function Ψ is the solution of the equation, which represents the states of the system for which the equation is written. We will be using the terms of the Schrödinger wave equation and the Schrödinger equation interchangeably; but they refer to the same equation.

Equation 6.1 is a very general equation. All the problems do not need this form of the equation. The problems are usually modeled, for example, to one spatial dimension, in which case the time-dependent Schrödinger equation would take the form shown in Study Checkpoint 6.1. So, various special cases from this general form of Eq. 6.1, such as the time-independent Schrödinger equation, can be derived in a straightforward way to represent the problem at hand.

STUDY CHECKPOINT 6.1

From Eq. 6.1, write the time-dependent Schrödinger equation in one spatial dimension, x.

Solution:

$$i\hbar\frac{\partial\Psi}{\partial t}+\frac{\hbar^2}{2m}\left(\frac{\partial^2\Psi}{\partial x^2}\right)-V(x,t)\Psi(x,t)=0$$

Note: While you are working your way through this book, try to realize how Eq. 6.1, the heart of quantum mechanics, is the basis for understanding the discipline of chemistry. Pardon, how this equation can be useful in biology.

PROBLEM 6.1

Show that one of the solutions of the time-dependent Schrödinger equation in one dimension is

$$\Psi = A\,e^{i(kx-\omega t)}$$

where A is a constant, and

$$k=\frac{p}{\hbar}$$

$$\omega=\frac{E}{\hbar}$$

Solution:

$$\Psi = A\,e^{i(kx-\omega t)}$$

which implies

$$\frac{\partial \Psi}{\partial t} = -i\omega A e^{i(kx-\omega t)} = -i\omega \Psi$$

$$\frac{\partial \Psi}{\partial x} = ikAe^{i(kx-\omega t)} = ik\Psi$$

$$\frac{\partial^2 \Psi}{\partial x^2} = i^2 k^2 A e^{i(kx-\omega t)} = -k^2 \Psi$$

By substituting the values of these derivatives into the time-dependent Schrödinger equation:

$$i\hbar \frac{\partial \Psi}{\partial t} + \frac{\hbar^2}{2m}\left(\frac{\partial^2 \Psi}{\partial x^2}\right) - V\Psi = 0$$

we obtain

$$\hbar\omega\Psi - \frac{\hbar^2 k^2}{2m}\Psi - V\Psi = 0$$

which implies

$$\left(\hbar\omega - \frac{\hbar^2 k^2}{2m} - V\right)\Psi = 0$$

But, $E = \hbar\omega$; therefore by substituting this, we obtain

$$\left(E - \left(\frac{p^2}{2m} + V\right)\right)\Psi = 0$$

Also realizing that kinetic energy ($p^2/2m$) plus potential energy is equal to the total energy E, we obtain

$$(E - E)\Psi = 0$$

which implies

$$0 = 0$$

Therefore, the given solution satisfies the time-dependent Schrödinger equation.

Here, we will demonstrate how from Eq. 6.1 you can derive the time-independent Schrödinger equation in one spatial dimension. It is straightforward to rewrite Eq. 6.1 in just one spatial dimension by eliminating the terms involving y and z:

$$i\hbar\frac{\partial \Psi(x,t)}{\partial t}+\frac{\hbar^2}{2m}\left(\frac{\partial^2 \Psi(x,t)}{\partial x^2}\right)-V(x,t)\Psi(x,t)=0 \tag{6.2}$$

> *Note*: In addition to the physical argument made for separation of variables in the body of the chapter, mathematically speaking, separation of variables is a technique used to solve some differential equations. If a proposed solution satisfies the differential equation (Schrödinger equation in our case), it is a solution then, regardless of how we came up with this; separation of variable or any other technique.

This represents the motion of a particle in one-dimension (1D) when the motion varies with time. However, in a great many situations, the forces that act upon a particle vary only with the position (spatial variables) of the particle and not with time. Electrons in an atom bound to the protons in the nucleus and atoms in a crystal lattice are some examples. Under these forces, called stationary forces, the potential energy of the particle, V, does not vary explicitly with time. In such cases, the space and time dependence of the wave function can be separated into two factors, space dependent and time dependent, as shown here:

$$\Psi(x,t)=\Psi(x)\,f(t) \tag{6.3}$$

By inserting this expression for $\Psi(x, t)$ into Eq. 6.2 and dividing both sides by $f(t)$, we obtain

$$\frac{i\hbar}{f(t)}\frac{df}{dt}\Psi(x)+\frac{\hbar^2}{2m}\left(\frac{d^2\Psi(x)}{dx^2}\right)-V(x)\Psi(x)=0 \tag{6.4}$$

> **STUDY CHECKPOINT 6.2**
>
> Why have we used total instead of partial derivatives in Eq. 6.4?
>
> **Solution:**
>
> Because f and Ψ in this equation are functions of only one variable: f is a function of t and Ψ is a function of x.

As shown in Problem 6.2, the first term (from the left) in Eq. 6.4 is equivalent to the energy of the system, E. Therefore, Eq. 6.4 can be rewritten as:

$$\frac{\hbar^2}{2m}\left(\frac{d^2\Psi(x)}{dx^2}\right)+(E-V(x))\Psi(x)=0 \tag{6.5}$$

In three dimensions (3D), it will be

$$\frac{\hbar^2}{2m}\left(\frac{\partial^2\Psi}{\partial x^2}+\frac{\partial^2\Psi}{\partial y^2}+\frac{\partial^2\Psi}{\partial z^2}\right)+(E-V(x,y,z))\Psi=0 \tag{6.6}$$

where Ψ is a function of x, y, and z.

STUDY CHECKPOINT 6.3

As shown in Problem 6.1, $A\ e^{i(kx-\omega t)}$ is an acceptable solution for the time-dependent Schrödinger equation in 1D. Use this fact to calculate $f(t)$ in Eq. 6.3.

Solution:

Equation 6.3

$$\Psi(x,t)=\Psi(x)\,f(t)=A\ e^{i(kx-\omega t)}=A\ e^{i(kx)}e^{-i\omega t}=\Psi(x)\ e^{-i\omega t}$$

which implies

$$f(t)=e^{-i\omega t}$$

Equations 6.5 and 6.6 are known as the time-independent Schrödinger equations in one and three spatial dimensions respectively. It is also called the steady-state Schrödinger equation because it represents a particle whose potential energy, V, does not depend on time. Therefore, the time variations of all functions of the particle have effectively the same forms as that of an unrestricted particle.

As mentioned in Chapter 5, the Schrödinger equation cannot be derived from fundamental principles because the equation itself is considered to be a fundamental principle. We hold this principle to be fundamentally true. Although, like any other scientific theory or principle, it needs to be continuously tested against the experimental results and observations. Since its invention, the Schrödinger equation has withstood experimental tests. To start with, it explains all the experimental results discussed in the previous chapters that classical physics failed to explain. All the quantum laws and hypothesis discussed in these chapters proposed to explain those results are embedded in the Schrödinger equation with the right interpretation of the wave function. For example, Problem 6.9 demonstrates how you can derive the de Broglie hypothesis from the Schrödinger equation. In this and forthcoming chapters, we will continue not only to explore how this equation can be applied to the situations described in the previous chapters, but also to further explore the anatomy and physiology of atoms and molecules.

PROBLEM 6.2

As shown in Study Checkpoint 6.3, the time-dependent part of the wave function can be written as:

$$f(t) = e^{-i\omega t} = e^{-i(E/\hbar)t} \qquad (1)$$

Use this to prove the following:

$$\frac{i\hbar}{f(t)}\frac{df}{dt} = E$$

Solution:

Because we are looking at a time-independent situation;

$$\frac{i\hbar}{f(t)}\frac{df}{dt} = C \qquad (2)$$

where C is independent of time. By integrating this equation, we obtain

$$i\hbar \ln(f) = Ct + c$$

where c is an integration constant that can be set to 0 for simplicity, which implies

$$f(t) = e^{-i(C/\hbar)t}$$

Comparing this result with Eq. 1, we realize that $C = E$
Combining this with Eq. 2, which implies

$$\frac{i\hbar}{f(t)}\frac{df}{dt} = E$$

In the following sections, we will apply the Schrödinger equation to some simple situations in order to demonstrate the power and usefulness of this simple looking equation.

6.3 THE SCHRÖDINGER EQUATION FOR A FREE PARTICLE

The simplest physical system to be explained with quantum mechanics is the system that has zero potential energy V everywhere; that is, a particle not subjected to any external forces. Because no forces are acting on this system, it is called a free particle. For simplicity, let us consider a free particle only in 1D, say x. Such a particle can be represented by Eq. 6.7 below, which is obtained by substituting $V = 0$ in the Schrödinger wave Eq. 6.5:

$$\frac{\hbar^2}{2m}\left(\frac{d^2\Psi(x)}{dx^2}\right)+E\Psi(x)=0 \tag{6.7}$$

This can be rewritten as:

$$\frac{d^2\Psi(x)}{dx^2}+\frac{2mE}{\hbar^2}\Psi(x)=0 \tag{6.8}$$

For brevity, let us define a constant k:

$$k \equiv \sqrt{\frac{2mE}{\hbar^2}} \tag{6.9}$$

With this substitution, Eq. 6.8 can be rewritten as:

$$\frac{d^2\Psi(x)}{dx^2}+k^2\Psi(x)=0 \tag{6.10}$$

STUDY CHECKPOINT 6.4

From Eq. 6.9, show that the momentum p under zero potential energy is given by

$$p=\hbar k$$

Solution:

From Eq. 6.9:

$$E=\frac{\hbar^2k^2}{2m}$$

For $V = 0$,

$$E=T=\frac{1}{2}mv^2=\frac{p^2}{2m}=\frac{\hbar^2k^2}{2m}$$

which implies

$$p=\hbar k$$

Problem 6.10 shows that the following are three of the possible solutions for Eq. 6.10:

$$\Psi(x) = A \begin{cases} \sin kx \\ \cos kx \\ e^{\pm ikx} \end{cases} \tag{6.11}$$

where A is a constant. Note that according to this solution, the particle can have any non-negative amount of energy. In other words, the energy for a free particle is not quantized even in quantum mechanics. However, the particle due to its wave nature will act like a wave under the right conditions. For example, if an electron is to pass through a partition with two slits on it, the result would be a wavelike diffraction pattern, as you saw in Chapter 5.

> *Note*: According to trigonometry, both solutions $\sin kx$ and $\cos kx$ can be represented in terms of $e^{\pm ikx}$, as shown in the following equation:
>
> $$\sin kx = \frac{1}{2i}\left(e^{ikx} - e^{-ikx}\right)$$
>
> $$\cos kx = \frac{1}{2}\left(e^{ikx} + e^{-ikx}\right)$$

In this section, we have discussed the simplest possible case: a free particle. Note that the wave function of a free particle and the energy are not quantized. In the real world, however, most of the systems are bounded. Think of an electron bounded inside an atom, two atoms bounded together to make a molecule, molecules of life in living cells in your body, the blood cells flowing in your veins, you bound to the Earth by gravitational force, and the Earth bound to the Sun by the same force, and so on. So, next we discuss the case of a bounded system.

6.4 SCHRÖDINGER EQUATION FOR A PARTICLE IN A BOX

A simple yet nontrivial situation is presented by a particle bounded (restricted) inside a box. Again, for the sake of simplicity, we will discuss the problem only in 1D, x. We are discussing this system with three goals in mind:

1. Get a feel of how the Schrödinger equation is applied to a bounded system; that is, a system that has some restriction on its motion.

2. Learn some characteristics of the solutions of the Schrödinger equation.
3. Learn how the quantum approach makes different predictions than the classical approach.

So, let us see how we can set up and solve the Schrödinger equation for this system.

6.4.1 Setting Up and Solving the Schrödinger Equation

The system of a particle bounded in a 1D box is illustrated in Figure 6.1.

Here is how we model this situation:

1. The particle is restricted to traveling along the x-axis from $x = 0$ to $x = a$.
2. The potential energy of the particle V is a constant inside, that is, between $x = 0$ and $x = a$. Let us assume for convenience that this constant is zero, that is, $V = 0$.
3. The potential V outside of the region from 0 to a is infinite.

This finding is the equivalent of having a potential well with walls at $x = 0$ and $x = a$. Inside the well, the particle can move freely and bounce back and forth from the walls without losing any energy, that is, E is constant. Because it cannot have infinite energy, it cannot exist outside the well. As you will see

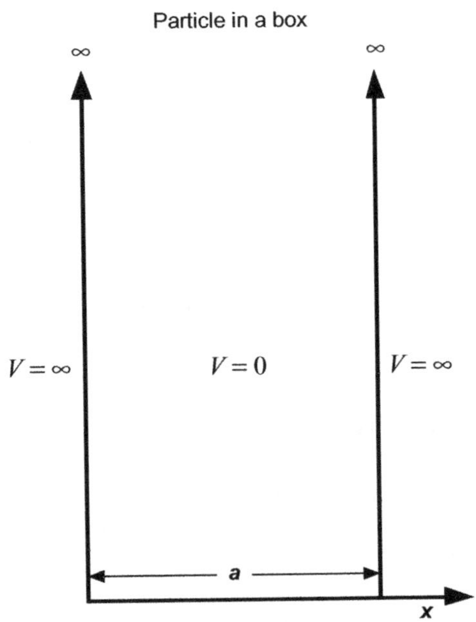

Figure 6.1. A particle confined to a potential box of width a.

later in this section, it also means that the wave solution has nodes at the boundaries of the box. Any solution needs to satisfy that condition.

Within the well, the particle is free and therefore can be represented by the time-independent Schrödinger equation with $V = 0$. In other words, the particle inside the well can be represented by

$$\frac{d^2\Psi(x)}{dx^2} + k^2\Psi(x) = 0 \tag{6.10}$$

As described earlier, two possible solutions for this equation are

$$\Psi(x) = A\sin kx$$

$$\Psi(x) = B\cos kx$$

where A and B are constants. As shown in Problem 6.11, a more general solution to Eq. 6.10 can be written as a combination (superposition) of the two solutions $A\sin kx$ and $B\cos kx$:

$$\Psi(x) = A\sin kx + B\cos kx \tag{6.12}$$

So far, we have seen no difference in this situation and the previous situation of a free particle because inside the box, the particle behaves as a free particle. However, the results will become more interesting when we consider the fact that the potential at the walls is infinite.

6.4.2 Here Comes the Energy Quantization

Here is the exciting part: the appearance of quantization. Because the potential at the walls of the box is infinite, the particle cannot exist outside the box; that is, the wave function outside the box should be zero. In other words, we have two boundary conditions:

1. The wave function should be zero at $x = 0$:
   which implies

$$\Psi(0) = A\sin 0 + B\cos 0 = 0$$

   which implies

$$B = 0$$

2. The wave function should be zero at $x = a$:

$$\Psi(a) = A\sin ka + B\cos ka = 0$$

Because $B = 0$, therefore,

$$A \sin ka = 0$$

which implies

$$ka = n\pi \tag{6.13}$$

where n must be a nonzero integer because if it is 0, then $k = 0$, and therefore Ψ becomes zero everywhere; that is, a particle does not exist anywhere, which is absurd. Substituting the value of k from Eq. 6.9 into Eq. 6.13, we obtain

$$\sqrt{\frac{2mE}{\hbar^2}}\,a = n\pi$$

which implies

$$E_n = n^2 \frac{\hbar^2 \pi^2}{2ma^2} = n^2 \frac{h^2}{8ma^2} \qquad n = 1, 2, 3, \ldots \tag{6.14}$$

These solutions are illustrated in Figure 6.2.

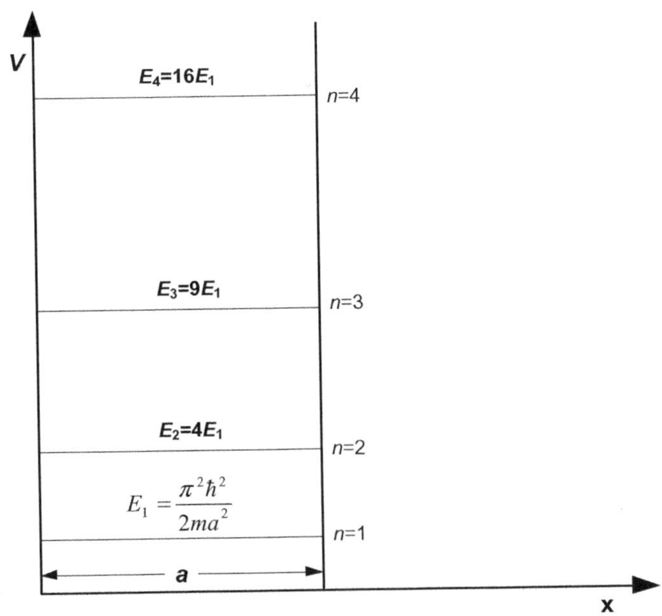

Figure 6.2. Energy quantization of a particle in the box.

Two important characteristics of the solution here are to be noted: Energy of the particle is quantized due to the integer values of n, and the particle always has a nonzero positive energy because $n = 0$ is not allowed, as discussed before. These two characteristics are the quantum mechanical results that have no counterpart in classical mechanics, where all values of energies including zero are permissible.

This is an amazing thing: Even through the particle is free in the box, it can only assume a discrete set of energy states. The particle is constrained to have only certain discreet values of energy, that is, the energy is quantized. Also interesting is that the energy of the particle cannot be zero: It can never rest. Now, if you go to the classical limit, that is, the box gets big and therefore the available quantum numbers n become very large, Eq. 6.14 shows that the solution becomes continuous within the accuracy of classical measurements.

STUDY CHECKPOINT 6.5

Will the energy states corresponding to $n = 3$ and -3 have different energies?

Solution:

No.

Because energy depends on the square of n.

Note that in Eq. 6.14, n is the same quantum number we talked about in the previous chapters, but now we obtained it from the Schrödinger equation. Also note that we obtained the quantized energy states as a characteristic of a bound system; that is, a particle confined in a finite region in space. The state of the particle with lowest energy is called the ground state and all other energy states are called excited states. Remember, the ground state is not the zero-energy state.

There is a straightforward, but very important implication of energy quantization: The particle whose energy is quantized (i.e., the particle is bound) can absorb or emit only certain allowed discrete amounts of energies corresponding to certain discrete frequencies. If the energy is the EM energy, we are talking about absorption and emission spectra. See how the spectra discussed in the previous chapters are naturally predicted by this simple example: the power of the Schrödinger equation.

Note: The results and conclusions, such as nonzero and quantized energy drawn from the scenario of a particle in a box, are very general and do not depend on the fact that we only considered one spatial dimension and not three.

STUDY CHECKPOINT 6.6

What is the energy of a particle in a 1D box of width a in its ground state?

Solution:

As we argued before from Eq. 6.13, n must be a nonzero integer because if it is 0, then $k = 0$, and therefore Ψ, becomes zero everywhere; that is, a particle does not exist anywhere, which is absurd.

The ground state corresponds to minimum energy and that is not zero. Remember $n = 0$ is not allowed. So the minimum energy the particle can have corresponds to $n = 1$.

Therefore the ground-state energy = $E_1 = h^2/8ma^2$.

Now that you have broken the ice with the Schrödinger equation, and you are excited with its predictive power displayed by the solution of a very simple situation, let us further explore the solutions of this equation, and see what we can discover.

6.4.3 Exploring the Solutions of the Schrödinger Equation

Equation 6.12 presented a solution of the Schrödinger equation that involved two constants A and B. We have already shown how the boundary conditions lead to $B = 0$; now we will determine the constant A, because to write the general solution we need the values for both A and B. By substituting $B = 0$ into Eq. 6.12, we obtain the solution for the problem at hand:

$$\Psi(x) = A \sin kx \tag{6.15}$$

By using Eq. 6.13 for k:

$$\Psi(x) = A \sin \frac{n\pi x}{a} \tag{6.16}$$

In general, the probability that the particle will be found somewhere is 1. In our case, because the particle cannot exist outside the well, the probability of finding the particle somewhere in the well is 1. Therefore, the normalization condition introduced in Chapter 5 becomes:

$$1 = \int_{-\infty}^{\infty} |\Psi|^2 \, dx = \int_0^a |\Psi|^2 \, dx = A^2 \int_0^a \sin^2 \frac{n\pi x}{a} \, dx = A^2 \frac{a}{2}$$

which implies

$$A = \sqrt{\frac{2}{a}}$$

Substitute the value of A back into Eq. 6.16 to obtain the solution we were after:

$$\Psi_n(x) = \sqrt{\frac{2}{a}} \sin \frac{n\pi x}{a} \tag{6.17}$$

The case of $n = 0$ has already been ruled out as unphysical because it leads to an absurd result: the particle does not exist. As shown in Problem 6.12, both positive and negative values of n represent the same physical quantitative meaning of the solution. The probability of finding the particle somewhere in the box is represented by $|\Psi_n(x)|^2$. The solutions presented in Eq. 6.17 are illustrated for a few values of n in Figure 6.3, which demonstrates that while $\Psi_n(x)$ could be positive or negative, $|\Psi_n(x)|^2$ is always positive. Also, the probability density $|\Psi_n(x)|^2$ of finding the particle inside the box is not the same everywhere. In every case, it is 0 at $x = 0$ and at $x = a$, and varies inside the box. For a given point in the box, the probability density depends on the quantum number n. For example, for $n = 1$, the particle is most likely to be

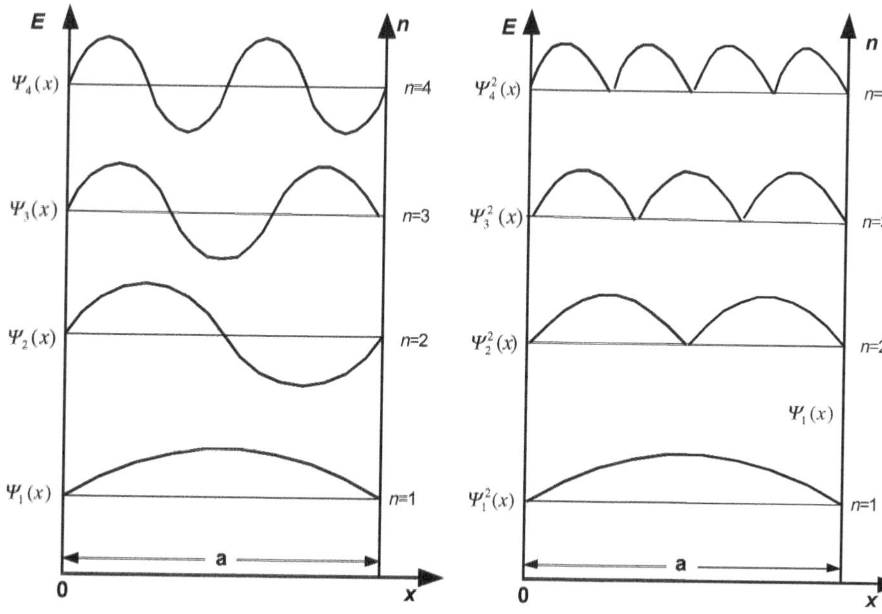

Figure 6.3. Wave functions and probability densities of a particle confined to a box with rigid walls. Read $\Psi^2(x)$ as $|\Psi(x)|^2$.

found in the middle of the box (maximum probability density), whereas for $n = 2$, the particle will least likely be found there because the probability density there is minimum. The variations of the probability over n can be understood in terms of energy. For higher values of E, the particle is more excited, and therefore bounces back and forth from the walls, consequently creating more fluctuations in the probability of its location in the box.

PROBLEM 6.3

According to classical physics, calculate the probability of finding the particle anywhere inside the box when there is an infinite potential outside the box.

Solution:

Infinite potential outside the box means the total probability of finding the particle inside the box is 1. According to classical physics the probability is continuously distributed; that is, the particle can be found anywhere inside the box with equal probability, say P.
which implies

$$1 = \int_0^a |\Psi|^2\, dx = \int_0^a P dx = Pa$$

which implies

$$P = \frac{1}{a}$$

The variation of probability for existence within the box is another result that is in contrast with classical physics, which predicts the probability will be $1/a$ everywhere in the box, as shown in Problem 6.3. Talking about classical physics, let us review the correspondence principle, first discussed in Section 4.8, now in the context of this problem.

6.4.4 The Uncertainty and Correspondence Principles: Revisited

The uncertainty and correspondence principles were discussed in Chapters 3 and 4, respectively. Let us see how we can realize these principles in the context of a simple problem of a particle in a box, which we are currently exploring.

As mentioned earlier, the variation of the probability of finding the particle with the width of the box, a, and the quantum number, n, is a salient feature of quantum mechanics, whereas in classical mechanics there is no concept of quantum number. In classical mechanics, the probability of finding the particle anywhere in the box will be the same, that is, equal to $1/a$. This classical result can be obtained from the quantum solution with a limit of $n \to \infty$, that is, for large values of n. As you can see from Figure 6.3, for large values of n there will be a large number of oscillations or fluctuations in the probability density within the box. Under this situation, every point in the box will have an average probability density, which is $1/a$, the same as predicted by classical physics. Therefore *classical physics meets with quantum physics at large values of n*; the correspondence principle.

As shown in Study Checkpoint 6.6, the minimum energy for this trapped particle (in a box) is nonzero, and is given by

$$E_1 = \frac{h^2}{8ma^2} \tag{6.18}$$

The corresponding system in classical physics will have a minimum energy equal to zero. Therefore, this minimum nonzero energy and the corresponding momentum can be considered to be a result of a quantum mechanical as uncertainty, or expressed as an uncertainty. Because the potential energy inside the box is zero, the corresponding uncertainty in momentum can be calculated as:

$$\Delta p = \sqrt{2mE_1} = \frac{h}{2a}$$

Here we have used the relation $E_1 = p^2/2m$ and Eq. 6.18.

The uncertainty in location of the particle is the size of the box:

$$\Delta x = a$$

which implies

$$\Delta p \Delta x = \frac{h}{2}$$

which is in accordance with Heisenberg's uncertainty principle introduced in Chapter 3:

$$\Delta p \Delta x \geq \frac{\hbar}{2}$$

So far, we have been assuming that the potential outside the box is infinite and therefore the value of the wave function outside the box is zero; that is, the particle cannot exist outside the box. However, the more realistic situation would be when the potential outside the box is not infinite, it is finite but way $>E$ of the particle. In classical physics, this does not change the situation because the particle will still not exist outside the box. But, what does quantum mechanics says about this situation? Assume we deposit a particle into the box with finite walls, that is, a finite potential outside the box. In quantum mechanics, we can ask the question What is the probability that the particle is still in the box after a certain time? This very question implies that after a certain time the particle will leave the box. It is called quantum tunneling, which we explore next in Section 6.4.5.

6.4.5 Quantum Mechanical Tunneling

Assuming there is an infinite potential outside the box starting from $x = 0$ on one end and from $x = a$ on the other, it should mean there are rigid walls that the particle can never jump over. The reason is because its energy inside the box with $V = 0$ is always finite, and hence always $<V$ at the walls. However, the assumption of an infinite potential is not very practical, and the world is filled with practical examples of finite potentials. Electrons in an atom are a practical example. So, let us say the potential V is very large, but not infinite at the walls and outside the box. Under this condition, the wave function $\Psi(x)$ will not vanish at the boundaries ($x = 0$ and $x = a$). In other words, walls of the box will not be rigid. As illustrated in Figure 6.4, the wave function will vanish around the walls, but it does not vanish exactly at the wall. This can also be understood in terms of the uncertainty principle:

$$\Delta p \Delta x \geq \frac{\hbar}{2}$$

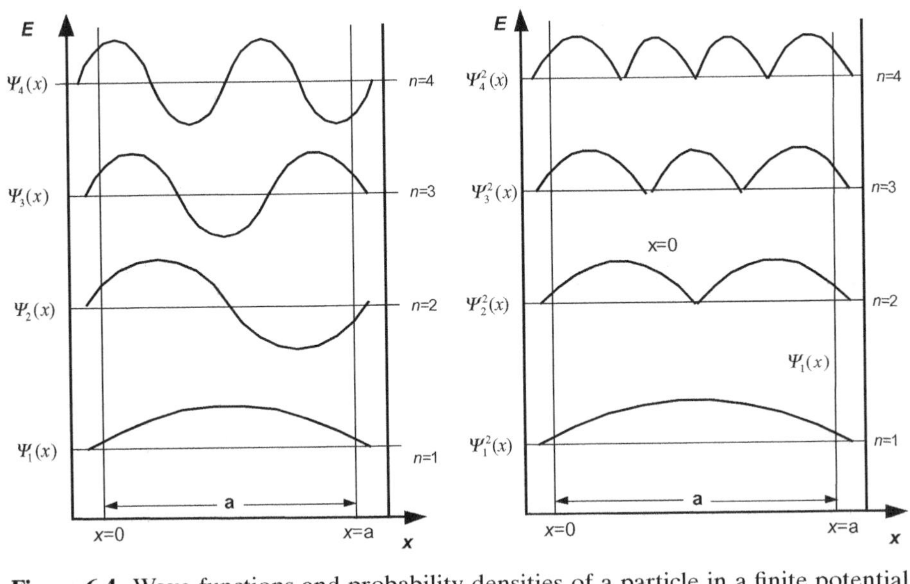

Figure 6.4. Wave functions and probability densities of a particle in a finite potential well. Read $\Psi^2(x)$ as $|\Psi(x)|^2$.

It means that the particle has to have infinite uncertainty in momentum in order to have zero uncertainty in its location. In other words, the particle has to have infinite momentum, and hence infinite energy in order for you to say for sure that it cannot be found at or around $x = 0$ or $x = a$. This means at finite energy E, even when it is $<V$, a particle can exist outside the walls, as shown in Figure 6.4.

Congratulations, you have just realized another salient feature of quantum mechanics called the tunneling effect: The particle has tunneled through the walls even if it did not have enough energy to jump over the potential walls. This obviously has no analogue in classical physics. Note from Figure 6.4 that as the value of a quantum number n increases, more of Ψ can be found outside the box. Increasing n means increasing the wavenumber according to $ka = n\pi$, and hence increasing momentum.

After these qualitative arguments for tunneling, let us see how the tunneling effect arises naturally from the analytic treatment of the Schrödinger equation for a particle in a box. We rewrite the time-independent Schrödinger equation in 1D, Eq. 6.5, in the following form:

$$\frac{\hbar^2}{2m}\left(\frac{d^2\Psi(x)}{dx^2}\right)+(E-V(x))\Psi(x)=0 \tag{6.5}$$

Now, consider the case in which V is $>E$ outside of the box starting from the walls:

$$V = V_0$$

where $V_0 > E$

Under this scenario, Eq. 6.5 can be rewritten as:

$$\frac{d^2\Psi(x)}{dx^2} = \frac{2m}{\hbar^2}(V_0 - E)\Psi(x) \tag{6.19}$$

Now, outside the box starting from $x = 0$ and $x = a$, $V_0 - E$ is positive because the potential outside the box is greater than the energy of the particle inside the box. Classically, this means that the particle stays trapped inside the box. But pay close attention to Eq. 6.19. It tells us that the second derivative of the wave function $\Psi(x)$ has the same sign as $\Psi(x)$ itself. This means that the wave function, that is, the solution of Eq. 6.19, is not a periodic function; it is exponential. You can verify that the following is one possible solution to this equation:

$$\Psi(x) = Ae^{-\alpha x} \tag{6.20}$$

As Problem 6.4 demonstrates:

$$\alpha = \frac{1}{\hbar}\sqrt{2m(V_0 - E)} \tag{6.21}$$

PROBLEM 6.4

Assuming that Eq. 6.20 is a solution to Eq. 6.19, calculate the value of α in Eq. 6.20.

Solution:

$$\Psi(x) = Ae^{-\alpha x}$$

$$\frac{d\Psi(x)}{dx} = -\alpha Ae^{-\alpha x}$$

$$\frac{d^2\Psi(x)}{dx^2} = \alpha^2 Ae^{-\alpha x} = \alpha^2\Psi(x)$$

By substituting these values in Eq. 6.19, we obtain

$$\alpha^2\Psi(x) = \frac{2m}{\hbar^2}(V_0 - E)\Psi(x)$$

$$\alpha^2 = \frac{2m}{\hbar^2}(V_0 - E)$$

which implies

$$\alpha = \frac{1}{\hbar}\sqrt{2m(V_0 - E)}$$

According to the solution in Eq. 6.20, the wave function decays exponentially in the region with $V = V_0$ and not periodically. That means there is a nonzero probability that the particle can exist outside of the box even when the potential there is greater than the energy of the particle. Now that we have the wave function, let us calculate this probability. The probability density $dP(a)$ of finding the particle at $x = a$ is

$$dP(a) = |\Psi(x)|^2_{x=a} = A^2 e^{-2\alpha a}$$

The probability density $dP(a + \Delta x)$ of finding the particle outside the box at $x = a + \Delta x$:

$$dP(a + \Delta x) = |\Psi(x)|^2_{x=a+\Delta x} = A^2 e^{-2\alpha a} e^{-2\alpha \Delta x}$$

The probability of finding the particle outside the box at $x = a + \Delta x$ as compared to finding it at the wall of the box at $x = a$:

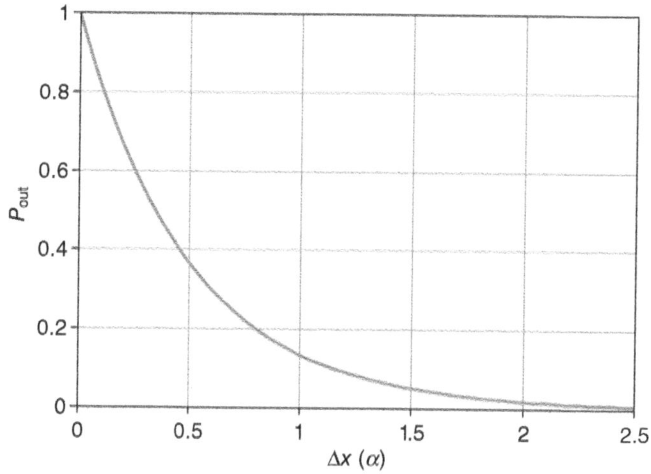

Figure 6.5. Tunneling probability for a particle trapped in a box to be found at a distance Δx away from the wall of the box shown in units of α.

$$P_{\text{out}} = \frac{dP(a+\Delta x)}{dP(a)} = e^{-2\alpha\Delta x} \tag{6.22}$$

This probability may be small, but not zero. These small energies and probabilities are only small in the macroscopic world; in the microworld of atoms and molecules, these values are typical: They run the world. Figure 6.5 plots the tunneling probability given in Eq. 6.22.

So, you can see how a particle trapped inside a box can tunnel through it. As an example, Problem 6.13 shows that P_{out} is equal to 5.00×10^{-7} for $\Delta x = 2\,\text{nm}$ when the potential is 2 eV higher than the energy of the particle trapped in the box. These numbers are significant in the microworld. For example, $\Delta x = 2\,\text{nm}$ can hold several atoms and is about the diameter of a deoxyribonucleic acid (DNA) molecule. It only takes ~13.6 eV to rip an electron off the hydrogen atom.

To summarize, we have discussed the Schrödinger equation for simple cases, such as a free particle and a particle in a box. In exploring the solutions of this equation for a particle confined to a box, we have seen how some salient features of quantum mechanics, such as energy quantization, ground state with nonzero energy, and quantum tunneling, arise naturally. Yet to some of us all this may seem like a nice intellectual exercise. What does it have to do with the real world?, you ask. Well, here are some relevant scenarios from the real world. You can find confinement everywhere: in the macroscopic world and in the microscopic world. Earth revolving around the Sun; that is, confined to it by the gravitational force; you confined to the Earth by the same force; Moon and satellites revolving around the Earth are some examples of confinement at the macroscale. At the microscale, some examples of confinement are DNA molecules confined to a nucleus in a cell, protons and neutrons confined in the nucleus of an atom due to the nuclear force, electrons in an atom confined to a region around the nucleus by the EM force, and electrons in a molecule confined in a region around multiple nuclei by an EM force.

Some of these electrons in molecules confined to a region around the nuclei may move freely along the bonds if the bonds are weak enough. But they are still confined to the molecule. Such a situation may be described by a particle in a box. We explore one such example in the context of what is called a free electron model (see Section 4.9.1).

The results obtained from a particle in a 1D box are very general and easily extendable to a particle in a 3D box, which we discuss next.

6.5 A PARTICLE IN A THREE-DIMENSIONAL BOX

In Section 6.4, we explored a particle in a 1D box. Even this simple case can be used to model some simple bound (or constrained) atomic and molecular systems in biology, chemistry, and nanotechnology to understand some structures and phenomena and to solve some problems. However, in a great many situations, it will be more practical to treat these atomic and molecular systems as 3D systems. The good news is that the results derived from the study of a particle in a 1D box are quite general and can easily be extended to 3Ds, as we show in this section.

Recall that the time-independent (steady-state) Schrödinger equation in 3D is given by

$$\frac{\hbar^2}{2m}\left(\frac{\partial^2\Psi}{\partial x^2}+\frac{\partial^2\Psi}{\partial y^2}+\frac{\partial^2\Psi}{\partial z^2}\right)+(E-V(x))\Psi=0 \tag{6.6}$$

where Ψ is a function of x, y, and z. Just like a particle in a 1D box, we assume that the potential V is a constant, say zero, inside the box and infinite outside the box. Let us further assume that the size dimensions of the box are a_1, a_2, and a_3 in the x, y, and z directions.

Recall that in the case of a 1D box, we started with the following solution:

$$\Psi(x)=A\sin kx+B\cos kx \tag{6.12}$$

and concluded that due to the boundary condition, $B = 0$, and therefore the solution was reduced to $A \sin kx$. The natural extension of this solution to 3Ds will be:

$$\Psi=A\sin(k_1x)\sin(k_2y)\sin(k_3z) \tag{6.23}$$

Therefore, the wave function becomes:

$$\Psi = A\sin\left(\frac{n_1\pi}{a_1}x\right)\sin\left(\frac{n_2\pi}{a_2}y\right)\sin\left(\frac{n_3\pi}{a_3}z\right) \tag{6.24}$$

By taking second-order partial derivatives of Ψ in Eq. 6.24 and substituting them into Eq. 6.6, where $V = 0$, we obtain

$$E = \frac{\hbar^2\pi^2}{2m}\left(\frac{n_1^2}{a_1^2} + \frac{n_2^2}{a_2^2} + \frac{n_3^2}{a_3^2}\right) \tag{6.25}$$

Note that the energy is quantized in terms of three quantum numbers n_1, n_2, and n_3 instead of one quantum number in the case of a 1D box. Nevertheless, it is a simple extension.

STUDY CHECKPOINT 6.7

A. What is the ground-state energy of a particle of mass m trapped in a cubical box of dimension a.

B. How does it compare to the 1D case?

Solution:

A. Given

$$a_1 = a_2 = a_3 = a$$

Also, in the ground state:

$$n_1 = n_2 = n_3 = 1$$

which implies

$$E = \frac{3\hbar^2\pi^2}{2ma^2} = \frac{3h^2}{8ma^2}$$

B. For the 1D case from Study Checkpoint 6.6,

$$E = \frac{h^2}{8ma^2}$$

Therefore, the ground-state energy in 3Ds is three times larger than that in 1D.

So far, we have been assuming that the potential does not vary over space: It is constant and finite (we assumed zero) inside the box and infinite outside. For some physical problems, these assumptions will be valid. However, there are some situations in nature where the potential is not zero and is not constant: It varies over spatial dimensions. A great many of these situations can be represented by a harmonic oscillator that we discuss next in Section 6.6.

6.6 HARMONIC OSCILLATOR

A harmonic oscillator is a system that vibrates about a position of equilibrium. Note the relationship between harmonic motion and equilibrium. Such a motion is called a harmonic motion, and such a system has a universal significance for two reasons. First, the situation of equilibrium is of great interest because there is an underlying physical law that systems tend to be in equilibrium and if disturbed they tend to restore their equilibrium. This physical law transcends itself in many laws and principles in physics and chemistry; and molecular biology cannot escape this principle either. Second, there are countless real-world situations in the microscopic world (just as in the macroscopic world) that can be represented by a harmonic oscillator. An atom in a crystal lattice and a diatomic molecule are two of the many categories of examples.

First, let us ask the question What is harmonic motion?

6.6.1 Understanding Harmonic Motion

Harmonic motion occurs due to the presence of some kind of restoring force F that continually works to bring the system back to its equilibrium configuration or state when it is disturbed. Due to some kind of inertia, such as mass, the system always overshoots the equilibrium point, and therefore oscillates around the equilibrium point indefinitely if there is no dissipative process.

If you took an introductory physics course, you may remember an illustration of harmonic motion in terms of a mass m oscillating in 1D, x, on the end of a spring, as shown in Figure 6.6. The farther away the spring is pulled from its equilibrium position at $x = 0$, the greater will be the restoring force. This relationship can be expressed in Hook's law:

$$F = -\kappa x \tag{6.26}$$

where κ is the proportionality constant called a spring constant.

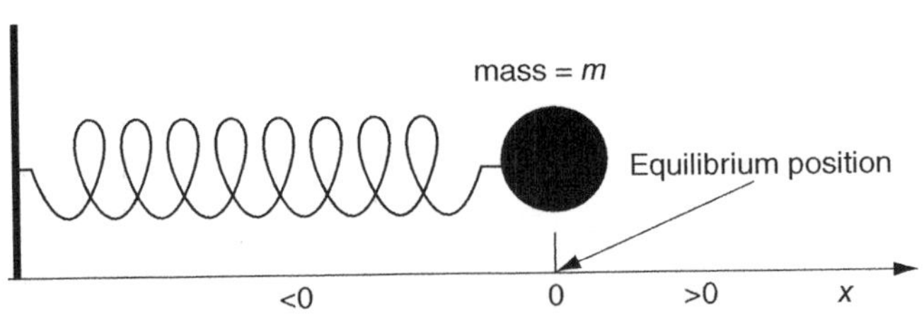

Figure 6.6. A spring mass system to produce a harmonic motion if disturbed from its equilibrium position.

From this equation, and using Newton's second law of motion ($F = ma$), you can derive the following differential equation:

$$\frac{d^2x}{dt^2} + \frac{\kappa}{m}x = 0 \tag{6.27}$$

where m is the mass attached to the spring, and κ is the spring constant. The derivation of this equation is given in Problem 6.5.

PROBLEM 6.5

Use classical mechanics to derive the following differential equation for a mass m vibrating on the end of a spring:

$$\frac{d^2x}{dt^2} + \frac{\kappa}{m}x = 0$$

Solution:

According to classical mechanics:
Acceleration a along the x-axis is given by

$$a = \frac{d^2x}{dt^2}$$

and force F is given by

$$F = ma$$

Therefore,

$$F = m\frac{d^2x}{dt^2}$$

By substituting this value of F in Eq. 6.26, we obtain

$$m\frac{d^2x}{dt^2} = -\kappa x$$

which is the same as:

$$\frac{d^2x}{dt^2} + \frac{\kappa}{m}x = 0$$

You can verify by differentiating and substituting back into Eq. 6.27, that the following is a general solution of Eq. 6.27:

$$x = Ae^{i\sqrt{(\kappa/m)}t}$$

which is a combination of the sine and cosine functions. Out of various ways of writing the solutions of Eq. 6.27, the convenient one is

$$x = A\sin(2\pi ft) \tag{6.28}$$

where A is a constant, which is equal to the maximum amplitude of oscillations and f is the frequency of oscillation. As is shown in Problem 6.14, f is given by

$$f = \frac{1}{2\pi}\sqrt{\frac{\kappa}{m}} \tag{6.29}$$

The potential energy $V(x)$ is given by the work needed to bring the particle from the equilibrium position at $x = 0$ to x:

$$V(x) = -\int_0^x F(x)dx = \kappa\int_0^x xdx = \frac{1}{2}\kappa x^2$$

Therefore the potential energy is given by

$$V(x) = \frac{1}{2}\kappa x^2 \tag{6.30}$$

So, how about the total energy of the oscillating system? We can calculate it without knowing the kinetic energy, if we put the following facts together:

1. The total energy E is given by the kinetic plus the potential energy:

$$E = T + V$$

2. Kinetic energy is zero when $x = A$, that is, when the oscillator is at its maximum displacement from the equilibrium point.
3. Total energy is conserved.

This, using Eq. 6.30 means

$$E = V(x)|_{x=A} = \frac{1}{2}\kappa A^2$$

Therefore,

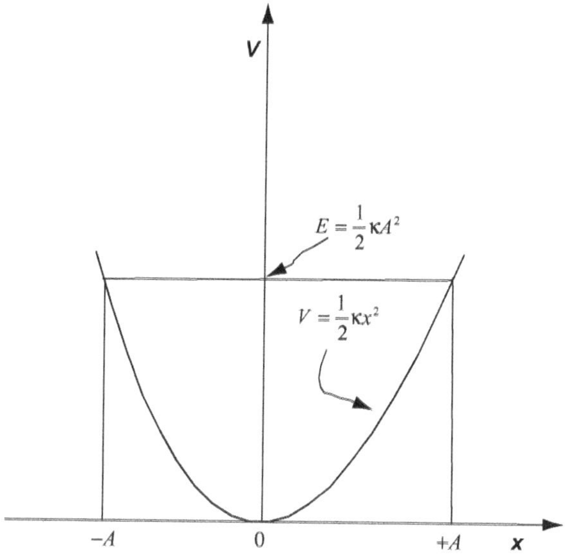

Figure 6.7. Potential energy of a harmonic oscillator.

$$E = \frac{1}{2}\kappa A^2 \tag{6.31}$$

The potential and total energy of a harmonic oscillator is illustrated in Figure 6.7. But the energy is not quantized. Of course not, because our treatment of harmonic motion so far has been within the framework of classical mechanics.

Now, let us ask a question and see how classical mechanics answers it before we ask the same question in quantum mechanics. The question is When the particle is oscillating about its equilibrium positions, what is the probability density of finding the particle in a specific region along the oscillation path? First, convince yourself that the slower the particle is passing through a given region, the higher is the probability of finding it there, because the more time the particle spends in a region the higher is the probability of finding it in that region. Hence, we can say that the probability dP of finding the particle somewhere between x and $x + dx$ is inversely proportional to its speed, v:

$$\frac{dP}{dx} = \frac{C}{v}$$

where C is the proportionality constant whose value we can determine from the normalization condition; that is, the total probability of finding the particle somewhere along the oscillating path is 1. But first, let us see how the speed varies along the oscillation path by writing it in terms of displacement x.

From Eq. 6.28

$$x = A\sin(2\pi ft)$$

Therefore,

$$v = \frac{dx}{dt} = A2\pi f\cos(2\pi ft) = A2\pi f\sqrt{1-\sin^2(2\pi ft)} = A2\pi f\sqrt{1-\left(\frac{x}{A}\right)^2}$$

Now, as said earlier, the probability of finding the particle somewhere along the oscillating path (from $x = -A$ to $x = +A$) is unity. This statement can mathematically be written as:

$$1 = P = \int_{-A}^{A}\frac{dP}{dx}dx = C\int_{-A}^{A}\frac{1}{v}dx = \frac{C}{A2\pi f}\int_{-A}^{A}\frac{1}{\sqrt{1-\left(\frac{x}{A}\right)^2}}dx = \left(\frac{C}{A2\pi f}\right)(A\pi) = \frac{C}{2f}$$

Therefore,

$$C = 2f$$

By using your calculus skills, you can convince yourself that the value of the integral involved in the above equation is actually $A\pi$.

Substituting the value of C and v back into the original equation for dP/dx, we obtain

$$\frac{dP}{dx} = \frac{2f}{v} = \frac{1}{A\pi\left(\sqrt{1-\left(\frac{x}{A}\right)^2}\right)} \tag{6.32}$$

The probability density is plotted in Figure 6.8 as a function of x in units of A from $-A$ to $+A$. As you can see from the plot, the probability of finding the particle at the end points is maximum because the velocity there is around zero.

Let us see how the things change when we switch to quantum mechanics.

6.6.2 Harmonic Motion in Quantum Mechanics

Question: How do we solve the harmonic motion problem in quantum mechanics? The obvious answer is by solving the Schrödinger equation for the system, such as a particle or an atom of mass m moving along the x dimension with the potential energy $V(x) = \frac{1}{2}\kappa x^2$:

$$\frac{\hbar^2}{2m}\left(\frac{d^2\Psi(x)}{dx^2}\right) + \left(E - \frac{1}{2}\kappa x^2\right)\Psi(x) = 0 \tag{6.33}$$

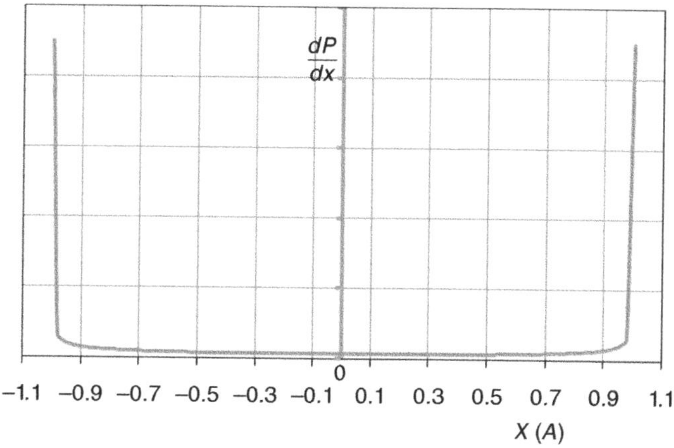

Figure 6.8. Probability density *dP*/*dx* of a classical harmonic oscillator as a function of displacement *x* in units of the maximum displacement *A*.

The solution of Eq. 6.33 turns out to be

$$\Psi_n(x)=\left(\frac{\sqrt{\alpha}}{2^n n!\sqrt{\pi}}\right)^{1/2} H_n\left(\sqrt{\alpha}x\right)e^{-(\alpha x^2/2)} \tag{6.34}$$

where

$$\alpha^2=\frac{m\kappa}{\hbar^2} \tag{6.35}$$

or

$$\alpha=\frac{\sqrt{m\kappa}}{\hbar}$$

Let us take the mystery out of the solutions, shown in Eq. 6.34. The first term depends on the quantum number n and also involves the constant α. The last term in Eq. 6.34 is an exponential function of displacement x and constant α. The term H_n represents a polynomial of degree n called a Hermite polynomial, named after a nineteenth century French mathematician Charles Hermite. Note from Eq. 6.34 that the Hermite polynomial is a function of the displacement x of the oscillating particle from its equilibrium position and the constant α, which depends on the mass of the particle and the spring constant. The Hermite polynomial is given by

$$H_n\left(\sqrt{\alpha}x\right)=(-1)^n e^{\alpha x^2}\frac{d^n}{dx^n}e^{-\alpha x^2} \tag{6.35a}$$

The quantum number n in the Hermite polynomial and in Eq. 6.34 represents the fact that there are many solutions: One corresponding to each value (0 or any positive integer) of n. As an example, Problem 6.6 determines $H_n(\sqrt{\alpha}x)$, and therefore $\Psi_n\left(\sqrt{\alpha}x\right)$ for $n = 0$, and $n = 1$:

$$\Psi_0(x)=\left(\frac{\alpha}{\pi}\right)^{1/4} e^{-(\alpha x^2/2)}$$

$$\Psi_1(x)=\left(\frac{\alpha}{\pi}\right)^{1/4} \left(\sqrt{2\alpha}\,x\right)e^{-(\alpha x^2/2)}$$

PROBLEM 6.6

Determine the wave functions of the harmonic oscillator for $n = 0$ and $n = 1$.

Solution:

From Eq. 6.35a:
for $n = 0$

$$H_0(x)=(-1)^0 e^{x^2}\frac{d^0}{dx^0}e^{-x^2}=1\,e^{x^2}\,e^{-x^2}=1$$

Therefore, from Eq. 6.34

$$\Psi_0(x)=\left(\frac{\sqrt{\alpha}}{2^0 0!\sqrt{\pi}}\right)^{1/2} H_0\left(\sqrt{\alpha}x\right)e^{-(\alpha x^2/2)}=\left(\frac{\sqrt{\alpha}}{\sqrt{\pi}}\right)^{1/2} e^{-(\alpha x^2/2)}=\left(\frac{\alpha}{\pi}\right)^{1/4} e^{-(\alpha x^2/2)}$$

Similarly, for $n = 1$:

$$H_1\left(\sqrt{\alpha}x\right)=(-1)^1 e^{\alpha x^2}\frac{d}{dx}e^{-\alpha x^2}=-1\,e^{\alpha x^2}\,(-2\sqrt{\alpha}x)e^{-\alpha x^2}=2\sqrt{\alpha}x$$

$$\Psi_1(x)=\left(\frac{\sqrt{\alpha}}{2^1 1!\sqrt{\pi}}\right)^{1/2} H_1\left(\sqrt{\alpha}x\right)e^{-(\alpha x^2/2)}=\left(\frac{\sqrt{\alpha}}{2\sqrt{\pi}}\right)^{1/2} 2\sqrt{\alpha}xe^{-(\alpha x^2/2)}=\left(\frac{\alpha}{\pi}\right)^{1/4}\left(\sqrt{2\alpha}\,x\right)e^{-(\alpha x^2/2)}$$

The acceptable solutions are limited by the condition that $n = 0$ or a positive integer. Also, it can be shown that for the wave function given by Eq. 6.34, the corresponding energy E in Eq. 6.33 is given by

$$E_n=\left(n+\frac{1}{2}\right)\hbar\omega \tag{6.36}$$

where

$$\omega = \sqrt{\frac{\kappa}{m}} \qquad (6.37)$$

In Eqs. 6.34 and 6.36 $n = 0, 1, 2 \ldots$

STUDY CHECKPOINT 6.8

Write the equation for E_n in terms of frequency f.

Solution:

From Eq. 6.29

$$f = \frac{1}{2\pi}\sqrt{\frac{\kappa}{m}}$$

By using Eq. 6.37:
which implies

$$\omega = \sqrt{\frac{\kappa}{m}} = 2\pi f$$

Substituting this in Eq. 6.36 yields:

$$E_n = \left(n + \frac{1}{2}\right)\hbar\omega = \left(n + \frac{1}{2}\right)\hbar 2\pi f = \left(n + \frac{1}{2}\right)hf$$

Note from Eq. 6.36 that although $n = 0$ is allowed, the ground-state energy or minimum energy of a harmonic oscillator is not zero; it is $\frac{1}{2}\hbar\omega$, a guest appearance of the uncertainty principle, as shown in Problem 6.7. Ground-state energy in this case is the energy at the equilibrium position around which the system oscillates or vibrates.

PROBLEM 6.7

Prove that the minimum energy allowed by the uncertainty principle for a harmonic oscillator is $\frac{1}{2}\hbar\omega$.

Solution:

According to the uncertainty principle there is always (even in the ground state) an uncertainty of Δx and an uncertainty of Δp in the simultaneous measurement of position and momentum, respectively.

Therefore minimum kinetic energy = $\Delta p^2/2m$

Minimum potential energy = $\kappa\Delta x^2/2$

You also know from your introductory physics course that average kinetic energy of a harmonic oscillator over the range of motion is equal to the average potential energy.

Therefore the minimum total energy is given by

$$E_0 = \frac{\Delta p^2}{2m} + \frac{\kappa\Delta x^2}{2} = \frac{\kappa\Delta x^2}{2} + \frac{\kappa\Delta x^2}{2} = \kappa\Delta x^2 \tag{1}$$

Also, this means

$$\frac{\kappa\Delta x^2}{2} = \frac{\Delta p^2}{2m}$$

which implies

$$\Delta x = \frac{\Delta p}{\sqrt{m\kappa}} \tag{2}$$

According to the uncertainty principle:

$$\Delta x\Delta p \geq \frac{\hbar}{2} \tag{3}$$

By using Eqs. 1–3:

$$E_0 = \kappa\Delta x^2 = \kappa\Delta x\Delta x = \kappa\frac{\hbar}{2\Delta p}\frac{\Delta p}{\sqrt{m\kappa}} = \frac{\hbar}{2}\sqrt{\frac{\kappa}{m}} = \frac{1}{2}\hbar\omega$$

Hence, the proof.

In the atomic and molecular world, this solution is a reality and not just a theoretical play of quantum mechanics. For example, atoms in a molecule vibrate and as a result the bond length between them periodically increases and decreases around an equilibrium value. As we will explore in Chapters 9 and 10, molecules also have a vibrational energy corresponding to these vibrations in addition to the translational and rotational energies corresponding to translational and rotational motion. As another example, the nonzero ground-state energy for ${}^4_2He$ is large enough so that liquid ${}^4_2He$ does not freeze at atmospheric pressure regardless of how cold the system gets. This is because of the harmonic oscillation nature of these molecules. Only quantum mechanics can explain the nonzero ground state, and hence why helium does not freeze.

Caution: Note a difference in the quantum numbers of harmonic oscillators from those of a particle in a box: $n = 0$ is allowed for a harmonic oscillator and not allowed for a particle in a box. However, in either case the ground energy is a nonzero energy.

As shown in Study Checkpoint 6.9, Eq. 6.36 has Planck's law of radiation embedded in it. Although this exhibits the power of the Schrödinger equation, it should not come as a surprise because emission of radiation at both the atomic and at the molecular level can be considered as an act of oscillation: each oscillator representing a mode of EM wave of specific frequency.

STUDY CHECKPOINT 6.9

A. Calculate the difference between the successive energy eigenvalues.
B. Can you relate the observation in part A above to Planck's law discussed in Chapter 2?

Solution:

A.

$$\Delta E = E_{n+1} - E_n = \left[\left(n+\frac{3}{2}\right)-\left(n+\frac{1}{2}\right)\right]\hbar\omega = \hbar\omega = hf$$

B.
According to Planck's law, radiation, E, is emitted in energy steps of hf:

$$E_n = nhf$$

This is exactly what we proved in part A.

As you learned in Chapter 5, the wave function of a system contains all the information about the system. So, let us take a closer look at the wave functions of a harmonic oscillator that we have obtained from the wave equation of the oscillator.

6.7 UNDERSTANDING THE WAVE FUNCTIONS OF A HARMONIC OSCILLATOR

In Section 6.6, we obtained the following wave functions as solutions to the wave equation governing a harmonic oscillator:

$$\Psi_n(x) = \left(\frac{\sqrt{\alpha}}{2^n n!\sqrt{\pi}}\right)^{1/2} H_n\left(\sqrt{\alpha}x\right)e^{-(\alpha x^2/2)} \qquad (6.38)$$

In Section 6.6, we explained the different terms involved in the solutions. Now, let us take a closer look at the wave functions represented by these solutions.

First, these wave functions and the associated energy (presented in Section 6.6) qualify to be called eigenfunctions and energy eigenvalues, discussed in Chapter 5 (e.g., Section 5.7). The energy eigenvalues and eigenfunctions for the first few quantum numbers (n) are shown in Table 6.1. As demonstrated in Problem 6.8 for $n = 0$, these functions are normalized to unity, that is, the probability of finding the oscillator somewhere along the oscillating path is 1.

Note from Table 6.1 that the wave function for $n = 0$ is a simple exponential function of the square of the displacement of the particle from its equilibrium positions. However, for higher quantum states, that is, higher values of n, the wave functions get more and more sophisticated, and hence more interesting. To help you visualize these wave functions, Figures 6.9–6.11 present the plots of these for $n = 0$–5. Note from these figures that as the value of n increases, the wave function gets more spread out. You can imagine that as n approaches infinity (the classical limit), the fluctuations in the wave function will vanish and the quantum mechanical oscillator will behave like a classical oscillator; the correspondence principle.

PROBLEM 6.8

Show that the harmonic oscillator function $\Psi_0(x)$ in Table 6.1 is normalized.

Solution:

$$\int_{-\infty}^{+\infty} \Psi_0(x)^* \Psi_0(x)dx = \left(\frac{\alpha}{\pi}\right)^{1/4} \left(\frac{\alpha}{\pi}\right)^{1/4} \int_{-\infty}^{+\infty} e^{-\alpha x^2} dx = \left(\frac{\alpha}{\pi}\right)^{1/2} 2\int_{0}^{+\infty} e^{-\alpha x^2} dx$$

$$= \left(\frac{\alpha}{\pi}\right)^{1/2} 2\left(\frac{1}{2}\sqrt{\frac{\pi}{\alpha}}\right) = 1$$

In Figures 6.9–6.11, corresponding to each wave function Ψ_n, $|\Psi_n|^2$ is also plotted, which represents the probability of finding the particle at a given displacement x. All these functions are plotted against the displacement x of the oscillator from its equilibrium position shown along the x-axis in units of $1/\sqrt{\alpha}$. This is just another way of saying that the wave function (and probability densities) are plotted as a function of $\sqrt{\alpha}x$.

Note from these plots that when the oscillator is in its ground state, that is, the lowest energy state ($n = 0$), the probability of finding it at its equilibrium position ($x = 0$) is maximum. This is contradictory to classical physics, which predicts, as shown in Figure 6.8, that the probability of finding the oscillator is maximum at the end points of oscillation where the particle moves slowly, and

TABLE 6.1. Energy Eigenvalues and Eigenfunctions for a First Few Quantum Numbers of a Harmonic Oscillator

Quantum Number n	Energy Eigenvalue E_n	$H_n(\sqrt{\alpha}x)$	Energy Eigenfunction $\Psi_n(x) = \left(\frac{\sqrt{\alpha}}{2^n n! \sqrt{\pi}}\right)^{1/2} H_n(\sqrt{\alpha}x) e^{-(\alpha x^2/2)}$
0	$\frac{1}{2}\hbar\omega$	1	$\left(\frac{\alpha}{\pi}\right)^{1/4} e^{-(\alpha x^2/2)}$
1	$\frac{3}{2}\hbar\omega$	$2\sqrt{\alpha}x$	$\left(\frac{\alpha}{\pi}\right)^{1/4} (\sqrt{2\alpha}x) e^{-(\alpha x^2/2)}$
2	$\frac{5}{2}\hbar\omega$	$4(\sqrt{\alpha}x)^2 - 2$	$\left(\frac{\alpha}{\pi}\right)^{1/4} \frac{1}{\sqrt{2}}[2(\sqrt{\alpha}x)^2 - 1] e^{-(\alpha x^2/2)}$
3	$\frac{7}{2}\hbar\omega$	$8(\sqrt{\alpha}x)^3 - 12\sqrt{\alpha}x$	$\left(\frac{\alpha}{\pi}\right)^{1/4} \frac{\sqrt{\alpha}x}{\sqrt{3}}[2(\sqrt{\alpha}x)^2 - 3] e^{-(\alpha x^2/2)}$
4	$\frac{9}{2}\hbar\omega$	$16(\sqrt{\alpha}x)^4 - 48(\sqrt{\alpha}x)^2 + 12$	$\left(\frac{\alpha}{\pi}\right)^{1/4} \frac{1}{2\sqrt{6}}[4(\sqrt{\alpha}x)^4 - 12(\sqrt{\alpha}x)^2 + 3] e^{-(\alpha x^2/2)}$
5	$\frac{11}{2}\hbar\omega$	$32(\sqrt{\alpha}x)^5 - 160(\sqrt{\alpha}x)^3 + 120\sqrt{\alpha}x$	$\left(\frac{\alpha}{\pi}\right)^{1/4} \frac{\sqrt{\alpha}x}{2\sqrt{15}}\left[4(\sqrt{\alpha}x)^4 - 20(\sqrt{\alpha}x)^2 + 15\right] e^{-(\alpha x^2/2)}$

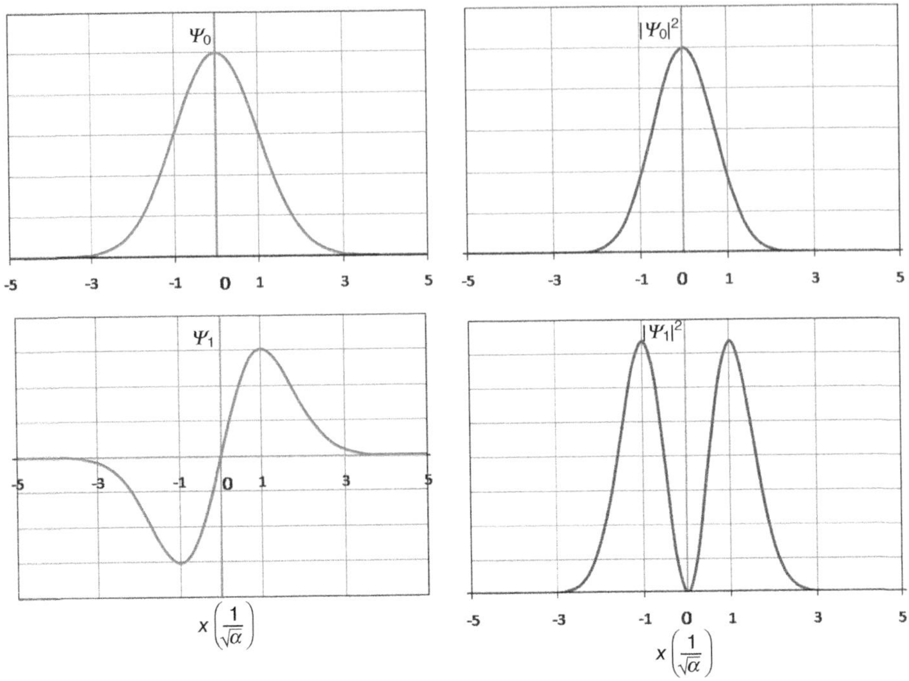

Figure 6.9. Harmonic oscillator wave functions and corresponding probability densities for $n = 0$ and 1 plotted against the displacement x shown along the x-axis in units of $1/\sqrt{\alpha}$.

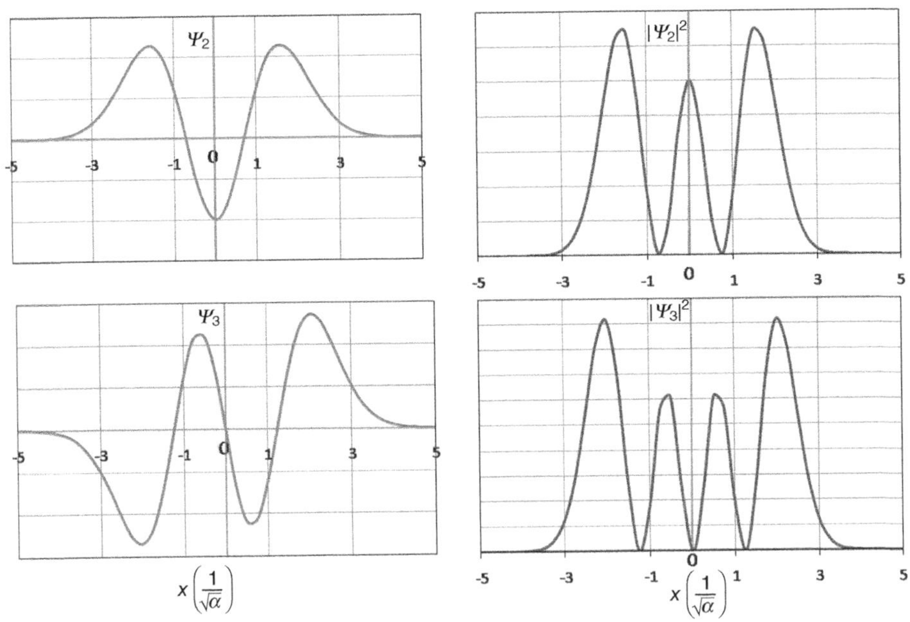

Figure 6.10. Harmonic oscillator wave functions and corresponding probability densities for $n = 2$ and 3 plotted against the displacement x shown along the x-axis in units of $1/\sqrt{\alpha}$.

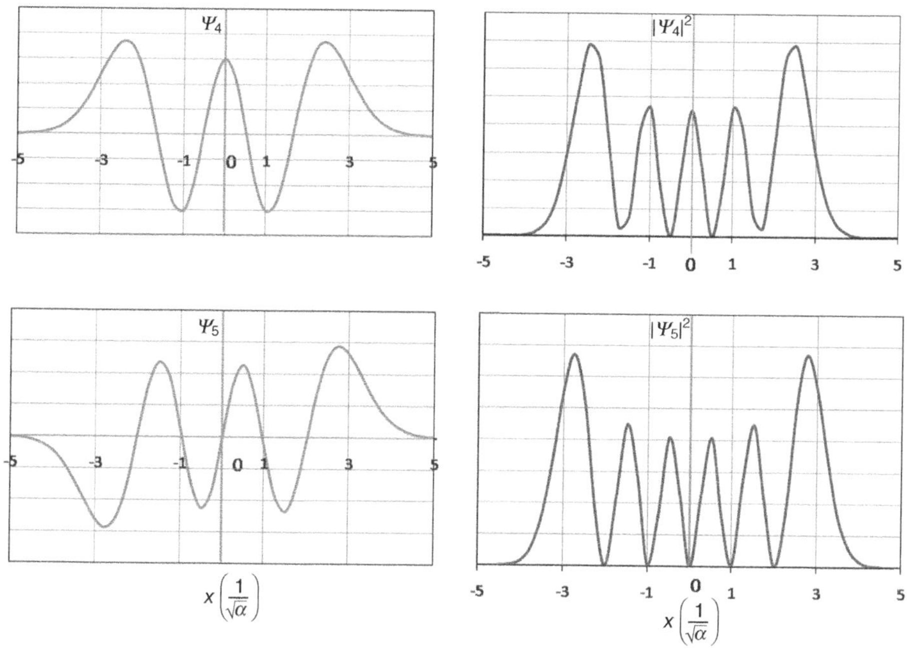

Figure 6.11. Harmonic oscillator wave functions and corresponding probability densities for $n = 4$ and 5 are plotted against the displacement x shown along the x-axis in units of $1/\sqrt{\alpha}$.

is least likely around the equilibrium position ($x = 0$) through which it passes quickly. However, as the plots demonstrate, this disagreement with classical physics becomes less and less marked as the value of quantum number n increases, again in accordance with the correspondence principle.

Next, in Section 6.8, we make a more detailed comparison of the quantum mechanical oscillator with the classical oscillator in order to realize the correspondence principle.

6.8 COMPARING QUANTUM MECHANICAL OSCILLATOR WITH CLASSICAL OSCILLATOR

Let us clarify the terminology. First, let us make it clear that the classical oscillator refers to the description and prediction of an oscillator by classical mechanics, whereas quantum mechanical oscillator refers to the description and predictions of the same oscillator by quantum mechanics. In this section, we compare the probability densities of both classical and quantum oscillators with each other.

From Eq. 6.38, you see that the scaling constant involved in the quantum mechanical probability density is α, whereas in the constant involved in the classical probability density A is the maximum displacement from the equilibrium position, which as already shown, is related to P by the following equation:

$$\frac{dP}{dx} = \frac{1}{A\pi\left(\sqrt{1-\left(\frac{x}{A}\right)^2}\right)} \tag{6.32}$$

As shown in Problem 6.17, α and A are related as follows:

$$A = \sqrt{\frac{2}{\alpha}\left(n+\frac{1}{2}\right)} = \sqrt{\frac{2n+1}{\alpha}}$$

Substituting it back into Eq. 6.32:

$$\frac{dP}{dx} = \frac{1}{A\pi\left(\sqrt{1-\left(\frac{x}{A}\right)^2}\right)} = \frac{1}{\pi}\sqrt{\frac{\alpha}{2n+1}}\frac{1}{\sqrt{1-\frac{\alpha x^2}{2n+1}}}$$

Now, we are in a position to plot both dP/dx and $|\Psi_n|^2$ for a given value of n against $\sqrt{\alpha}x$ for a meaningful comparison.

STUDY CHECKPOINT 6.10

Write the expression for the probability density dP/dx for the classical oscillator corresponding to $n = 1$.

Solution:

$$\frac{dP}{dx} = \frac{1}{\pi}\sqrt{\frac{\alpha}{2n+1}}\frac{1}{\sqrt{1-\frac{\alpha x^2}{2n+1}}}$$

for $n = 1$

$$\frac{dP}{dx} = \frac{1}{\pi}\sqrt{\frac{\alpha}{2+1}}\frac{1}{\sqrt{1-\frac{\alpha x^2}{2+1}}} = \frac{1}{\pi}\sqrt{\frac{\alpha}{3}}\frac{1}{\sqrt{1-\frac{\alpha x^2}{3}}}$$

$$\frac{dP}{dx} = \frac{1}{\pi}\sqrt{\frac{\alpha}{3}}\frac{1}{\sqrt{1-\frac{\alpha x^2}{3}}}$$

As Figure 6.12 demonstrates, the classical and quantum approach are pretty much in opposition to eachother. When the quantum mechanical oscillator is in its ground state, that is, the lowest energy state ($n = 0$), the probability of finding it at its equilibrium position ($x = 0$) is maximum. This is contradictory to the classical approach that predicts the probability of finding the oscillator is maximum at the end points of oscillation, where the particle moves slowly, and it is least around the equilibrium position ($x = 0$) through which it passes quickly. However, as the plots in Section 6.7 demonstrate, this disagreement with classical physics becomes less and less marked as the value of the quantum number n increases, again in accordance with the correspondence principle.

Figure 6.13 for $n = 10$ compared to Figure 6.12 for $n = 0$ clearly shows how the disagreement between the classical and quantum approaches is getting less and less marked. The probability density $|\Psi_{10}|^2$ when smoothed is indeed approaching dP/dx. As n increases the fluctuations in the quantum probability density getting more closely spaced, and hence their effect is getting weaker and their experimental observation is getting more difficult. From this, you can realize that for a large value of quantum number n, quantum physics yields the same result as classical physics; the correspondence principle.

Also, note in Figures 6.12 and 6.13 that the classical probabilities just drop off to zero at the end points, whereas the quantum probabilities have exponential tails there: a demonstration of tunneling. Also note that the exponential tails are getting smaller as n increases and the classical and quantum approaches proceed toward agreement in the macroscopic world.

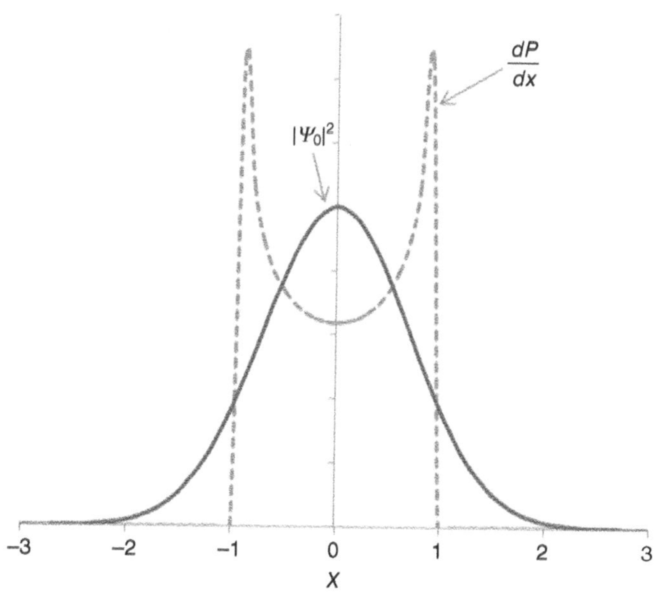

Figure 6.12. Probability density dP/dx of a classical oscillator and the probability density $|\Psi_0|^2$ of a quantum mechanical oscillator for $n = 0$ at the same energies.

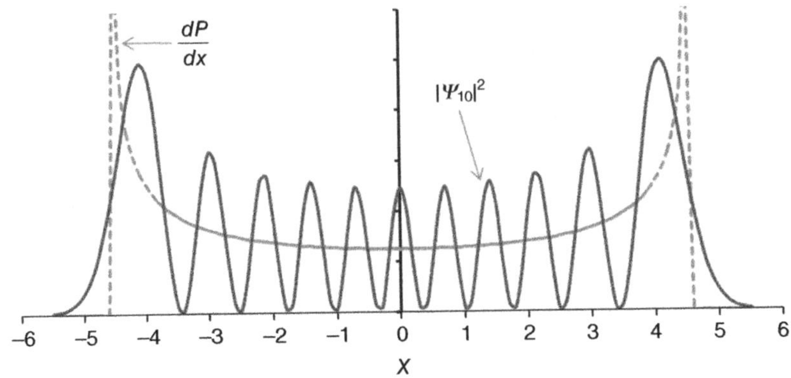

Figure 6.13. Probability density *dP*/*dx* of a classical oscillator and the probability density $|\Psi_{10}|^2$ of a quantum mechanical oscillator for $n = 10$ at the same energies.

Diatomic molecules, such as hydrochloric acid (HCl) and carbon monoxide (CO), are good examples of systems that can be modeled to a harmonic oscillator. The molecule is in equilibrium when the two atoms bonding together in the molecule are at a certain distance from each other where they have a minimum energy. A small displacement from their equilibrium positions increases the molecules energy and the atoms experience a restoring force that tries to push them back to the equilibrium position and harmonic oscillation begins. We will come back to this topic for molecules in Chapter 9.

6.9 LIVING IN THE QUANTUM WORLD

The most rigorous test of a model or a theory is to test its predictions. One of the most daring results of the predictions of the quantum theory of atomic structure was the transistor, which was invented at Bell Laboratories in 1947. Due to its low power consumption and small size, the transistor replaced the vacuum tube and made possible the sophisticated electrical circuits in microprocessors, which started the electronics age, including the computer revolution.

To manipulate material and create devices at the microscale or macroscale level, by using the knowledge of what is happening at the nanoscale level, has been the basis of electronic revolution. By applying the laws of quantum physics, which is the physics of the microscale and nanoscale, we can even manipulate entities at the nanoscale level, such as molecules and atoms. This is also proof of the validity of quantum physics. If you can manipulate atoms a few at a time, it opens the door to a whole new era of electronics and other applications. Some examples are making circuits with atoms and creating special materials with special properties, such as extreme strength or conductivity.

Figure 6.14. Electron corrals created and viewed by manipulating atoms on the Cu surface with a STM. Courtesy of Crommui, Lutz & Eigler, IBM Corporation.

A specialized microscope called a scanning tunneling microscope (STM) uses the quantum mechanical effect called quantum tunneling, discussed in Section 6.4.5, to observe and manipulate atoms. It means you can view a surface with atomic details to an unprecedented resolution of 0.1 nm, about the size of an atom, and move around the atoms at the surface. As an example, Figure 6.14 presents the electron corrals, a circle of ripples created by the electrons on the copper (Cu) surface. The IBM researchers built this corral by manipulating and positioning individual iron (Fe) atoms on a Cu surface by using STM. The atoms are arranged into a ring and the ripples in the ring are the wave patterns of the electrons that were trapped in the corral.

The point is that by understanding and using the laws of quantum physics, researchers and technologists are now able to manipulate matter atom by atom and molecule by molecule. This is what has given birth to the brave new world of nanotechnology.

STUDY CHECKPOINT 6.11

Question:

What is the underlying quantum mechanics principle in the functioning of an electron STM?

Answers:

Quantum tunneling

6.10 SUMMARY

The basic principles of quantum mechanics are built into the Schrödinger equation, which governs the dynamics of systems, such as electrons, atoms, and molecules. Almost all the salient features of quantum mechanics can be logically understood from the probabilistic nature of the solution of the Schrödinger equation. A particle in a box and a harmonic oscillator are two general situations to which an enormous number of structures and systems correspond in the real physical microscopic world. In this chapter, by applying the Schrödinger equation to these two situations, we witnessed some of the features of quantum mechanics naturally arising from solving the equation, such as uncertainty, and quantized states when the particle is confined.

To some students, a harmonic oscillator and a particle in a box may still come across as rather abstract situations. But you will see in the forthcoming chapters how real physical systems can be modeled to correspond to these situations in order to solve problems. You have gained some experience in this chapter with the Schrödinger equation.

In Chapter 7, we will demonstrate the power of the Schrödinger equation by applying it to a very simple, but real, physics system: The hydrogen atom that requires the full complexity of the 3D Schrödinger equation.

6.11 ADDITIONAL PROBLEMS

Problem 6.9 The standing wave of displacement on a string discussed in Section 1.17 can be represented by the following equation:

$$\frac{d^2\Psi}{dx^2} + \frac{4\pi^2}{\lambda^2}\Psi = 0$$

where Ψ depends only on one spatial dimension, x.

A. Compare this equation with the time-independent Schrödinger equation in one spatial dimension in order to determine the value of the wavelength λ.

B. Compare the expression for wavelength obtained in part A above with the de Broglie wavelength.

Problem 6.10 Show that $\sin kx$, $\cos kx$, and $e^{\pm ikx}$ are solutions of the Schrödinger equation, Eq. 6.10, representing a free particle.

Problem 6.11 Show that $A \sin kx + B \cos kx$ is a solution of the Schrödinger equation, Eq. 6.10, representing a free particle inside a potential well.

Problem 6.12 In applying the Schrödinger equation to a particle bounded in a box, we have shown that the value of n appearing in energy and wave function expressions cannot be zero, but it can be negative. Will the solutions to the Schrödinger equation for this problem be different for + and $-n$ in any meaningful way?

Problem 6.13 Consider an electron trapped in a 1D potential box. Assume that the potential outside the box is 2 eV higher than the energy of the electron inside the box. Calulate the probability of finding the electron 2 nm away and outside the box as compared to the probability of finding it at the boundaries of the box.

Problem 6.14 Consider the differential equation describing the harmonic oscillator in classical mechanics:

$$\frac{d^2x}{dt^2}+\frac{\kappa}{m}x=0 \qquad (1)$$

with a solution: $x = A\sin(2\pi ft)$. Caluclate the value of f.

Problem 6.15 A hydrogen atom consists of one proton confined in a nucleus of ~1.6×10^{-15} m. By modeling this situation to a particle confined in a 1D box, calculate the ground-state energy of the proton.

Problem 6.16

A. Write down the wave function for the ground state of a harmonic oscillator.

B. From the wave function in A, calculate:

$$\Delta x=\sqrt{\langle x^2\rangle-\langle x\rangle^2}$$

$$\Delta p=\sqrt{\langle p^2\rangle-\langle p\rangle^2}$$

C. Calculate the value of C in the following:

$$\Delta x\Delta p=C$$

Can you relate it to the Heisenberg uncertainty principle?

Problem 6.17 Prove that the maximum amplitude A of the classical oscillator is related to the constant α in the quantum oscillator as:

$$A=\sqrt{\frac{2}{\alpha}\left(n+\frac{1}{2}\right)}$$

QUANTUM MECHANICS OF AN ATOM 7

An animal that embarks on forming states without greatly restricting egoism will perish.

Erwin Schrödinger

Figure 7.0. *Old Woman Dozing* by Nicolaes Maes (1656), Oil on canvas. Royal Museums of Fine Arts, Brussels.

Quantum Physics for Scientists and Technologists, by Paul Sanghera

Atoms and molecules with unpaired electrons, called free radicals, are responsible for many diseases including old age. Their motion is governed by quantum mechanics.

Although a cell (the smallest unit of life) largely consists of four types of molecules (nucleic acids, proteins, lipids, and carbohydrates), atoms and ions (atoms that have gained or lost one or more electrons) play a crucial role inside an organism. For example, many chemical reactions in biological systems involve a proton (H^+) transfer, and therefore their rates depend on the concentrations of H^+. Because the correct rate of these reactions is crucial to maintain life, the wrong H^+ concentration can result in illness and even death.

A free radical is another example. It is any atom or molecule that has a single unpaired electron in an orbital of its outer shell. An orbital needs two electrons to be complete, and the atom experiences instability due to the lack of a missing electron. In order to gain stability, the atom (free radical) has a very strong tendency to rip an electron from neighboring atoms and molecules. So, most biologically relevant free radicals are highly reactive, and therefore they can participate in unwanted side reactions resulting in cell damage and even deoxyribonucleic acid (DNA) damage. The free radical theory of aging states that organisms age because cells accumulate damage caused by free radical over time. Furthermore, damage caused by free radicals has also been linked to a large number of diseases, including Alzheimer's disease, arthritis, atherosclerosis, cancer, diabetes, heart attack, and Parkinson's disease.

In order to understand and control free radicals, you need to understand that their motion is governed by quantum mechanics.

In This Chapter

7.1 INTRODUCTION

The origin of the name atom comes from the Greek word atoms, which means uncuttable or indivisible: something that cannot be divided further. Ancient Greek philosophers, such as Democritus (460–370 BC), believed that matter must be made up of some fundamental stuff they called atoms. It took centuries after that to scientifically prove what we know today, that all known mater in the universe (living or dead) is made up of atoms, and the atom is made up of a dense central nucleus that contains positively charged protons and possibly electrically neutral neutrons, surrounded by a cloud of negatively charged electrons. Because an atom itself is made up of these subatomic particles, it is not indivisible. However, it remains true that an atom is the smallest building block of a chemical substance.

This explanation shows how important it is to understand an atom in order to understand the universe and its physical systems including living organisms. Quantum mechanics has played an epochal role in the understanding of the universe by providing us insight into the atom. Starting with replacing our classical approach, which did not work, with a quantum approach to atomic phenomena, led us to understand other related phenomena as well, such as the explanation of the periodic table of elements, how the atoms combine to make molecules and crystals, and how the substances made of molecules develop certain electric, magnetic, and mechanical characteristics.

In Chapter 4, you learned how Bohr's model explained the line spectra emitted by atoms. Even though it was the first quantum mechanical (or pseudoquantum mechanical) model, it did not explain the existence of stationary states and hence the quantum number; it just assumed them. Now that we have the Schrödinger equation, we will see how stationary states naturally arise from it.

The hydrogen atom, being the smallest atom, and hence the simplest to study, has been used for experimental studies more than any other atom. So, it should not come as a surprise that we begin this chapter with the hydrogen atom. Yes, this means applying the Schrödinger equation to the hydrogen atom. In doing so, we will demonstrate how most of the postulates put forward by Bohr and others to explain the experimental results can be naturally derived from quantum theory, that is, the Schrödinger equation. For example, you will see how the spectral lines discussed in Chapter 4 are naturally predicted by quantization that results from solving the Schrödinger equation. In fact, it took quantum mechanics to explain why there could be atoms at all. Recall from Chapter 4 that in the classical model of the atom with electrons revolving around a positive nuclei, there is nothing to stop them from collapsing, since positive and negative charges attract.

You will see that three quantum numbers, which emerge naturally when we try to solve the Schrödinger equation for the hydrogen atom, define the state of the hydrogen atom. We will explore the physical meaning of those quantum numbers. We will also explore what quantum theory predicts about the location of an electron in the atom. So, the major question in this chapter is How can the quantum theory in the form of the Schrödinger equation be used to explore a simple atom? To answer that question, we will explore three avenues: applying the Schrödinger equation to the hydrogen atom, quantum numbers of an atom and their physical meanings, and locating an electron inside an atom.

7.2 APPLYING THE SCHRÖDINGER EQUATION TO THE HYDROGEN ATOM

In this section, we will set up the Schrödinger equation for the hydrogen atom. First realize that an atom of any element overall has a zero net charge because it has an equal number of protons in the nucleus and the electrons around the nucleus. When an atom acquires a positive or a negative charge by losing an electron or by gaining an electron, respectively, it is called an ion; a cation if it has acquired a positive charge, or an anion if it has acquired a negative charge.

To set up the Schrödinger equation for the hydrogen atom, we need to determine the form of the potential energy (or potential) term, V, in the equation. A hydrogen atom consists of a proton with electric charge $+e$ and an electron with electric charge $-e$. Therefore, the potential energy of this system is provided by the electrostatic force between the proton and the electron and is called the electrostatic potential. Because a proton is much heavier than an electron (the mass ratio of a proton to an electron is 2000) it can be considered stationary as compared to the electron, which is moving around in the vicinity of the proton. As shown in Problem 7.1, the potential of the hydrogen atom is given by

$$V(r) = -\frac{e^2}{4\pi\varepsilon_0 r} \tag{7.1}$$

where ε_0, the permittivity of vacuum, is a constant with the value of $8.854 \times 10^{-12}\,C^2/N\,m^2$; you have already met this constant in Section 1.13.

PROBLEM 7.1

Derive the equation for the potential shown in Eq. 7.1.

Solution:

Consider an electron with charge $-e$ at a distance r from a proton (nucleus of the hydrogen atom) that has a charge of $+e$. The electrostatic force between the two charges is give by

$$F(r) = -\frac{e^2}{4\pi\varepsilon_0 r^2}$$

Potential energy difference dV is the work done to move the electron by a distance dr away from the nucleus, which implies

$$dV = -\frac{e^2}{4\pi\varepsilon_0 r^2}dr$$

Work done to move it from r to infinity is given by

$$V = V(\infty) - V(r) = -\int_r^\infty \frac{e^2}{4\pi\varepsilon_0 r^2}dr = \frac{e^2}{4\pi\varepsilon_0 r}\Big|_{r=\infty} - \frac{e^2}{4\pi\varepsilon_0 r} = -\frac{e^2}{4\pi\varepsilon_0 r}$$

We assumed that the potential at infinity is zero, because at an infinite distance from the proton, the electron will be free of the proton, that is, no longer bound to it.

From Chapter 6, we know that the time-independent Schrödinger equation in three dimensions (3Ds) written in Cartesian coordinates is

$$\frac{\hbar^2}{2m}\left(\frac{\partial^2\Psi}{\partial x^2} + \frac{\partial^2\Psi}{\partial y^2} + \frac{\partial^2\Psi}{\partial z^2}\right) + (E - V)\Psi = 0 \tag{7.2}$$

This equation can be considered to represent the hydrogen atom with a potential V given by Eq. 7.1. However, this potential is a function of r that is part of the spherical polar coordinates, whereas the rest of the equation is in terms of Cartesian coordinates: x, y, z. To make the whole equation consistent, we can either express V in terms of Cartesian coordinates or transform the rest of the equation into polar coordinates. To take advantage of the radial symmetry of the hydrogen atom, and end up with simpler and more intuitive equations, we decide to transform Eq. 7.2 into spherical polar coordinates: r, θ, ϕ. To make this transformation, we need to know the relationship between the spherical polar coordinates and the Cartesian coordinates, which is illustrated in Figure 7.1, from which we write the following equations:

$$x = r\sin\theta\cos\phi \tag{7.3a}$$

$$y = r\sin\theta\sin\phi \tag{7.3b}$$

$$z = r\cos\theta \tag{7.3c}$$

From these equations, you can prove the following:

$$r = \sqrt{x^2 + y^2 + z^2} \tag{7.3d}$$

The following range of values will cover the 3D space:

$$r \geq 0,\ 0 \leq \theta \leq \pi, \text{ and } 0 \leq \phi \leq 2\pi.$$

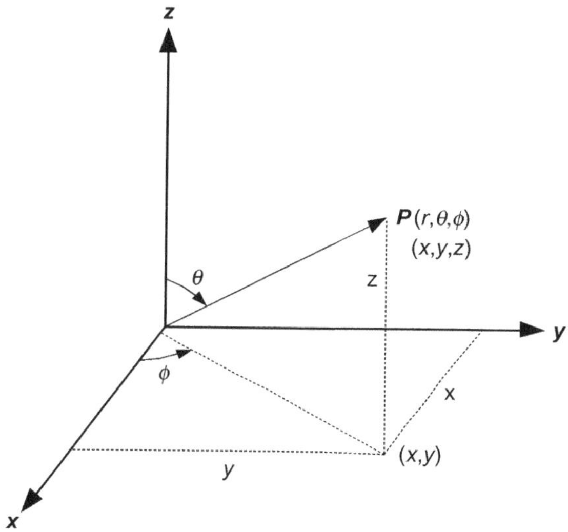

Figure 7.1. The relationship between the Cartesian and the spherical polar coordinates.

STUDY CHECKPOINT 7.1

Prove the relationship:

$$r = \sqrt{x^2 + y^2 + z^2}$$

Solution:

$$\begin{aligned} x^2 + y^2 + z^2 &= r^2 \sin^2\theta \cos^2\phi + r^2 \sin^2\theta \sin^2\phi + r^2 \cos^2\theta \\ &= r^2 \sin^2\theta(\cos^2\phi + \sin^2\phi) + r^2\cos^2\theta \\ &= r^2\sin^2\theta + r^2\cos^2\theta \quad (\text{because } \cos^2\phi + \sin^2\phi = 1) \\ &= r^2(\sin^2\theta + \cos^2\theta) \\ &= r^2 \quad (\text{because } \sin^2\theta + \cos^2\theta = 1) \end{aligned}$$

which implies

$$r = \sqrt{x^2 + y^2 + z^2}$$

By applying the transformation relationships presented in Eqs. 7.3, Eq. 7.2 becomes:

$$\frac{1}{r^2}\frac{\partial}{\partial r}\left(r^2\frac{\partial \Psi}{\partial r}\right) + \frac{1}{r^2 \sin\theta}\frac{\partial}{\partial \theta}\left(\sin\theta\frac{\partial \Psi}{\partial \theta}\right) + \frac{1}{r^2\sin^2\theta}\left(\frac{\partial^2\Psi}{\partial \phi^2}\right) + \frac{2m}{\hbar^2}(E - V)\Psi = 0 \tag{7.4}$$

where Ψ is now a function of r, θ, ϕ. Substituting the value of V from Eq. 7.1 into Eq. 7.4 and multiplying the whole equation by $r^2\sin^2\theta$, we obtain

$$\sin^2\theta\frac{\partial}{\partial r}\left(r^2\frac{\partial\Psi}{\partial r}\right)+\sin\theta\frac{\partial}{\partial\theta}\left(\sin\theta\frac{\partial\Psi}{\partial\theta}\right)+\frac{\partial^2\Psi}{\partial\phi^2}+\frac{2mr^2\sin^2\theta}{\hbar^2}\left(E+\frac{e^2}{4\pi\varepsilon_0 r}\right)\Psi=0 \tag{7.5}$$

Equation 7.5 represents the hydrogen atom. In other words, it represents the dynamics of the electron in the hydrogen atom. In this equation, Ψ is a function of r, θ, ϕ as in $\Psi(r, \theta, \phi)$, but we will denote it as Ψ for brevity.

So, we have set up the Schrödinger equation for the hydrogen atom. The predictions of the quantum theory (i.e., quantum mechanics) about the hydrogen atom will appear when you try to solve this Schrödinger equation.

7.3 SOLVING THE SCHRÖDINGER EQUATION FOR THE HYDROGEN ATOM

The solution to the Schrödinger equation written in polar coordinates will in general depend on the polar coordinates: r, θ, and ϕ. To solve this equation and to understand the solutions, it is convenient to use the method of separation of variables.

7.3.1 Separating the Variables in the Schrödinger Equation

In an attempt to solve the equation, we would like to separate the variables r, θ, ϕ. In other words, we want to look for the solution of this equation that can be written as a product of three functions, each of which depends on only one of the coordinates r, θ, ϕ, as shown in the following:

$$\Psi(r,\theta,\phi)=R(r)\Theta(\theta)\Phi(\phi) \tag{7.6}$$

The goal here is to separate the partial differential equation 7.5, into three separate differential equations, each depending only on one of the variables r, θ, ϕ. From Eq. 7.6, we obtain the following derivatives:

$$\frac{\partial\Psi}{\partial r}=\Theta\Phi\frac{\partial R}{\partial r}$$

$$\frac{\partial\Psi}{\partial\theta}=R\Phi\frac{\partial\Theta}{\partial\theta}$$

$$\frac{\partial^2\Psi}{\partial\phi^2}=R\Theta\frac{\partial^2\Phi}{\partial\phi^2}$$

Substituting these results into Eq. 7.5 and dividing the resulting equation by $R\Theta\Phi$, we obtain

$$\frac{\sin^2\theta}{R}\frac{\partial}{\partial r}\left(r^2\frac{\partial R}{\partial r}\right)+\frac{\sin\theta}{\Theta}\frac{\partial}{\partial\theta}\left(\sin\theta\frac{\partial\Theta}{\partial\theta}\right)+\frac{1}{\Phi}\frac{\partial^2\Phi}{\partial\phi^2}+\frac{2mr^2\sin^2\theta}{\hbar^2}\left(E+\frac{e^2}{4\pi\varepsilon_0 r}\right)=0 \tag{7.7}$$

By rearranging the terms in Eq. 7.7, we obtain

$$\frac{\sin^2\theta}{R}\frac{\partial}{\partial r}\left(r^2\frac{\partial R}{\partial r}\right)+\frac{\sin\theta}{\Theta}\frac{\partial}{\partial\theta}\left(\sin\theta\frac{\partial\Theta}{\partial\theta}\right)+\frac{2mr^2\sin^2\theta}{\hbar^2}\left(E+\frac{e^2}{4\pi\varepsilon_0 r}\right)=-\frac{1}{\Phi}\frac{\partial^2\Phi}{\partial\phi^2} \tag{7.8}$$

If you look closely at Eq. 7.8, you will notice that the right-hand side (rhs) of the equation only depends on the variable ϕ, and the left-hand side (lhs) depends only on r and θ. This means that the lhs will not change with the change in ϕ, and the rhs will not change with the change in r or θ; yet both sides are equal to each other. That means both sides are equal to a constant, that is, something independent of r, θ, ϕ. Let us call this constant m_l^2 for reasons that will become apparent later on in this chapter. Equating the rhs of Eq. 7.8 to m_l^2 and rearranging the terms, we obtain

$$\frac{d^2\Phi}{d\phi^2}+m_l^2\Phi=0 \tag{7.9}$$

We changed the partial derivative to ordinary derivative because Φ only depends on ϕ. This equation is also called the *azimuthal equation* because angle ϕ in spherical coordinates is also called the azimuthal angle. Similarly, equating the lhs of Eq. 7.8 to m_l^2 and rearranging the terms, we obtain

$$\frac{1}{R}\frac{\partial}{\partial r}\left(r^2\frac{\partial R}{\partial r}\right)+\frac{2mr^2}{\hbar^2}\left(E+\frac{e^2}{4\pi\varepsilon_0 r}\right)=\frac{m_l^2}{\sin^2\theta}-\frac{1}{\Theta\sin\theta}\frac{\partial}{\partial\theta}\left(\sin\theta\frac{\partial\Theta}{\partial\theta}\right)$$

Note that the rhs of this equation only depends on the variable θ, and the lhs only depends on the variable r. As argued earlier, both sides have to be equal to the same constant, which we call $l(l+1)$ for reasons that will become apparent further on in this chapter. This yields the following two equations:

Note: The reason for choosing these two constants in the form of m_l^2 and $l(l+1)$ will become apparent later in this chapter.

$$\frac{m_l^2}{\sin^2\theta} - \frac{1}{\Theta\sin\theta}\frac{\partial}{\partial\theta}\left(\sin\theta\frac{\partial\Theta}{\partial\theta}\right) = l(l+1)$$

$$\frac{1}{R}\frac{\partial}{\partial r}\left(r^2\frac{\partial R}{\partial r}\right) + \frac{2mr^2}{\hbar^2}\left(E + \frac{e^2}{4\pi\varepsilon_0 r}\right) = l(l+1)$$

By rearranging the terms in these two equations, we obtain

$$\frac{1}{\sin\theta}\frac{d}{d\theta}\left(\sin\theta\frac{d\Theta}{d\theta}\right) + \left[l(l+1) - \frac{m_l^2}{\sin^2\theta}\right]\Theta = 0 \tag{7.10}$$

$$\frac{1}{r^2}\frac{d}{dr}\left(r^2\frac{dR}{dr}\right) + \left[\frac{2m}{\hbar^2}\left(E + \frac{e^2}{4\pi\varepsilon_0 r}\right) - \frac{l(l+1)}{r^2}\right]R = 0 \tag{7.11}$$

Equation 7.10 is called the angular equation, and Eq. 7.11 is called the radial equation. We changed the partial derivatives to ordinary derivatives because R depends only on r and Θ depends only on θ. So, we have separated the Schrödinger equation into three separate differential equations, each depending only on one r, θ, ϕ. Note that only the equation for R (Eq. 7.11), called the radial equation, depends on the potential energy.

Now, let us examine the azimuthal, angular, and radial equations for their solutions.

7.3.2 Solution of the Azimuthal Equation

The azimuthal equation (Eq. 7.9), the simplest of the three equations (Eqs. 7.9–7.11), is rewritten here:

$$\frac{d^2\Phi}{d\phi^2} + m_l^2\Phi = 0 \tag{7.9}$$

Recognize from Section 6.6 that this is an equation of a harmonic oscillator, and therefore its solution can be written in sines and cosines or equivalently as an exponential function. We conveniently choose the following solution:

$$\Phi = Ae^{im_l\phi} \tag{7.12}$$

This solution involves two constants A and m_l that need to be determined. As Problem 7.4 demonstrates, the normalization condition requires that $A = \sqrt{1/2\pi}$. By substituting this into Eq. 7.12, we obtain the normalized solution:

$$\Phi = \frac{1}{\sqrt{2\pi}}e^{im_l\phi} \tag{7.13}$$

You can verify it by directly substituting this solution into Eq. 7.9. Let us look at the solution closely in order to determine the value of the constant m_l. As Study Checkpoint 7.2 shows, if $Ae^{im_l\phi}$ is the solution, as is $Ae^{im_l(\phi+2\pi)}$, and that

$$\Phi(\phi) = \Phi(\phi + 2\pi)$$

so we can write

$$e^{im_l(\phi+2\pi)} = e^{im_l\phi} \tag{7.14}$$

STUDY CHECKPOINT 7.2

Prove that $\Phi(\phi) = \Phi(\phi + 2\pi)$

Solution:

In order for it to represent a meaningful solution, $\Phi(\phi)$ has to have a single value at a given point in space; and ϕ and $(\phi + 2\pi)$ identify the same point. Therefore,

$$\Phi(\phi) = \Phi(\phi + 2\pi)$$

You can also see that from the identities:

$$\sin(\phi + 2\pi) = \sin\phi$$

$$\cos(\phi + 2\pi) = \cos\phi$$

which implies

$$e^{i(\phi+2\pi)} = e^{i(\phi)}$$

which implies

$$e^{im_l(\phi+2\pi)} = e^{im_l\phi}$$

if $m_l = 0, \pm1, \pm2, \pm3, \cdots$

For Eq. 7.14 to be true, the following has to be true:

$$m_l = 0, \pm1, \pm2, \pm3, \ldots$$

So, we have determined the possible values of the constant m_l. The constant m_l is called the magnetic quantum number. Note that solving the azimuthal part of the Schrödinger equation for a meaningful physical solution has given rise to certain permissible values of m_l, which are discrete values. In other words m_l, is a quantum number.

> *Note*: The magnetic quantum number, m_l, determines the energy shift of an atomic orbital due to an external magnetic field, hence the term *magnetic* in the name. You will learn more about it later in this book.

Now, let us turn our attention to the angular wave equation (Eq. 7.10).

7.3.3 Solutions of the Angular Equation

The angular part of the Schrödinger equation is given by Eq. 7.10:

$$\frac{1}{\sin\theta}\frac{d}{d\theta}\left(\sin\theta\frac{d\Theta}{d\theta}\right)+\left[l(l+1)-\frac{m_l^2}{\sin^2\theta}\right]\Theta=0 \tag{7.10}$$

Note that this equation contains both l and m_l. We showed how the requirement for a physical solution for the azimuthal equation gave rise to the quantization of m_l. It turns out that the angular equation (Eq. 7.10) will have a physical solution if l is an integer equal to or greater than the absolute value of m_l, $|m_l|$. This requirement puts a further constraint on the values of m_l, as shown in the following:

$$m_l = 0, \pm1, \pm2, \pm3, \ldots, \pm l \tag{7.15}$$

The constant l is called the orbital quantum number.

You can verify by differentiating and substituting back into the equation that $(\sqrt{3}/2)\sin\theta$ is a solution of Eq. 7.10 for a certain value of l and m_l, for example, 1 for l and 1 or −1 for m_l. Similarly, you can verify that $(\sqrt{6}/2)\cos\theta$ is a solution for $l = 1, m_l = 0$.

Now, let us consider Eq. 7.11 for the radial part, $R(r)$, of the wave function Ψ.

7.3.4 Solutions of the Radial Equation

The radial part of the Schrödinger equation is given by Eq. 7.11:

$$\frac{1}{r^2}\frac{d}{dr}\left(r^2\frac{dR}{dr}\right)+\left[\frac{2m}{\hbar^2}\left(E+\frac{e^2}{4\pi\varepsilon_0 r}\right)-\frac{l(l+1)}{r^2}\right]R=0 \tag{7.11}$$

Note that this is the equation that contains the total energy E and the expression for potential energy V. It also contains the term that involves the orbital quantum number l. If you have some experience in solving differential equations, you will guess that an exponential solution will satisfy this equation. You can verify by differentiating and substituting back into the equation that the following is a solution for this equation for $l = 0$:

$$R = Ae^{-(r/a_0)}$$

By differentiating

$$\frac{dR}{dr} = -\frac{1}{a_0}Ae^{-(r/a_0)} = -\frac{R}{a_0}$$

Similarly,

$$\frac{d^2R}{dr^2} = \frac{R}{a_0^2}$$

As shown in Problem 7.5, by substituting this solution and the derivatives back into Eq. 7.11, we obtain

$$\left(\frac{1}{a_0^2} + \frac{2m}{\hbar^2}E\right) + \frac{1}{r}\left(\frac{2me^2}{4\pi\varepsilon_0\hbar^2} - \frac{2}{a_0}\right) = 0$$

To satisfy this equation, both terms in the parenthesis must be separately equal to zero. Equating the second term to zero leads to

$$a_0 = \frac{4\pi\varepsilon_0\hbar^2}{me^2}$$

This is the Bohr's radius; naturally arising from the Schrödinger equation demonstrating the predictability of the Schrödinger equation. By equating the first term equal to 0, we obtain

$$E = -\frac{\hbar^2}{2ma_0^2} = -\frac{me^4}{32\pi^2\varepsilon_0^2\hbar^2} = E_0$$

Again, recall from Chapter 4 (Eq. 4.6a) that is equivalent to Bohr's result for the ground-state energy for the hydrogen atom. Notice, as we will discuss later, the results of Bohr's theory naturally arise from these simple solutions of the Schrödinger equation.

We have just shown how a simple solution to Eq. 7.11 gives rise to Bohr's radius. In general, it turns out that this equation will have a solution if the following two requirements are met.

1. Energy E either has a positive (that means the electron is free) or a negative value, as shown below in Eq. 7.16:

$$E_n = -\frac{me^4}{32\pi^2\varepsilon_0^2\hbar^2}\left(\frac{1}{n^2}\right) \tag{7.16}$$

where n is a positive integer, that is, $n = 1, 2, 3, \cdots$

2. The value of n must be $\geq(l + 1)$, where l is the orbital quantum number. In other words, l must be $<n$. This means the following:

$$l = 0, 1, 2, 3, \ldots, (n-1) \tag{7.17}$$

The integer n is called the principal quantum number.

> *Note*: The relationships between the three quantum numbers can be summarized as:
>
> $$n > 0$$
>
> $$l < n \quad \text{and} \quad l \geq 0$$
>
> $$|m_l| \leq l$$
>
> So, for a given value of n, you can determine the possible values of l and m_l.

A negative value of E_n means the electron is bound to the nucleus in the atom. Note that Eq. 7.16 for energy E is identical with that obtained from Bohr's atomic model, covered in Chapter 4. Also note that the Schrödinger equation has yielded three quantum numbers from 3Ds of motion. As mentioned above, the values of these quantum numbers are constrained by the following:

Principal quantum number: $n = 1, 2, 3, \cdots$
Orbital quantum number: $l = 0, 1, 2, 3, \cdots, (n - 1)$

Magnetic quantum number: $m_l = 0, \pm1, \pm2, \pm3, \cdots, \pm l$

> *Note*: Note that quantum numbers n, l, and m_l appeared naturally from the condition that the electron is bound to the atom. Again, a reminder that quantum numbers appear for particles confined to specific regions in space.

Although for sake of simplicity we have not shown the details for solving the Schrödinger equation, we have shown what the simple solutions for the azimuthal, angular, and radial parts look like and how they lead to important results. Let us take a look at what the general solutions look like.

7.3.5 Solutions of the Schrödinger Equation for the Hydrogen Atom: Putting It All Together

To show the dependence of solutions R, Θ, Φ on the quantum numbers n, l, m_l, we rewrite Eq. 7.6:

$$\Psi(r, \theta, \phi) = R_{nl}\Theta_{lm_l}\Phi_{m_l} \tag{7.18}$$

We have presented and discussed some simple solutions for certain values of these quantum numbers. As it is sufficiently complicated, we do not present the details of solving this equation here for $\Psi(r, \theta, \phi)$. However, the functions R_{nl}, Θ_{lm_l}, Φ_{m_l} are presented in Table 7.1 for first few values of n, l, m_l. Corresponding to each row in this table, you can write a function $\Psi(r, \theta, \phi)$ from Eq. 7.18, for example, for $(n, l, m_l) = (2, 1, 0)$:

$$\Psi(r, \theta, \phi) = \frac{1}{4\sqrt{2\pi}a_0^{3/2}}\frac{r}{a_0}e^{-r/2a_0}\cos\theta$$

where

$$a_0 = \frac{4\pi\varepsilon_0\hbar^2}{me^2}$$

is the Bohr radius.

STUDY CHECKPOINT 7.3

In Table 7.1, why are the expressions for $R(r)$ identical for $(n, l, m) = (3, 1, 0)$ and $(3, 1, 1)$?

Solution:

Because R varies with only n and l; and (n, l) values for both of the given cases are identical.

Instead of getting overwhelmed with the functions presented in Table 7.1, let us have fun by making sense of them. First, note that the azimuthal and angular solutions are rather simple, whereas the radial solutions start of being simple for $(n, l) = (1, 0)$, and quickly become more sophisticated for higher

TABLE 7.1. Radial, Angular, and Azimuthal Wave Functions for the Hydrogen Atom for the First Few Values of *n*

(n, l, m_l)	$R(r)$	$\Theta(\theta)$	$\Phi(\phi)$
(1, 0, 0)	$\frac{2}{a_0^{3/2}} e^{-r/a_0}$	$\frac{1}{\sqrt{2}}$	$\frac{1}{\sqrt{2\pi}}$
(2, 0, 0)	$\frac{1}{2\sqrt{2}a_0^{3/2}}\left(2-\frac{r}{a_0}\right)e^{-r/2a_0}$	$\frac{1}{\sqrt{2}}$	$\frac{1}{\sqrt{2\pi}}$
(2, 1, 0)	$\frac{1}{2\sqrt{6}a_0^{3/2}}\frac{r}{a_0}e^{-r/2a_0}$	$\frac{\sqrt{6}}{2}\cos\theta$	$\frac{1}{\sqrt{2\pi}}$
(2, 1, ±1)	$\frac{1}{2\sqrt{6}a_0^{3/2}}\frac{r}{a_0}e^{-r/2a_0}$	$\frac{\sqrt{3}}{2}\sin\theta$	$\frac{1}{\sqrt{2\pi}}e^{\pm i\phi}$
(3, 0, 0)	$\frac{2}{81\sqrt{3}a_0^{3/2}}\left(27-18\frac{r}{a_0}+2\frac{r^2}{a_0^2}\right)e^{-r/3a_0}$	$\frac{1}{\sqrt{2}}$	$\frac{1}{\sqrt{2\pi}}$
(3, 1, 0)	$\frac{4}{81\sqrt{6}a_0^{3/2}}\left(6-\frac{r}{a_0}\right)\frac{r}{a_0}e^{-r/3a_0}$	$\frac{\sqrt{6}}{2}\cos\theta$	$\frac{1}{\sqrt{2\pi}}$
(3, 1, 1)	$\frac{4}{81\sqrt{6}a_0^{3/2}}\left(6-\frac{r}{a_0}\right)\frac{r}{a_0}e^{-r/3a_0}$	$\frac{\sqrt{3}}{2}\sin\theta$	$\frac{1}{\sqrt{2\pi}}e^{\pm i\phi}$
(3, 2, 0)	$\frac{4}{81\sqrt{30}a_0^{3/2}}\frac{r^2}{a_0^2}e^{-r/3a_0}$	$\frac{\sqrt{10}}{4}(3\cos^2\theta-1)$	$\frac{1}{\sqrt{2\pi}}$
(3, 2, ±1)	$\frac{4}{81\sqrt{30}a_0^{3/2}}\frac{r^2}{a_0^2}e^{-r/3a_0}$	$\frac{\sqrt{15}}{2}\sin\theta\cos\theta$	$\frac{1}{\sqrt{2\pi}}e^{\pm i\phi}$
(3, 2, ±2)	$\frac{4}{81\sqrt{30}a_0^{3/2}}\frac{r^2}{a_0^2}e^{-r/3a_0}$	$\frac{\sqrt{15}}{4}\sin^2\theta$	$\frac{1}{\sqrt{2\pi}}e^{\pm 2i\phi}$

values of the quantum numbers. Keeping things in perspective, these solutions represent various quantum states of the electron in a hydrogen atom. Being more interesting, the radial wave functions (solutions) are plotted in Figure 7.2. Figure 7.2a presents the wave function corresponding to the 1s state of the electron, that is, $n = 1$ and $l = 0$, whereas Figure 7.2b presents the wave functions for the 2s ($n = 2, l = 0$), and 2p ($n = 2, l = 1$) states. The wave functions for the 3s ($n = 3, l = 0$), 3p ($n = 3, l = 1$), and 3d ($n = 3, l = 2$) states are presented in Figure 7.2c. All wave functions are plotted against the atomic radius r in units of the Bohr radius a_0.

Note from Figure 7.2 that the wave function R is maximum at $r = 0$, that is at the nucleus, for all s ($l = 0$) states, whereas its value is zero at $r = 0$ for p, d, and f states, which have a nonzero value for l, and hence for angular momentum.

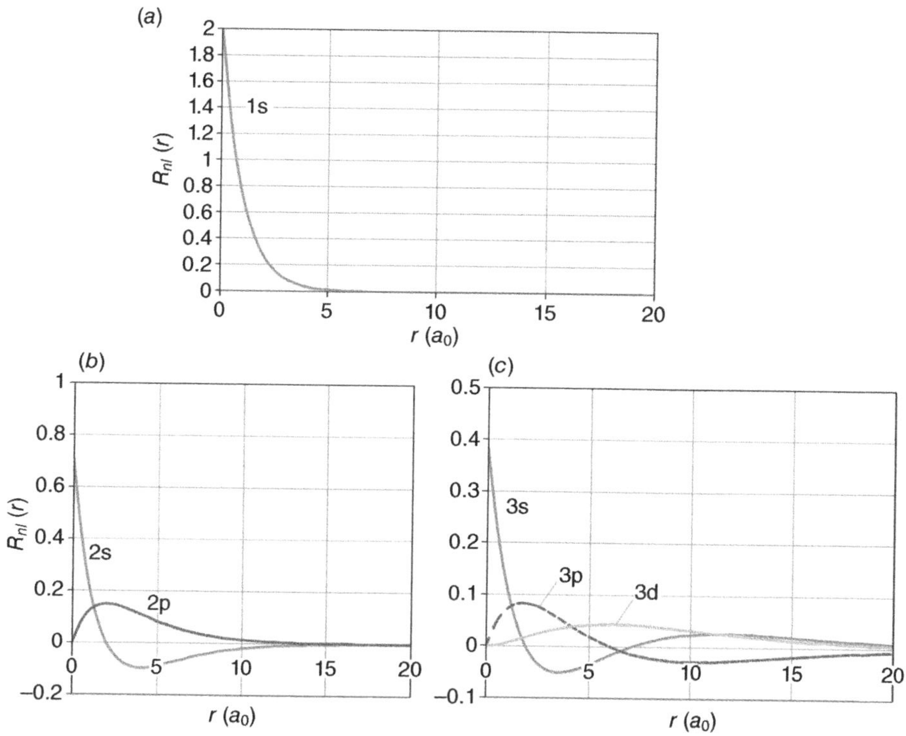

Figure 7.2. The radial wave function of the electron in a hydrogen atom as a function of the radial distance from the nucleus shown along the x-axis in units of the Bohr radius a_0. The wave function is shown for various quantum states of the electron.

STUDY CHECKPOINT 7.4

The radial wave function R is maximum at $r = 0$ for the s state. Because the wave function represents the probability of finding the particle, does it mean that the electron in the hydrogen atom can be found in the nucleus?

Solution:

The answer is no.

Yes, we usually say that the wave function of a particle represents the probability of finding the particle. However, strictly speaking, the probability density for finding the particle is proportional to the square of the mod of the function, that is, $|R|^2$ in this case.

You will see in Section 7.4 that the probability density for finding the electron along the atomic radius is actually $r^2|R|^2$.

Now that these functions make sense, let us get a bit more ambitious in understanding them. You know that the wave function of a particle represents the probability of finding the particle. So, let us ask a question What do these functions tell us about where the electron is in the hydrogen atom?

7.4 FINDING THE ELECTRON

As you learned in Chapter 4, according to the Bohr's model, electrons in an atom orbit around the nucleus with a well-defined radius r. For example, the radius of the innermost orbit is a_0, where the electron in a ground-state hydrogen atom will be found. However, in quantum theory, you can only talk about the probability of finding the electron and no definite values of r, θ, and ϕ about the location of the electron can be predicted: an expression of the wave nature of the electron and the uncertainty principle. The probability density of finding the electron with wave function Ψ is given by $|\Psi|^2$.

Furthermore, contrary to Bohr's model, we cannot think of an electron as revolving around the nucleus in an atom in any conventional sense because $|\Psi|^2$ is independent of time and varies over space.

We already have a wealth of wave functions for the electron in the hydrogen atom that we examined in Table 7.1. Let us see how we can use these wave functions to examine the probabilities of finding the electron at various places inside the atom.

Let us start with looking for the electron around the z-axis, that is, along the azimuthal direction by varying the azimuthal angle ϕ. Using the azimuthal wave function from Section 7.3.5, the azimuthal probability density can be written as:

$$|\Phi|^2 = Ae^{im_l\phi}Ae^{-im_l\phi} = A^2$$

where A is a constant.

Therefore, the probability of finding the electron at any azimuthal angle ϕ is a constant. In other words, the electron's probability density is symmetric around the z-axis regardless of the quantum state of the electron. The electron is equally likely to be found at any azimuthal angle. This result is interesting, indeed.

Now let us try to find the electron along the radius of the atom. In other words, how does the probability of finding the electron change as we move away from the nucleus of the atom. In general, the probability dP of finding the electron in a differential volume element dV is given by

$$dP = |\Psi|^2 \, dV = \Psi^* \Psi dV = |\Psi(r,\theta,\phi)|^2 \, dV \tag{7.19}$$

where

$$\Psi(r,\theta,\phi) = R(r)\Theta(\theta)\Phi(\phi) \tag{7.6}$$

and dV in spherical polar coordinates is given by

$$dV = r^2 \sin\theta \, dr \, d\theta \, d\phi \tag{7.20}$$

Now combining these equations with Eq. 7.19, we can write the probability $P(r)dr$ of finding the electron somewhere in the spherical shell between r and $r + dr$ as:

$$P(r)dr = \left(\int |\Psi(r,\theta,\phi)|^2 \, dV\right) dr = r^2 |R|^2 \, dr \int_0^{\pi} |\Theta|^2 \sin\theta \, d\theta \int_0^{2\pi} |\Phi|^2 \, d\phi = r^2 |R|^2 \, dr$$

Therefore,

$$P(r)dr = r^2 |R|^2 \, dr \tag{7.21}$$

In during Eq. 7.21 we have used the fact that $\Theta(\theta)$ and $\Phi(\phi)$ are normalized functions, that is, there integrals are equal to 1. Note from Eq. 7.21, that although the electron wave function for s states, as we noted in Section 7.3.5, is maximum at the nucleus ($r = 0$), the probability density is still zero here due to the term r^2. Hence, the Schrödinger equation saves the electron from collapsing into the nucleus. To help you visualize where this electron spends most of its time in the atom around the nucleus, we have plotted the probability density presented in Eq. 7.21 in Figure 7.3 for all the quantum states for which we plotted the wave functions in Section 7.3.5. Note from Figure 7.3 that the probability density, P, for the s states is zero at the nucleus and is maximum when it is closer to the nucleus, but at a certain distance from it. Precisely speaking, the maximum P for the 1s state is exactly at $r = a_0$, the Bohr radius. Note that the probability distributions are different for different quantum states, as they are epected to be.

As an example, consider the hydrogen atom in its ground state. From Table 7.1, the wave function of the electron in this atom is given by

$$\Psi = \frac{1}{\sqrt{\pi} a_0^{3/2}} e^{-r/a_0} \tag{7.22}$$

Therefore the average (or expectation) value of $1/r$ of an electron in the ground-state hydrogen atom is

$$\left\langle \frac{1}{r} \right\rangle = \int_0^{\infty} \frac{1}{r} |\Psi|^2 \, dV = \frac{1}{\pi a_0^3} \int_0^{\infty} r e^{-2r/a_0} \, dr \int_0^{\pi} \sin\theta \, d\theta \int_0^{2\pi} d\phi$$

$$= \frac{1}{\pi a_0^3} \frac{a_0^2}{4} (2) 2\pi = \frac{1}{a_0}$$

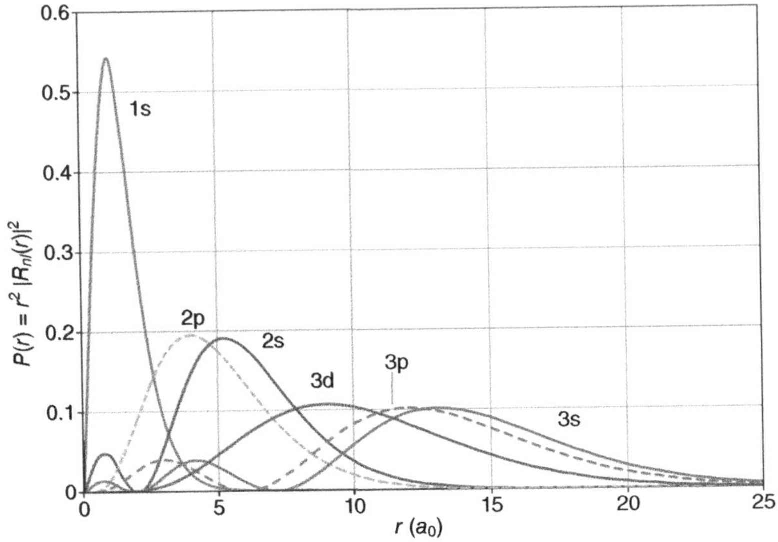

Figure 7.3. The probability density of finding the electron in a hydrogen atom along the radial distance from the nucleus shown along the x-axis in units of the Bohr radius a_0. The probability distributions are shown for various quantum states of the electron.

Recall that a_0 is the Bohr radius with the value of 0.529 Å.

As Problem 7.9 shows the total energy of the electron will be equal to its potential energy at $r = 2a_0$. Therefore, thinking in terms of classical mechanics, this is the maximum radius an electron in the hydrogen atom can have. In other words, the electron cannot exist outside the sphere of radius $2a_0$ with a proton at the center. However, as Problem 7.9 demonstrates, according to quantum mechanics there is a significant probability that an electron can be found outside this sphere. Although we are using the hydrogen atom as an example here, this result is very general. This is one of the very important points that distinguishes quantum mechanics from classical mechanics: You can find particles where Newtonian physics says they cannot go. Think of the implications of this while trying to understand the various phenomena in chemistry and biology at the molecular and particle level. For example, think of ions passing through membranes to carry information through the nervous system of our bodies, electrons and protons in motion in the body cells facilitating necessary processes, such as formation of energy to keep us alive and moving. And the motion of these particles is governed by quantum mechanics. So you can see that, if Newtonian physics were governing in the microworld instead of quantum mechanics, we would be living in a very different macroworld.

The beauty of the Schrödinger equation is that it generates quantum numbers naturally. As you have seen when we solve the Schrödinger equation for the hydrogen atom, we obtain three quantum numbers that are integers and have a relationship with each other. But, what do these quantum numbers really represent? In other words, what is their physical significance?

7.5 UNDERSTANDING THE QUANTUM NUMBERS

As you saw in Section 7.3, three quantum numbers n, l, and m_l arise naturally while attempting to solve the 3D Schrödinger equation for the hydrogen atom. What these numbers mean is that the atom can only be in a state that is defined by this triplet. In other words, the set of states for the hydrogen atom is determined by the set of triplet quantum numbers (n, l, m_l). Furthermore, the solutions to the Schrödinger equation require that the three quantum numbers are integers and their values are constrained by the following conditions:

$$\begin{aligned}
&\text{Principal quantum number: } n = 1, 2, 3, \ldots \quad n > 0 \\
&\text{Orbital quantum number: } l = 0, 1, 2, 3, \ldots, (n-1) \quad l < n \\
&\text{Magnetic quantum number: } m_l = -l, -(l+1), \ldots, 0, 1, 2, \ldots, \\
&\quad (l-2), (l-1), +l \quad |m_l| \le l
\end{aligned} \tag{7.23}$$

Put in this form, note that the principal quantum number n is the only quantum number that does not depend on any other quantum number. So we begin our discussion with this quantum number.

7.5.1 The Principal Quantum Number and Energy Radiations

As mentioned earlier, the principal quantum number n arises when we try to solve the radial wave equation for the hydrogen atom, Eq. 7.11. As in the case of a particle in a box and a harmonic oscillator, the potential energy that represents the confinement of the particle also quantizes its energy. One of the conditions for the radial equation to have a solution is

$$E_n = -\frac{me^4}{32\pi^2\varepsilon_0^2\hbar^2}\left(\frac{1}{n^2}\right) \tag{7.24}$$

Presenting a detailed method to solve the radial wave equation is beyond the scope of this book. However, we can verify that Eq. 7.17 is true for the ground state that is $n = 1$ and $l = 0$. As shown in Table 7.1, the solution for the radial wave equation for the ground state ($n = 1$, $l = 0$) is

$$R(r) = \frac{2}{a_0^{3/2}} e^{-r/a_0} \tag{7.25}$$

As demonstrated in Problem 7.5, this solution can be substituted into the radial wave equation to yield

$$E_n = -\frac{me^4}{32\pi^2\varepsilon_0^2\hbar^2}$$

This is exactly Eq. 7.24 for $n = 1$.

So, this formula for quantized energy that Bohr obtained from the classical approach by postulating the principal quantum number n, arises naturally from the Schrödinger equation. The Bohr model also postulates that as long as an electron is in this state (value of n), it does not radiate any energy, and it radiates or absorbs energy when it jumps from one state to another. Can we derive this postulate from the Schrödinger equation as well? The answer is yes, and we demonstrate it next.

In Section 6.2, we showed that the following will be a valid solution for the time-dependent Schrödinger equation:

$$\Psi_n(x,t) = \psi_n(x)e^{-i\omega t} = \psi_n(x)e^{-i(E_n/\hbar)t} \tag{7.26}$$

Therefore,

$$\Psi_n(x,t)^* = \psi_n(x)^* e^{+i(E_n/\hbar)t} \tag{7.27}$$

For simplicity, we are considering the motion of the electron only in one dimension, the x dimension. The electron can be located at x with the following expectation value:

$$\begin{aligned}\langle x\rangle &= \int_{-\infty}^{+\infty} \Psi_n(x,t)^* x\Psi_n(x,t)dx = \int_{-\infty}^{+\infty} x\psi_n(x)^* \psi_n(x)e^{-i(E_n/\hbar)t+i(E_n/\hbar)t}dx \\ &= \int_{-\infty}^{+\infty} x\psi_n(x)^* \psi_n(x)dx \end{aligned} \tag{7.28}$$

Equation 7.28 shows that the expectation value of x is independent of time. This means that the electron in a specific state does not oscillate, and therefore does not radiate. This is a postulate of Bohr's model that comes here naturally from quantum theory. Now consider the case that the electron transitions from state n to the excited state m and then jumps back to state n. During the time the electron is transitioning from state m to the state n, it can be found in either of these two states with some probability. Therefore, the wave function ψ of the electron can be written as follows:

$$\Psi = a\Psi_m + b\Psi_n$$

where a^*a (modules of a^2) is the probability that the electron is in state m and b^*b (modules of b^2) is the probability that it is in state n.

The expectation value of x during this transition can be written as follows:

$$\langle x\rangle=\int_{-\infty}^{+\infty}(a^*\Psi_m^*+b\Psi_n^*)x(a\Psi_m+b\Psi_n)dx$$

$$=a^2\int_{-\infty}^{+\infty}x|\Psi_m|^2\,dx+b^2\int_{-\infty}^{+\infty}x|\Psi_n|^2\,dx+a^*b\int_{-\infty}^{+\infty}x\Psi_m^*\Psi_n dx+b^*a\int_{-\infty}^{+\infty}x\Psi_n^*\Psi_m dx$$

$$=a^2\int_{-\infty}^{+\infty}x|\psi_m|^2\,dx+b^2\int_{-\infty}^{+\infty}x|\psi_n|^2\,dx+a^*b\int_{-\infty}^{+\infty}x\psi_m^*\psi_n e^{(it/\hbar)(E_m-E_n)}dx+$$

$$b^*a\int_{-\infty}^{+\infty}x\psi_n^*\psi_m e^{-(it/\hbar)(E_m-E_n)}dx$$

(by using Eqs. 7.27 and 7.28)

By using

$$e^{i\theta}=\cos\theta+i\sin\theta$$

and

$$e^{-i\theta}=\cos\theta-i\sin\theta$$

the time-dependent part of the above equation becomes:

$$\langle x\rangle|_t=\cos\left(\frac{E_m-E_n}{\hbar}\right)t\int_{-\infty}^{+\infty}(a^*b\psi_m^*\psi_n+b^*a\psi_n^*\psi_m)xdx+$$
$$i\sin\left(\frac{E_m-E_n}{\hbar}\right)t\int_{-\infty}^{+\infty}(a^*b\psi_m^*\psi_n-b^*a\psi_n^*\psi_m)xdx$$

Considering only the real part <x> varies with time as:

$$\cos\left(\frac{E_m-E_n}{\hbar}\right)t=\cos 2\pi\left(\frac{E_m-E_n}{h}\right)t=\cos 2\pi ft$$

Therefore the electron oscillates with a frequency:

$$f=\left(\frac{E_m-E_n}{h}\right) \tag{7.29}$$

This means that when the electron is in the process of transitioning from state m to state n, it oscillates with frequency f given by Eq. 7.29. This oscillating electron acts like an electric dipole and therefore emits EM radiation of frequency f. So quantum theory naturally predicts the following:

1. An electron in the hydrogen atom in a specific state does not radiate any energy.
2. An electron transitioning from an excited state m to lower state n radiates energy given by the following:

$$E = hf = E_m - E_n$$

You saw in Section 4.6.1 that these were the postulates of the Bohr model to explain the experimental results. However, as you have seen, the quantum theory predicts these results in a natural fashion.

STUDY CHECKPOINT 7.5

What will be the energy difference between the following (n, l, m_l) states of the hydrogen atom?

(2, 0, 0), (2, 1, 0), (2, 1, −1), (2, 1, 1)

Solution:

The energy for all these states will be the same because the energy only depends on the principal quantum number n, which is the same for all of these states.

So, the principal quantum number defines the stationary states of the electron in the atom by determining and quantizing its energy. What is the orbital quantum number good for?

7.5.2 The Orbital Quantum Number

As you witnessed earlier in this chapter, the orbital quantum number l arises from the radial and angular part of the Schrödinger equation and its wave function: $R(r)$ and $\Theta(\theta)$. We will show that this quantum number is related to the angular momentum. This angular momentum of the atom exists as the oppositely charged electron and proton in the hydrogen atom move around each other.

7.5.2.1 Physical Meaning of the Orbital Quantum Number. Let us draw the physical meaning of the orbital quantum number from the radial Schrödinger equation of the hydrogen atom:

$$\frac{1}{r^2}\frac{d}{dr}\left(r^2\frac{dR}{dr}\right)+\left[\frac{2m}{\hbar^2}\left(E+\frac{e^2}{4\pi\varepsilon_0 r}\right)-\frac{l(l+1)}{r^2}\right]R=0 \tag{7.11}$$

The energy E appearing in this equation is the total energy that consists of kinetic energy due to radial motion T_r, kinetic energy due to orbital motion T_o, and the potential energy V:

$$E = T_r + T_o + V = T_r + T_o - \frac{e^2}{4\pi\varepsilon_0 r}$$

Substituting this expression for E into Eq. 7.11 and rearranging terms, we obtain

$$\frac{1}{r^2}\frac{d}{dr}\left(r^2\frac{dR}{dr}\right)+\frac{2m}{\hbar^2}\left[T_r+T_o-\frac{\hbar^2 l(l+1)}{2mr^2}\right]R=0 \tag{7.30}$$

The kinetic energy of the electron due to its orbital motion can be written as:

$$T_o=\frac{1}{2}mv_o^2$$

where v_o is the orbital speed of the electron. The magnitude of angular momentum of the electron due to this motion can be written as:

$$L=mv_o r$$

which implies

$$v_o=\frac{L}{mr}$$

Substituting this into the expression for T_o:
which implies

$$T_o=\frac{L^2}{2mr^2} \tag{7.31}$$

Equation 7.30 is a radial equation, therefore the nonradial energy must be zero. This means

$$T_o-\frac{\hbar^2 l(l+1)}{2mr^2}=0 \tag{7.32}$$

From Eqs. 7.31 and 7.32, we obtain

$$T_o=\frac{\hbar^2 l(l+1)}{2mr^2}=\frac{L^2}{2mr^2}$$

Therefore,

$$L=\hbar\sqrt{l(l+1)} \tag{7.33}$$

Because l is an integer, the angular momentum of the electron is quantized just like the energy. Again we have obtained a principle from quantum mechanics that Bohr's model postulated to explain experimental results: orbital momentum is quantized. However, in this case, quantum mechanics disagrees with Bohr's model on the detail. In Bohr's model, the quantization is given by $L = n\hbar$, whereas quantum theory predicts $L = l(l + 1)\hbar$. This is a demonstration of how quantum theory goes beyond Bohr's semiclassical planet-like model of electrons orbiting the nucleus. As shown in Problem 7.2, for an electron with $l = 3$, $L = 3.67 \times 10^{-34}$ J s. Compare it with 2.66×10^{40} J s as the angular momentum of Earth around the Sun, as calculated in Problem 7.6.

PROBLEM 7.2

Calculate the angular momentum of a particle in a state of $l = 3$.

Solution:

$$\begin{aligned} L &= \hbar\sqrt{l(l+1)} \\ &= \hbar\sqrt{3(3+1)} \\ &= \hbar\sqrt{12} \end{aligned}$$

Given $\hbar = 1.06 \times 10^{-34}$ J s

$$L = 3.67 \times 10^{-34} \text{ J s}$$

Compare it with 2.66×10^{40} J s as the angular momentum of Earth around the Sun.

You must have already seen these quantum numbers in chemistry.

7.5.2.2 *Making the Connection with Chemistry.* It is common, especially in chemistry, to denote angular quantum number values in letters: s for $l = 0$, p for $l = 1$, d for $l = 2$, and so on. Table 7.2 presents the atomic electron states for the first few (n, l) values in the terminology used in chemistry. Because the energy of an electron in the atom only depends on n, it is also called the principal energy level, and accordingly l states are called sublevels. Note that the number of sublevels is equal to the principal quantum number. The sublevels that appear in bold in Table 7.2 are not needed by the elements that are presently known. We will continue this exploration in Chapter 8 when we discuss the periodic table of elements in the context of quantum mechanics.

Note: Our discussion of quantum numbers here is not limited to the hydrogen atom. This applies to electrons in any atom, and also to any particle or body that is confined in a certain region by a potential and has a radial and orbital motion.

TABLE 7.2. Electron States in an Atom

n	*l*	Sublevels[a]	Number of Sublevels
1	0	1s	1
2	0, 1	2s, 2p	2
3	0, 1, 2	3s, 3p, 3d	3
4	0, 1, 2, 3	4s, 4p, 4d, 4f	4
5	0, 1, 2, 3, 4	5s, 5p, 5d, 5f, **5g**	5
6	0, 1, 2, 3, 4, 5	6s, 6p, 6d, **6f, 6g, 6h**	6
7	0, 1, 2, 3, 4, 5, 6	7s, **7p, 7d, 7f, 7g, 7h, 7i**	7

[a]The sublevels that appear in boldface are not needed by the elements that are presently known.

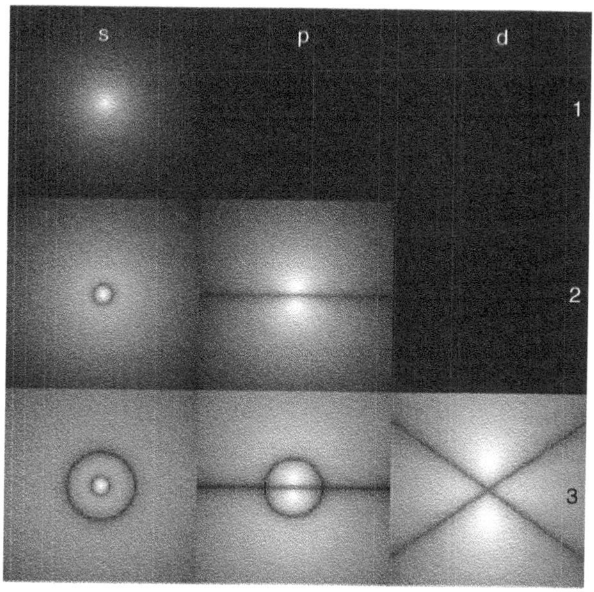

Figure 7.4. Probability density, represented by the degree of brightness, of finding the electron in a hydrogen atom in various orbits and orbitals. Courtesy of Michael Diederich and Falcorian.

For now, however, realize that these quantum numbers n, and s, p, d, f for a given n represent the quantum states (wave functions) of the electrons in the atom. As you already know that the square of the magnitude of the wave function represents the probability of finding the particle in the state represented by the wave function. Figure 7.4 illustrates the probability densities of an electron in a hydrogen atom. Brighter areas correspond to higher probability density. Also, the electron in the atom possesses discrete energy levels with energy increasing from the top of the figure to the bottom with increasing $n = 1, 2, 3 \cdots$. Angular momentum increases from left to right in the figure.

So, the orbital quantum number determines the magnitude of the angular momentum of the electron and quantizes it. However, because angular momentum is a vector, it has a direction. Is the direction of an electron in the atom quantized too? To answer this question, we will explore magnetic quantum number.

7.5.3 Magnetic Quantum Number

The magnetic quantum number m_l arises from solving the azimuthal part of the Schrödinger equation of the hydrogen atom to determine the wave function $\Phi_{m_l}(\phi)$. The angle ϕ is a measure of rotation of the electron along the z-axis. It turns out that the solution $\Phi_{m_l}(\phi)$ of the Schrödinger equation requires that m_l is an integer and satisfies the following conditions:

$$L_z = m_l \hbar \tag{7.34}$$

where L_z is the z component of the angular momentum and

$$m_l = 0, \pm 1, \pm 2, \ldots, \pm l$$

Condition 7.34 is called space quantization. This means the possible values of L_z corresponding to a value of l range from $-l$ to $+l$ in steps of 1, for a total of $2l + 1$ allowed values.

STUDY CHECKPOINT 7.6

What are the possible values of the magnetic quantum number m_l for an electron in an atom whose orbital quantum number l is 3?

Solution:

$m_l = -3, -2, -1, 0, 1, 2, 3$

We can derive Eq. 7.34 by knowing that the operator for the angular momentum in the z direction is

$$\hat{L}_z = -i\hbar \frac{\partial}{\partial \phi} \tag{7.35}$$

Also, in Section 7.3.2, we solved the azimuthal part of the Schrödinger equation to find

$$\Phi = \frac{1}{\sqrt{2\pi}} e^{i m_l \phi}$$

Therefore, the expectation value of L_z can be calculated as:

$$\langle L_z \rangle = \int_0^{2\pi} \phi^* \hat{L}_z \phi d\phi = \frac{-i\hbar}{2\pi} \int_0^{2\pi} e^{-im_l\phi} \frac{\partial}{\partial\phi}(e^{im_l\phi}) d\phi = \hbar m_l$$

Also,

$$\hat{L}_z \Phi = -i\hbar \frac{\partial \Phi}{\partial \phi} = \frac{-i\hbar}{\sqrt{2\pi}} \frac{\partial}{\partial \phi}(e^{im_l\phi}) = \frac{\hbar m_l}{\sqrt{2\pi}}(e^{im_l\phi}) = \hbar m_l \Phi$$

This means

$$L_z = m_l \hbar \tag{7.34}$$

and Φ is an eigenfunction of the angular momentum operator $\hat{L}_z$. It means that the angular momentum is quantized along one direction in space. The choice of z to be in that direction is a matter of convenience. Regardless of which direction we choose for quantization, the result will be $m_l\hbar$. If the atom is inside a magnetic field, the z-axis is chosen to be along the direction of that field. That is why m_l is called the magnetic quantum number.

PROBLEM 7.3

Would it be possible for L_z to be equal to L?

Solution:

Note the following facts:

$$L = \hbar\sqrt{l(l+1)}$$

$$L_z = m_l \hbar$$

$$|m_l| \le l$$

$$\text{and } l \ge 0$$

This means L_z is always $<L$. Therefore, $L = L_z$ only when $L = 0$

In a nutshell, we have obtained three quantum numbers of the electron in a hydrogen atom in an attempt to solve the Schrödinger equation for it.

In this chapter, we have used the hydrogen atom to demonstrate the power of the Schrödinger equation. Just because the hydrogen atom is the simplest atom does not diminish its importance. Although we used the hydrogen atom in this chapter to demonstrate some general features of quantum mechanics, understanding the quantum mechanics of hydrogen by virtue of the hydrogen atom itself can be very useful as well, the reason is because hydrogen and its ions and bonds have enormous use in biology and chemistry.

7.6 THE SIGNIFICANCE OF HYDROGEN

An incredibly large amount of chemistry can be understood in terms of acid–base reactions, the reactions that involve the transfer of H^+ ions from one substance to another. An acid is a substance, a molecule or an ion, that donates an H^+ ion and a base is a substance that accepts an H^+ ion. The beauty of the hydrogen atom is that its ion, H^+, is simply a proton.

Acids and bases play an important role in numerous chemical reactions or processes occurring almost everywhere: from industry to environment, to inside our body. For example, the ability of aquatic environment to support fish or plant life depends on the right degree of acidity in the environment. If the acidity in our body is not maintained within a certain narrow range, we cannot stay alive. Inside living organisms, many biological reactions involve proton transfer, and therefore the reaction rate depends on the concentration of H^+. To keep the organisms healthy and alive, these reactions must occur at certain rates. In other words, the wrong concentration of H^+ will cause illness or even death.

7.7 LIVING IN THE QUANTUM WORLD

So far in this book, you have studied the basic postulates of the theory of quantum mechanics. You also learned about the Schrödinger equation that describes the dynamics of a physical system at the microscopic level. In the microscopic world, quantum mechanics replaces the classical (Newtonian) mechanics. That underlies the importance of these postulates largely built into the Schrödinger equation. When you think of these quantum mechanic postulates or the Schrödinger equation, think of the DNA molecule, and think of the whole machinery at work in each cell of your body turning carbohydrate molecules into energy and amino acid molecules into protein molecules following the instructions coded in the DNA molecule (illustrated in Fig. 7.5): The two strands of the DNA molecule are held together by hydrogen bonds. Think of the neurons carrying the message from your eyes to your brain while you are reading this book. One may argue that these molecules are big enough to be treated as classical objects. Then, how about the anatomy and inner dynamics of these molecules? How about the fundamental particles directly involved in helping these molecules perform their jobs. The point is that quantum mechanics is at work around this area.

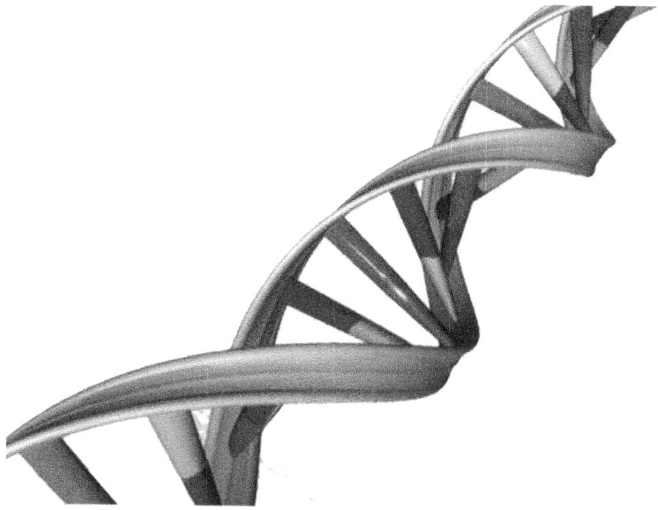

Figure 7.5. Illustration of paired and coiled strands forming the double-helix structure of the DNA molecule. Courtesy of genome.gov.

The DNA molecule, discussed in more detail in Chapter 11, contains the code according to which proteins in the cell are made. Let us explore an example related to the topic at hand. The DNA mutations, the errors in the sequence of DNA, are mostly generated during DNA replication, as illustrated in Figure 7.6. These mutations are sources of many diseases and provide the raw material for evolution. An enzyme called DNA polymerase, makes a copy of the already existing DNA. Occasionally, during the process of making the new DNA strand, DNA polymerase makes a mistake, such as insert the wrong base, miss to insert a base, or insert an additional base. There can be several reasons for these errors including exposure to radiation and hazardous chemicals. We can do something about sources of errors, such as radiation and chemicals, but there is nothing we can do about errors or mutations arising from the quantum nature of the DNA code, at least not until we understand them in their terms: the quantum terms. These errors arise from the uncertainty principle at action during replication. It is believed that these mutations may be initiated by motion of a proton from one site in the DNA molecule to another, a process called tautomerism. The proton will exist as a superposition of both (initial and final) position states, and this uncertainty will result in mutation.

Just as in the macroscopic world, the benefits of understanding the laws at work in the microscopic world of life are enormous. By using quantum mechanics, we can understand the functioning of life at a molecular level by probing the structure and dynamics of the molecules of life, such as nucleic acids (DNA and RNA), proteins, carbohydrates, and lipids. The knowledge obtained from this understanding can have an endless array of applications in the fields of medicine, biotechnology, bioengineering, and can even give rise to new fields.

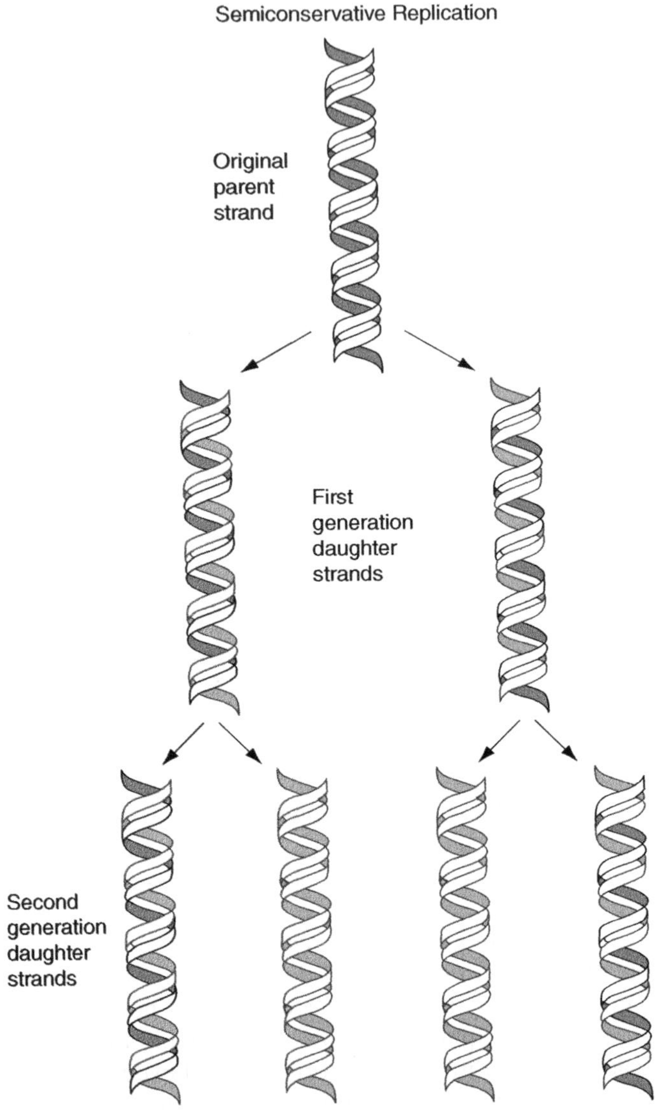

Figure 7.6. The DNA replication process according to the semiconservative model. Courtesy of Lizanne Koch.

After the computer revolution and the Internet revolution, do you want another revolution? Fuse quantum mechanics with the existing molecular biology.

7.8 SUMMARY

In this chapter, we have demonstrated how the Schrödinger equation can be used to apply quantum mechanics to atoms, such as the hydrogen atom. By using the case of the hydrogen atom, we have shown how quantum mechanics naturally predicts some principles that were empirically postulated by Bohr and others to explain the experimental data in the absence of the theory. The requirement of having meaningful solutions for the equation gives rise to three quantum numbers: the principal quantum number n, the orbital quantum number l, and the magnetic quantum number m_l. This theory predicts that these quantum numbers are integers and meet the following conditions:

$$\begin{aligned}
&\text{Principal quantum number: } n = 1, 2, 3, \ldots \qquad n > 0 \\
&\text{Orbital quantum number: } l = 0, 1, 2, 3, \ldots, (n-1) \qquad l < n \\
&\text{Magnetic quantum number: } m_l = -l, -(l+1), \ldots, 0, 1, 2, \ldots, \\
&\quad (l-2), (l-1), +l \quad |m_l| \leq l
\end{aligned} \tag{7.23}$$

The principal quantum number n appears from the radial part of the Schrödinger equation, which must satisfy the following conditions to have a solution:

$$E_n = -\frac{me^4}{32\pi^2 \varepsilon_0^2 \hbar^2}\left(\frac{1}{n^2}\right)$$

where E_n is the energy of the electron in the atom with quantum number n. The electron with a specific quantum number n is said to be in a stationary energy level E_n. It is called the stationary energy state because quantum theory predicts that the electron will not emit any radiation while it is in this energy state. Furthermore, quantum theory predicts that when the electron jumps from a higher state m to a lower state n, it emits radiation that is equal to $E_m - E_n$.

The orbital quantum number l arises from the radial and angular part of the Schrödinger equation and it determines and quantizes the angular momentum of the atom (electron proton system), as shown in the equation:

$$L = \hbar\sqrt{l(l+1)}$$

The magnetic quantum number that arises from solving the azimuthal part of the Schrödinger equation determines and quantizes the angular momentum in one dimension, conveniently chosen to be the z dimension as below:

$$L_z = m_l \hbar$$

Finally, the location of the electron inside the atom cannot be exactly determined. You can only calculate the probability that the electron will be found at a certain point in the atom by using the wave function of the electron.

7.9 ADDITIONAL PROBLEMS

Problem 7.4 The azimuthal part of the Schrödinger equation for the hydrogen atom is

$$\frac{d^2\Phi}{d\phi^2} + m_l^2\Phi = 0$$

A solution for this equation is

$$\Phi = Ae^{im_l\phi}$$

Normalize this wave function to find the value of A, the constant.

Problem 7.5 Consider the ground state ($n = 1, l = 0$) of the hydrogen atom.

A. Write down the equation for the radial wave function.

B. Substitute the radial wave function into the radial equation to demonstrate:

$$\left(\frac{1}{a_0^2} + \frac{2m}{\hbar^2}E\right) + \frac{1}{r}\left(\frac{2me^2}{4\pi\varepsilon_0\hbar^2} - \frac{2}{a_0}\right) = 0$$

and

$$E = -\frac{me^4}{32\pi^2\varepsilon_0^2\hbar^2}$$

Problem 7.6 Use classical mechanics to calculate the angular momentum of the Earth.

Problem 7.7 Write down all the possible combinations of (n, l, m_l) for the $n = 5$ state of the hydrogen atom.

Problem 7.8 Calculate the average value of distance r of the electron from the nucleus in a hydrogen atom in its ground state.

Problem 7.9 Consider a hydrogen atom in its ground state, that is, the 1s state.

A. Calculate the maximum radius of the electron in the hydrogen atom assuming that the maximum radius will be reached when the total energy of the electron becomes equal to the potential energy.

B. Use quantum mechanics to calculate the probability that the electron can exist at a radius greater than the maximum calculated in part A.

QUANTUM MECHANICS OF MANY-ELECTRON ATOMS 8

About seven years later I was given a book about the periodic table of the elements. For the first time I saw the elegance of scientific theory and its predictive power.
Albert Einstein

Figure 8.0. A bus wearing the periodic table of elements to promote Oxford Science Park.

The periodic table of elements is a well-known scientific fact that is one of a very few items in science that has obtained a wide acceptance and popularity inside and outside the scientific community. It is a prime example and a public display of how scientific research is based on finding patterns and trends in the data (observations) and then trying to explain those patterns and trends in terms of hypotheses, models, and theories. This is also an example of how quite often scientists make models based on empirical

Quantum Physics for Scientists and Technologists, by Paul Sanghera

research and the true scientific explanation comes later by a fundamental physical law or theory, a topic that we discussed in Sections 5.1 and 5.2.

The periodic table of elements is a tabular display of chemical elements organized in such a way as to illustrate trends in the observed properties of the elements. Although precursors to this table in one or another form exist, its invention is widely credited to Russian chemist Dmitri Mendeleev in 1869. Since then, the layout of the table has been refined and extended over time to accommodate newly discovered elements and improved scientific understanding of it. The fundamental explanation and understanding of the periodic table came from quantum mechanics, the theory that did not even exist when the periodic table was invented.

In This Chapter

8.1 INTRODUCTION

In chapter 7, we explored the quantum mechanics of the atom by applying the Schrödinger equation to the hydrogen atom. We demonstrated how quantum mechanics explains the anatomy and physiology of the hydrogen atom in such an elegant way and we were also able to derive some salient features of quantum mechanics in the process. An obvious question to ask is Can we apply the Schrödinger equation to other atoms, which have multiple electrons? In theory, we should be able to do that by adding more electrons, that is, potentials, and solving the equation for the wave function. In practice, however, we run into formidable mathematical problems in formulating and solving the equation. In many-electron atoms, not only do we have many electrons around the nucleus, we also have many protons in the nucleus. Furthermore, we also have a repulsive force among similarly charged electrons making the potential even more complex.

This problem is solved by taking a different approach. It turns out that the anatomy and physiology of many-electron atoms can be better understood if quantum mechanics includes another principle, called the Pauli's exclusion principle, also referred to as the exclusion principle. This principle uses an additional quantum number called the spin quantum number that you will learn about in this chapter.

This approach was able to explain the periodic table of elements toward the end of the 1920s, which was considered to be a significant achievement of quantum mechanics. We will see how the qualitative understanding of atomic structure with the help of quantum mechanics allows us to understand some of the physical and chemical properties of the elements. These properties include atomic sizes, ionization energies, magnetic moments, spectroscopic behavior, and chemical behavior. In the context of spectroscopic behavior, the electron spin helps explain the split of spectral lines exhibited by atoms in an external magnetic field, an effect called the Zeeman effect. So, the central theme of this chapter is to explain the behavior of many-electron atoms using quantum mechanics. To this end, we will explore three avenues: introducing the concept of the spin quantum number, explaining the periodic table of the elements with quantum mechanics, and explaining the Zeeman effect with quantum mechanics.

8.2 TWO CHALLENGES TO QUANTUM MECHANICS: THE PERIODIC TABLE AND THE ZEEMAN EFFECT

In Chapter 7, we applied quantum mechanics to the simplest atom, the hydrogen atom, which has one proton in its nucleus and one electron around it. More complex atoms contain multiple protons and possibly one or more neutron in their nuclei with multiple electrons around it. Now that we are talking about multiple-electron atoms, it is time to define some related terms. The number of protons in the nucleus of atoms of a given element is called its atomic number. The number of electrons in a neutral atom is equal to the number of protons in its nucleus. An atom that has lost or gained one or more electrons, and thus has acquired an electric charge, is called an ion. An ion with a positive charge is called a cation. An ion with a negative charge is called an anion. Isotopes are the different forms of the same element that differ in the number of neutrons in their nuclei.

Quantum mechanics described so far could not explain the behavior of many-electron atoms well. The general behavior and trends of many-electron atoms, however, is well expressed in a periodic table. Therefore, explaining the general behavior of many-electron atoms is equivalent to discovering the physical basis of the periodic table. On the spectroscopic side, the so-called Zeeman effect, the splitting of spectral lines when an atom is subjected to a magnetic field, was a challenge for the growing field of quantum mechanics. In this section, we will introduce the periodic table and the Zeeman effect.

8.2.1 The Periodic Table of Elements

As part of the quest to understand matter, the universe, and its origin, physicists and chemists have been studying the properties of elements and atoms for centuries, long before the development of quantum mechanics. In 1869, a chemist named Dmitry Mendeleev arranged the just over 60 elements known at the time into a table based on their known chemical properties. The elements in this table, known as a periodic table of elements, were primarily arranged in order of their atomic masses. Mendeleev noticed that a certain arrangement of rows in the table gave rise to a pattern that made sense. For example, elements of the same column had similar properties, for example, all the elements had metal-like properties, but were different from the elements of another column that had gas-like nonreactive properties. However, there were some vacancies in Mendeleev's table. While skeptics used these vacancies to doubt the validity of the table, Mendeleev used these vacancies to predict some unknown elements. The periodic table was widely accepted after the discovery of three of these unknown elements: gallium in 1875, scandium in 1879, and germanium in 1886. Recall from Chapters 2 and 3 that classical physics was running into great trouble during these times. Mendeleev's periodic table added another challenge for physicists: Explain the physical basis of the periodic table.

Dmitri Ivanovich Menedeleev (1834–1907).

Dmitri Menedeleev was born in a village called Verhnie Aremzyani in Siberia on February 8, 1834. He joined the University of Saint Petersburg, Russia and completed his doctorate in science. He stayed in and around Saint Petersburg, Russia for most of his life. In 1866, Mendeleev was appointed as professor of Chemistry at Saint Petersburg University. Three years later he published the first version of the periodic table of elements. Mendeleev died on February 2, 1907. The element of atomic number 101 is called mendelevium in his honor.

Note: The elements in the modern periodic table are arranged according to atomic numbers, whereas the elements in the first periodic table published by Mendeleev were arranged according to atomic masses, as the concept of atomic number was unknown at the time. Recall that atomic mass directly depends on atomic number.

Also, simple quantum theory fails to explain a spectroscopic result called the Zeeman effect.

8.2.2 The Split Spectral Lines and the Zeeman Effect

Chapter 4 explained how the spectral series observed during the spectroscopic experiments posed a challenge to classical physics. The chapter also explained how these series were eventually explained by some ad hoc quantum principles or postulates, such as the Bohr atomic model. These principles were gradually integrated into quantum theory, as explained in the previous chapters.

The quantum theory developed so far cannot explain some experimental results from spectroscopic experiments. Many spectral lines discussed in Chapter 4 (e.g., Sections 4.6.2 and 4.7) actually consist of two separate lines close to each other. For example, the first line in the Balmer series of the hydrogen atom appears due to the transition of an electron between the $n = 3$ and 2 levels. As shown in Problem 8.1, quantum mechanics discussed so far predicts for one line of wavelength 656.3 nm. Watched closely in an experiment, there are actually two lines 0.14-nm apart. From a macroscopic viewpoint, this is indeed a small effect. However, in the microscopic world, this is

PROBLEM 8.1

Calculate the wavelength of the lowest frequency line in the Balmer series.

Solution:

Consider the Rydberg ($\Re$) formula:

$$\frac{1}{\lambda} = \Re\left(\frac{1}{l^2} - \frac{1}{n^2}\right)$$

For the Balmer series, $l = 2$

The highest frequency corresponds to $n = 3$.

By substituting these values of l and n in the Rydberg formula, we obtain

$$1/\lambda = 0.1389 \times \Re$$

which implies

$$\begin{aligned}\lambda &= 1/(0.139 \times \Re) = 1/(0.1389 \times 10.97\ \mu\text{m}^{-1}) \\ &= 0.6563\ \mu\text{m} \\ &= 656.3\ \text{nm}\end{aligned}$$

a sizable effect, and therefore a blow to quantum mechanics, the theory of the microscopic world. The other related blow to quantum mechanics discussed so far comes from the Zeeman effect, which refers to the experimental observation that a spectral line of an atom split into multiple lines when the atom is in a magnetic field.

In order to explain both the periodic table and the split in the spectral lines, physicists had to wait for the concept of electron spin to be discovered.

8.3 INTRODUCING THE ELECTRON SPIN

The split in spectral lines, called the fine structure in spectral lines, was finally explained by this quantum mechanical principle put forward by two Dutch graduate students, Samuel Goudsmit and George Uhlenbeck in 1925: Every electron has an intrinsic angular momentum S due to a property called the electron spin, s. The angular momentum would have a magnetic moment μ_s associated with it. In quantum mechanics, the electron's intrinsic angular momentum and corresponding magnetic moment are quantized; therefore its spin s is considered as the spin quantum number just like the orbital quantum number l corresponding to the orbital angular momentum L.

Because both S and L represent some type of angular momentum, the relationship between S and s is identical to the relationship between L and l, that is,

$$S = \sqrt{s(s+1)}\hbar \tag{8.1}$$

The major difference between s and l is that s can have only one value: $s = \frac{1}{2}$.

> *Note*: Each electron has a spin whose magnitude is always the same for all the electrons and is ½, which follows from spectral data and also from the relativistic quantum mechanics that we are not discussing in this book. Some particles have integral spin. For example, photons have a spin of 1.

Recall from Section 7.5.3 that the space quantization of angular momentum is represented by

$$L_z = m_l\hbar$$

where m_l is called the magnetic quantum number. The orbital angular momentum can have $2l + 1$ orientations corresponding to $2l + 1$ values of m_l given by

$$m_l = 0, \pm 1, \pm 2, \ldots, \pm l$$

Similarly, space quantization of spin angular momentum is represented by

$$S_z = m_s \hbar \tag{8.2}$$

where m_s is called the spin magnetic quantum number. The spin angular momentum can have $2s + 1$ orientations corresponding to $2s + 1$ values of m_s given by

$$m_s = +\frac{1}{2}, -\frac{1}{2} \tag{8.3}$$

As shown by Eq. 8.2, the spin magnetic quantum number m_s determines the component of the spin angular momentum along a magnetic field in the z direction. The magnetic moment due to spin is called the spin magnetic moment μ_s given by the following equation:

$$\mu_s = -\frac{e}{m} S \tag{8.4}$$

where m is the mass of the electron. Due to the space quantization of spin angular momentum S, spin magnetic moment can only have limited values along an axis. As shown in Problem 8.8, the possible values of μ_s along the z-axis are given by

$$\mu_{sz} = \pm \frac{e\hbar}{2m} = \pm \mu_B \tag{8.5}$$

where the constant μ_B is called the Bohr magneton.

STUDY CHECKPOINT 8.1

A. Is the spin angular momentum, like spin, always the same in all electrons?
B. What is the spin angular momentum of an electron with spin magnetic quantum number given by $m_s = -\frac{1}{2}$

Solution:

A. Because

$$S = \sqrt{s(s+1)}\hbar$$

and s is always ½
Therefore S is always the same for all electrons.

B. Regardless of the value of the magnetic spin, the electron spin is always given by

$$s = \frac{1}{2}$$

Therefore,

$$S = \sqrt{s(s+1)}\hbar$$

which implies

$$S = \frac{\sqrt{3}}{2}\hbar$$

When Goudsmit and Uhlenbeck proposed the concept of electron spin, what they had in mind was a charged sphere spinning on its axis: a classical picture. Scattering experiments show that an electron must be $<10^{-16}$ m across in size. All the experiments by particle physicists at the high energy labs, such as those at CERN, have not provided any evidence against the possibility that an electron is a point particle, that is, it has no structure. Even assuming that the electron is not a point particle and is a sphere with a diameter 10^{-16} m, it has to have an equatorial velocity greater than the speed of light to generate the spin angular momentum of $S = (\sqrt{3}/2)\hbar$, as shown in Problem 8.2. However, Einstein's theory of relativity does not allow the speed of any particle to be greater than the speed of light.

PROBLEM 8.2

Consider an electron as a charged sphere of diameter 10^{-16} m spinning around its axis. Determine the equatorial velocity that it will need to generate an angular momentum of $\frac{\sqrt{3}}{2}\hbar$

Solution:

The angular velocity of the spinning sphere is given by

$$\omega = \frac{v}{r}$$

where v is the equatorial velocity

and

$$r = d/2 = 0.5 \times 10^{-16}\ \text{m}$$

Angular momentum is given by

$$S = I\omega = \left(\frac{2}{5}mr^2\right)\left(\frac{v}{r}\right) = \frac{\sqrt{3}}{2}\hbar$$

which implies

$$v = \frac{5\sqrt{3}}{4}\frac{\hbar}{mr} = \frac{5\sqrt{3}}{4}\frac{197.3\dfrac{\text{eV.nm}}{c}}{\left(0.5110\dfrac{\text{MeV}}{c^2}\right)(0.5\times 10^{-16}\ \text{m})} = 1.67\times 10^4 c$$

which is greater than the speed of light, c.

So, you should know that this classical picture of electron spin lends itself to serious objections. However, objections themselves may be coming from our classical way of looking at the spin. One problem is that this argument treats an electron as a nonrelativistic particle. When an electron is treated as a relativistic particle by using Dirac's equation, the relativistic form of the Schrödinger equation, the spin quantum number is predicted in a natural way like the other three quantum numbers: l, m, and n. Regardless of how you look at it, the fact remains that the spin quantum number is endorsed by experimental observations and by its ability to help explain the periodic table and Zeeman effect, as you will see in this chapter.

How does the spin quantum number help explain the periodic table? For this answer, you need to understand Pauli's exclusion principle, a principle that takes spin into account.

8.4 EXCLUSION PRINCIPLE

As pointed out several times in this book, a system has a natural tendency to be in an equilibrium or stable state, which usually means in a state of minimum energy. For a hydrogen atom, the state of minimum energy is its ground state with $n = 1$. Now, extend your thinking to many-electron atoms. An obvious expectation will be that all the electrons in an atom will fall into the same ground state under normal conditions so that the atom goes into its stable state. However, by just looking at the known chemical properties of elements, we can tell that this is not the case.

Wolfgang Ernst Pauli (1900–1958). Courtesy of S. A. Goudsmit, AIP Niels Bohr Library.

Wolfgang Pauli was born on April 25, 1900 in Vienna, Austria. In 1922, he received his doctorate from the University of Munich. After performing some assignments at Göttingen, Copenhagen, and Hamburg, he became professor of physics at the Institute of Technology in Zurich, Switzerland, in 1928. Pauli spent most of his life at Zurich, except during the years of the World War II. He spent those years at U.S. universities including Princeton University. He discovered the Exclusion Principle in 1925, for which he was awarded the Nobel Prize in 1945. He is also known for postulating the existence of the particle neutrino to explain nuclear beta decay.

Throughout his life, Pauli had been struggling with the question of why the fine structure constant, a dimensionless fundamental constant, has a value nearly equal to 1/137. Pauli died on December 15, 1958, at the rather young age of 58, in a hospital room number 137 due to pancreatic cancer.

The electronic configuration of an atom largely determines its chemical properties. Therefore, if all the electrons in an atom jam into the ground state, the elements whose atoms differ only by a few electrons should have very similar chemical properties, and those whose atoms differ by a large number of electrons should have dissimilar chemical properties. However, reality is contrary to this expectation. For example, consider elements with atomic numbers 9–11 that differ only by one electron in their atoms, but their chemical properties differ drastically: Fluorine (9 electrons) is a chemically active halogen gas, neon (10 electrons) is an inert gas, and sodium (11 electrons) is a metal. On the other hand, helium (2 electrons) and argon (18 electrons) are both inert gases.

So, it was obvious that all the electrons in many-electron atoms are not in the same ground state, and perhaps cannot be in the same quantum state. With this in mind, physicist Wolfgang Pauli was looking at the atomic spectra. He knew that the various quantum states of an atom can be determined from the spectral lines it generates. Pauli noticed that in the spectra of many-electron atoms, certain spectral lines were missing that corresponded to transitions of electrons among states with certain combination of quantum numbers.

Note: Classically thinking, the ultimate stable state for the atom will be for the electron to fall into the nucleus. The introduction of simple quantum mechanics postulates and the principal quantum number helped explain why this does not happen. We needed another quantum number and the exclusion principle to explain why all the electrons do not fall into the ground state of $n = 1$.

It dawned on Pauli that he could use the spin quantum number to explain the missing spectral lines. He proposed what is known as Pauli's exclusion principle:

PAULI'S EXCLUSION PRINCIPLE

No two electrons in an atom can exist in the same quantum state uniquely defined by the combination of quantum numbers n, l, m_l, and m_s. In other words, each electron in an atom must have a unique set of quantum numbers n, l, m_l, and m_s different from all other electrons in the same atom. Note that the Pauli exclusion principle can be extended to other particles, such as neutrons. For example, Pauli's exclusion principle is used to explain the formation of neutron stars, and above a certain critical mass, the transition to a black hole, as the degeneracy pressure cannot fight off the force of gravity. A star collapsed under a gravitational field is called a black hole because its gravitational force is so strong nothing can leave the star, not even light.

Pauli was able to show that all the missing atomic spectral lines involved two or more electrons with an identical set of quantum numbers. The exclusion principle is just a declaration of this observation.

Note: There are two components to Pauli's exclusion principle:

1. A quantum state is defined by a set of quantum numbers $\{n, l, m_l, m_s\}$.
2. Each electron in an atom must have a unique quantum state different from those of all other electrons in the same atom.

STUDY CHECKPOINT 8.2

The helium atom has two electrons. Which following combination of quantum states $\{n, l, m_l, m_s\}$ is possible for these electrons to exist in?

A. $\{1, 0, 0, \frac{1}{2}\}$; $\{1, 0, 0, -\frac{1}{2}\}$
B. $\{1, 0, 0, \frac{1}{2}\}$; $\{1, 0, 0, \frac{1}{2}\}$
C. $\{2, 1, 0, \frac{1}{2}\}$; $\{2, 0, 0, \frac{1}{2}\}$
D. $\{2, 1, -1, \frac{1}{2}\}$; $\{1, 1, -1, \frac{1}{2}\}$

Solution:

All combinations are possible except for B.

So, Pauli's exclusion principle explains why all the electrons in many-electron atoms do not jam up into the ground state. But how exactly does it help us to understand the actual detailed electronic configuration of atoms, in other words, the periodic table? Let us explore it.

8.5 UNDERSTANDING THE ATOMIC STRUCTURE

Although we cannot exactly solve the Schrödinger equation for many-electron atoms, no radical changes in quantum mechanics theory are expected as we move from the hydrogen atom to more complex atoms. With the addition of a spin quantum number, the fundamentals of quantum mechanics still remain valid with the increase in atomic number of atoms: the same set of four quantum numbers $\{n, l, m_l, m_s\}$, the same quantum states defined by this set of quantum numbers, the same wave functions representing the probability of locating an electron in the atom, and the same general formula for energy absorption and radiation during the transition between two states, and so on.

A key to understanding the periodic table is to understand an underlying unity behind the diversity of all atoms: The electrons in all atoms live in orbitals (essentially the states) defined by a set of four quantum numbers: $\{n, l, m_l, m_s\}$. The orbitals exist in subshells defined by l, which in turn exist in shells defined by n. In order to understand the physical basis of the periodic table, it is important to understand the concepts of shells, subshells, and orbitals, and their relationship with one another.

8.5.1 Understanding Shells, Subshells, and Orbitals

A shell is defined by the principal quantum number n, the quantum number that quantizes the energy. Electrons with the same quantum number n (i.e., the same value of n) are roughly at the same average distance from the nucleus of the atom they are in. Therefore, they are essentially considered to be in the same shell. In this context, n is also referred to as a shell or shell number. Because the electron energy largely depends on n, the electrons in the same shell have roughly the same energy. Atomic shells, that is, the values of n, are also denoted by letters (e.g., K, L, M, N, and O) corresponding to the values of $n = 1, 2, 3, 4$, and 5, respectively, as summarized below:

Value of n	1	2	3	4	5
Letter Used	K	L	M	N	O

As mentioned earlier, the energy of electrons in a specific shell also depends to some extent on the quantum number l. The intervening shells of other electrons shield the nuclear charge from an electron and the degree to which the nuclear charge is shielded from the electron depends on its probability–

density distribution. An electron with a small l value is more likely to be found near the nucleus where it is poorly shielded by other electrons than an electron with a larger value of l is within the same shell. Nearer to the nucleus means higher binding energy and therefore lower total energy. In other words, the energy of electrons within the same shell increases with an increase in l. Electrons with the same value of l are said to belong to the same subshell. Because there are different values of l for the same value of n, there are multiple subshells in a shell. You know from Chapter 7 that l has n integral values corresponding to each n ranging from 0 to $n - 1$. Subshells, that is, the values of l, are also denoted by letters, for example, s, p, d, f, and g corresponding to the values of 0, 1, 2, 3, and 4, respectively, as summarized below:

Value of l	0	1	2	3	4
Letter Used	s	p	d	f	g

Note: Because the energy, E, of electrons within the same shell increases with an increase in l, the stability of an atom also decreases with the increase in l.

Each subshell is identified by its principal quantum number n followed by the letter corresponding to the value of l. For example, a subshell with $n = 2$ and $l = 0$ is denoted by 2s.

STUDY CHECKPOINT 8.3

Determine the number of orbitals and the number of electrons that can exist in subshells s, p, d, f, and g.

Solution:

Subshell	Value of l	Number of Orbitals = $2l + 1$	Number of Electrons = 2 × Number of Orbitals
s	0	1	2
p	1	3	6
d	2	5	10
f	3	7	14
g	4	9	18

Recall from Chapter 7 that corresponding to a specific value of l, there are $2l + 1$ values of m_l because m_l can have integral values ranging from $-l$ to $+l$ including zero. In other words, the total number of orbitals in a subshell l are

$2l + 1$: one corresponding to each value of m_l. The quantum number m_l describes the space orientation of the orbital in which the electron exists. Recall that an orbital is the region around the nucleus in which the electron can be found with a probability distribution. It is important to realize the difference between an orbital and an orbit in the Bohr's model discussed in Chapter 4. Bohr's orbit is a well-defined circular orbit around the nucleus with a certain radius in which the electron moves. This concept violates the uncertainty principle. An orbital, in accordance with the uncertainty principle, is a region about the nucleus in which an electron can be found by using the probability wave function.

STUDY CHECKPOINT 8.4

Explain exactly how Bohr's orbit violates the uncertainty principle.

Solution:

The orbit in Bohr's model is a well-defined circular orbit around the nucleus. In order to draw such an orbit in which the electron moves, you must know the location of the electron and its velocity, and hence the momentum at a given time simultaneously. This is not possible according to the uncertainty principle, which states:

$$\Delta x \Delta p \geq \frac{\hbar}{2}$$

Corresponding to each value of m_l, there are only two values of the magnetic spin quantum number m_s: 1/2 and −1/2. Therefore, according to Pauli's exclusion principle, an orbital can only have two electrons at maximum. As demonstrated in Problem 8.3, the maximum number of electrons that can reside in a shell n are $2n^2$, a fact that chemist take for granted.

As examples, Table 8.1 summarizes the possible values of all the quantum numbers n, l, m_l, and the maximum number of electrons in a shell corresponding to $n = 1$ through $n = 7$. The last column of the table lists the maximum number of electrons that can occupy a shell. Not all the elements have all the shells completed, that is, having maximum electrons; a factor that contributes to the diversity of elements. For example, hydrogen (H) has only one shell ($n = 1$), and even that is incomplete; it only has one electron; whereas a helium (He) atom has one shell ($n = 1$) and it is complete; that is, it contains the maximum number of electrons. The carbon (C) atom has its first shell complete, but has only four electrons in the second shell, while it can contain a maximum of eight electrons; whereas Neon (Ne) has two shells (n + 1 and +2) and both are complete.

PROBLEM 8.3

Calculate the maximum number of electrons a shell with principal quantum number n can have. Express your answer in terms of n only.

Maximum number of electrons in an orbital = 2

Maximum number of orbitals in a subshell = $2l + 1$

which implies

Maximum number of electrons in a subshell = $2(2l + 1)$

Maximum number of subshells in a shell = n; ranging from $l = 0$ to $n - 1$

Maximum number of electrons, N, in a subshell:

$$N = \sum_{l=0}^{l=n-1} 2(2l+1) = 2[1+3+5+\cdots+\{2(n-1)+1\}]$$

$$N = 2[1+3+5+\cdots+(2n-1)]$$

The quantity in brackets has n terms with an average value = $\frac{1}{2}[1+(2n-1)] = n$

Total value of the quantity in the bracket = $n \times n = n^2$.

Therefore,

$$N = 2n^2$$

Now that you understand how different quantum numbers define the concepts of shells, subshells, and orbitals, it is time to explore how these concepts help explain the distribution of electrons in the atom; the electron configuration of atoms.

8.5.2 Understanding the Electron Configuration of Atoms

The anatomy of atoms is determined by the way electrons are distributed among the various orbitals of an atom. This distribution is called electron configuration, which in turn is determined by the following three principles:

1. *Principle of Stability or Equilibrium.* Here comes the stability principle again: A system of particles tends to be in a stable state, which is the state of minimum energy. This statement implies that orbitals in an atom are filled with available electrons in order of increasing energy. Electron energy varies with quantum states; largely with n, but also somewhat with l. To be specific, energy increases with the increase in the value of n, and it also increases with the increase in the value of l. This idea is illustrated in Figure 8.1.
2. *Exclusion Principle.* A unique state defined by the set of quantum numbers $\{n, l, m_l, m_s\}$ can be occupied only by one electron in an atom. This means no two electrons in the same atom can have the same values of n, l, m_l, and m_s.

TABLE 8.1. Realtionship Among Values of Quantum Numbers n, l, m_l, and the Maximum Number of Electrons in Shells with n = 1–7

Shell n	Subshell l	Subshell Designation	Possible values of m_l	Number of Orbitals in a Subshell	Total Number of Orbitals in a Shell	Maximum Number of Electrons in a Shell
1	0	1s	0	1	1	2
2	0	2s	0	1	4	8
	1	2p	1,0,–1	3		
3	0	3s	0	1	9	18
	1	3p	1,0,–1	3		
	2	3d	2,1,0,–1,–2	5		
4	0	4s	0	1	16	32
	1	4p	1,0,–1	3		
	2	4d	2,1,0,–1,–2	5		
	3	4f	3,2,1,0,–1,–2,–3	7		
5	0	5s	0	1	25	50
	1	5p	1,0,–1	3		
	2	5d	2,1,0,–1,–2	5		
	3	5f	3,2,1,0,–1,–2,–3	7		
	4	**5g**	4,3,2,1,0,–1,–2,–3,–4	9		
6	0	6s	0	1	36	72
	1	6p	1,0,–1	3		
	2	6d	2,1,0,–1,–2	5		
	3	**6f**	3,2,1,0,–1,–2,–3	7		
	4	**6g**	4,3,2,1,0,–1,–2,–3,–4	9		
	5	**6h**	5,4,3,2,1,0,–1,–2,–3,–4,–5	11		
7	0	7s	0	1	49	98
	1	**7p**	1,0,–1	3		
	2	**7d**	2,1,0,–1,–2	5		
	3	**7f**	3,2,1,0,–1,–2,–3	7		
	4	**7g**	4,3,2,1,0,–1,–2,–3,–4	9		
	5	**7h**	5,4,3,2,1,0,–1,–2,–3,–4,–5	11		
	6	**7i**	6,5,4,3,2,1,0,–1,–2,–3,–4,–5,–6	13		

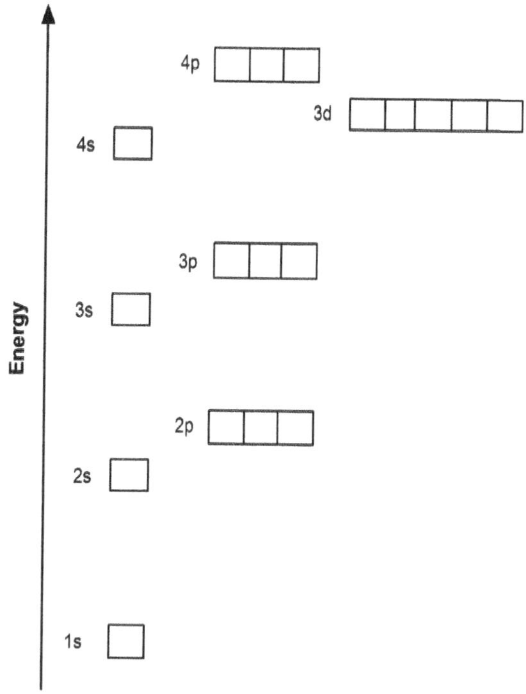

Figure 8.1. An example of orbital energy levels for $n = 1$–4 in many-electron atoms. The energies of the subshells in each shell increase in the order $ns < np < nd < nf$.

3. *Hund's Rule.* Hund's rule states that when electrons are added to orbitals at the same energy level, that is, the same principal quantum number, they will fill each of the available orbitals with only one electron before doubling up in any orbital. This rule is in accordance with the principle of stability because two electrons in the same orbital make the system less stable due to the increased repulsion between the two electrons.

STUDY CHECKPOINT 8.5

Give the increasing order of energy along the values of l for each of $n = 2, 3, 4$.

Solution:

For $n = 2$, the increasing order of energy is 2s < 2p
For $n = 3$, the increasing order of energy is 3s < 3p < 3d
For $n = 4$, the increasing order of energy is 4s < 4p < 4d < 4f

To reach stability, a lower subshell of a higher shell can be filled before filing the higher subshell of a lower shell. For example, consider the noble gas element argon (Ar), which marks the end of the row started by sodium (Na). The electron configuration of Ar, with atomic number 18 is, $1s^2 2s^2 2p^6 3s^2 3p^6$

You may think that the next element in the periodic table, that is the lement with atomic number 19, which is potassium (K), will have its outermost electron in the 3d shell. However, experimental results about the properties of K demonstrate that it is a member of the alkali metal group. This means that it puts its outermost electron, that is, the electron with highest energy, in the 4s subshell instead of putting it in the 3d subshell. As Figure 8.1 shows, 3d orbitals have higher energy than the 4s orbital, which implies that 4s will be filled before the 3d is filled. Why is this? Due to its greater penetration near the nucleus, the 4s orbital has a slightly larger binding energy, and hence a slightly less total energy than the 3d, and therefore makes the atom slightly more stable.

You can ask a question Is there any general visual order (or sequence) in which electrons fill the orbitals? The answer is yes: The general order is shown in Figure 8.2, which is true in general with only few exceptions. The figure is read as shown in the following:

1. Write the subshells corresponding to each shell (principal quantum number) in horizontal lines starting with s at the bottom.
2. Draw diagonal arrows over the sublevels, as shown in Figure 8.2*a*.
3. Start reading the sublevels along the arrows from the bottom up and left to right.
4. When you reach the last subshell, along an arrow, start from the beginning (bottom) of the next arrow.

Following this procedure, you can read the order in which the subshells in various shells will be filled. The results are shown in Figure 8.2*b*.

From what we have learned so far about how the shells and orbitals are filled, let us tackle a few examples of the electronic configuration of real elements with their atoms in the ground state.

A hydrogen atom contains only one electron, which obviously goes to the 1s subshell. In He, the second electron pairs with the first electron in the same 1s orbital. In spite of the repulsion between the two electrons in the same orbital, an atom has less energy (more stability) in this case than it would if the second electron in He went to the 2s orbital. According to Pauli's exclusion principle, the 1s orbital is full with two electrons, and therefore the third electron in the Li atom must go to the 2s orbital. The fourth electron in Be fills the 2s orbital, and the fifth electron in B goes to the first orbital (there are three orbitals in a p subshell as compared to one orbital in an s subshell) in the 2p orbital. Following Hund's rule, the sixth electron in C goes to the second 2p orbital and the seventh electron in N goes to the third 2p orbital. The next

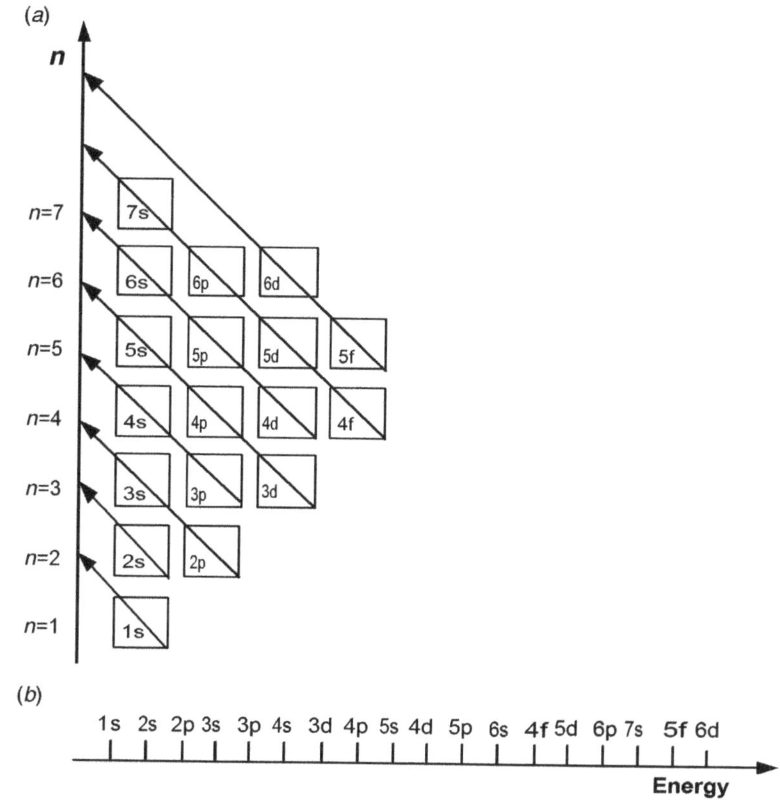

Figure 8.2. (*a*) The sequence of filling the atomic shells and subshells with electrons. (*b*) The sequence of filling subshells with electrons with increasing energy.

three elements O, F, and Ne fill the first, second, and third orbital of the 2p subshell, respectively. This filling of atomic orbitals with electrons, illustrated in Figure 8.3, covers atoms of the elements in the first two periods (rows) of the periodic table.

The filling of the 1s subshell, and hence the primary shell $n = 1$, is complete with the noble gas (or rare earth) element He, and the electron configuration of the next element, Li, can be presented in terms of helium, Li: [He]$2s^1$.

Similarly, the 2p subshell, and hence the primary shell $n = 2$, is complete with the noble gas element Ne, and the electron configuration of the next element sodium (Na), can be written in terms of sodium, Na: [Ne]$3s^1$.

This method of writing the electron configuration is called the condensed electron configuration. The inner-shell electrons represented by the symbol of a noble gas element are called core electrons and the electrons that follow are called the outer-shell electrons, which include the electrons that take part in atomic bonding and are called *valence electrons*.

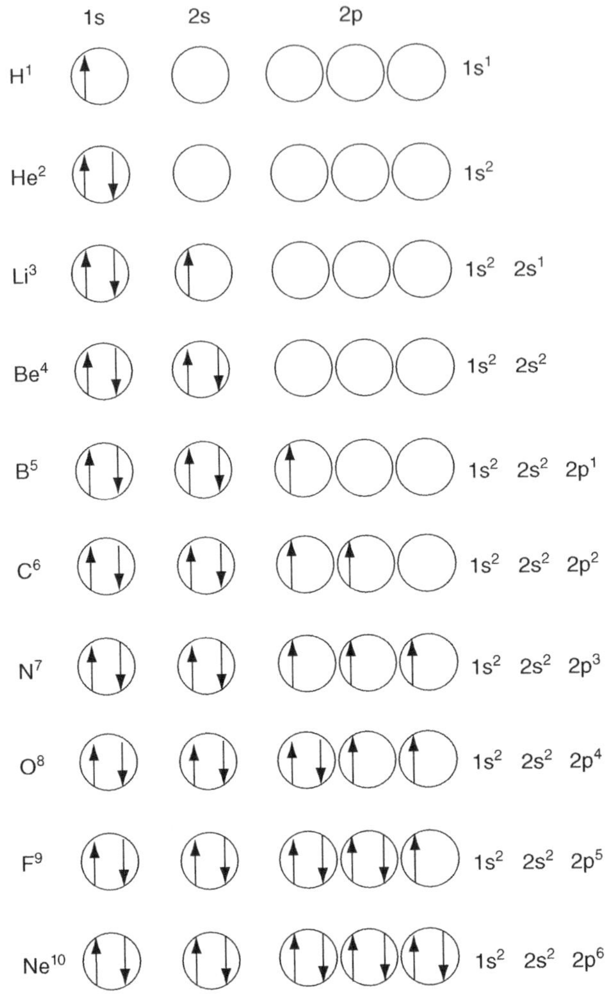

Figure 8.3. Filling of atomic orbitals with electrons for the elements in the first two periods of the periodic table.

Note: Valence electrons are the outermost electrons of an atom, which play a significant role in determining how the atom chemically reacts with other atoms. For example, atoms with a complete shell of valence electrons (corresponding to an electron configuration s^2p^6) tend to be chemically inert, whereas atoms with one or two valence electrons more than a closed shell are highly reactive. The reason is because the extra electrons are easily removed, leaving behind positive ions. Furthermore, atoms with one or two valence electrons less than a closed shell are also highly reactive because they have a strong tendency either to gain the missing electrons and form negative ions, or to share electrons with other atoms, and thereby form bonds called covalent bonds.

The general trend across the second period is the increase in ionization energy with an increase in atomic number, as each new electron is more tightly held to the nucleus because of the increased charge in the nucleus. This general tendency is true for other periods for the same reason with few exceptions due to other factors. Now, we are already talking about explaining the physical basis of the periodic table.

Note: Because the outermost electron in B is in the 2p orbital and unpaired as compared to the two outermost electrons in Be, which are paired in a 2s orbital, the outermost electron in B has more total energy and less potential (binding) energy. Therefore, the first ionization energy, that is, the energy it will take to knock out the outermost electron from the atom, is less than that of Be. This is an example of how the electron configuration of atoms influences the properties of the elements.

PROBLEM 8.4

Use Figure 8.2*a* to write down the electron configuration of the K^{19} atom.

Solution:

In Figure 8.2*a*:
1st arrow:
$1s^2$
2nd arrow:
$2s^2$
3rd arrow:
$2p^6\ 3s^2$
4th arrow:
$3p^6\ 4s^1$
which implies $1s^2\ 2s^2\ 2p^6\ 3s^2\ 3p^6\ 3d^0\ 4s^1$
Note that 4s is beginning to fill without filling 3d.

8.6 UNDERSTANDING THE PHYSICAL BASIS OF THE PERIODIC TABLE

From the three physical principles (stability principle, exclusion principle, and Hund's rule) discussed in Section 8.5.2, the electron configuration of atoms of the elements can be determined within the framework of quantum mechanics. This configuration is presented in Figure 8.4. In this section, we explore how this configuration (atomic structure) predicts the properties of the elements. Let us start with observing some general trends and understanding what those trends predict.

Transition Metals (groups 3–10)

1A 1	2A 2	3B 3	4B 4	5B 5	6B 6	7B 7	8B 8	8B 9	8B 10
1 **H** $1s^1$									
3 **Li** $2s^1$	4 **Be** $2s^2$								
11 **Na** $3s^1$	12 **Mg** $3s^2$								
19 **K** $4s^1$	20 **Ca** $4s^2$	21 **Sc** $3d^14s^2$	22 **Ti** $3d^24s^2$	23 **V** $3d^34s^2$	24 **Cr** $3d^54s^1$	25 **Mn** $3d^54s^2$	26 **Fe** $3d^64s^2$	27 **Co** $3d^74s^2$	28 **Ni** $3d^84s^2$
37 **Rb** $5s^1$	38 **Sr** $5s^2$	39 **Y** $4d^15s^2$	40 **Zr** $4d^25s^2$	41 **Nb** $4d^45s^1$	42 **Mo** $4d^55s^1$	43 **Tc** $4d^65s^1$	44 **Ru** $4d^75s^1$	45 **Rh** $4d^85s^1$	46 **Pd** $4d^{10}$
55 **Cs** $6s^1$	56 **Ba** $6s^2$	57 **La** $5d^16s^2$	72 **Hf** $4f^{14}5d^2$ $6s^2$	73 **Ta** $4f^{14}5d^3$ $6s^2$	74 **W** $4f^{14}5d^4$ $6s^2$	75 **Re** $4f^{14}5d^5$ $6s^2$	76 **Os** $4f^{14}5d^6$ $6s^2$	77 **Ir** $4f^{14}5d^9$	78 **Pt** $4f^{14}5d^9$ $6s^1$
87 **Fr** $7s^1$	88 **Ra** $7s^2$	89 **Ac** $6d^17s^2$	104 **Rf** $5f^{14}6d^2$ $7s^2$	105 **Db** $5f^{14}6d^3$ $7s^2$	106 **Sg** $5f^{14}6d^4$ $7s^2$	107 **Bh** $5f^{14}6d^5$ $7s^2$	108 **Hs** $5f^{14}6d^6$ $7s^2$	109 **Mt** $5f^{14}6d^7$ $7s^2$	110 **Ds** $5f^{14}6d^9$ $7s^1$

Lanthanides	58 **Ce** $4f^26s^2$	59 **Pr** $4f^36s^2$	60 **Nd** $4f^46s^2$	61 **Pm** $4f^56s^2$	62 **Sm** $4f^66s^2$	63 **Eu** $4f^76s^2$
Actinides	90 **Th** $6d^27s^2$	91 **Pa** $5f^26d^1$ $7s^2$	92 **U** $5f^36d^1$ $7s^2$	93 **Np** $5f^46d^1$ $7s^2$	94 **Pu** $5f^67s^2$	95 **Am** $5f^77s^2$

Figure 8.4. Periodic table with electron configurations.

1B 11	2B 12	3A 13	4A 14	5A 15	6A 16	7A 17	8A 18
							2 **He** $1s^2$
		5 **B** $2s^22p^1$	6 **C** $2s^22p^2$	7 **N** $2s^22p^3$	8 **O** $2s^22p^4$	9 **F** $2s^22p^5$	10 **Ne** $2s^22p^6$
		13 **Al** $3s^23p^1$	14 **Si** $3s^23p^2$	15 **P** $3s^23p^3$	16 **S** $3s^23p^4$	17 **Cl** $3s^23p^5$	18 **Ar** $3s^23p^6$
29 **Cu** $3d^{10}4s^1$	30 **Zn** $3d^{10}4s^2$	31 **Ga** $3d^{10}4s^2 4p^1$	32 **Ge** $3d^{10}4s^2 4p^2$	33 **As** $3d^{10}4s^2 4p^3$	34 **Se** $3d^{10}4s^2 4p^4$	35 **Br** $3d^{10}4s^2 4p^5$	36 **Kr** $3d^{10}4s^2 4p^6$
47 **Ag** $4d^{10}5s^1$	48 **Cd** $4d^{10}5s^2$	49 **In** $4d^{10}5s^2 5p^1$	50 **Sn** $4d^{10}5s^2 5p^2$	51 **Sb** $4d^{10}5s^2 5p^3$	52 **Te** $4d^{10}5s^2 5p^4$	53 **I** $4d^{10}5s^2 5p^5$	54 **Xe** $4d^{10}5s^2 5p^6$
79 **Au** $4f^{14}5d^{10} 6s^1$	80 **Hg** $4f^{14}5d^{10} 6s^2$	81 **Tl** $4f^{14}5d^{10} 6s^26p^1$	82 **Pb** $4f^{14}5d^{10} 6s^26p^2$	83 **Bi** $4f^{14}5d^{10} 6s^26p^3$	84 **Po** $4f^{14}5d^{10} 6s^26p^4$	85 **At** $4f^{14}5d^{10} 6s^26p^5$	86 **Rn** $4f^{14}5d^{10} 6s^26p^6$
111 **Rg** $5f^{14}6d^{10} 7s^1$	112 $5f^{14}6d^{10} 7s^2$						

64 **Gd** $4f^76s^2 5d^1$	65 **Tb** $4f^96s^2$	66 **Dy** $4f^{10}6s^2$	67 **Ho** $4f^{11}6s^2$	68 **Er** $4f^{12}6s^2$	69 **Tm** $4f^{13}6s^2$	70 **Yb** $4f^{14}6s^2$	71 **Lu** $4f^{14}5d^1 6s^2$
96 **Cm** $5f^76d^1 7s^2$	97 **Bk** $5f^86d^1 7s^2$	98 **Cf** $5f^{10}7s^2$	99 **Es** $5f^{11}7s^2$	100 **Fm** $5f^{12}7s^2$	101 **Md** $5f^{13}7s^2$	102 **No** $5f^{14}7s^2$	103 **Lr** $5f^{14}6d^1 7s^2$

8.6.1 General Trends Across Groups and Periods

Note in Figure 8.4 that reading from the top to the bottom, each new period has a new shell or principal level. In other words, the periods in the periodic table are distinguished by a different set of shells. For example, the first period contains only the K shell, the second period contains the K and L shells, the third period contains K, L, and M shells, and so on. The number of shells does not change within a given period, which corresponds to filling of subshells. Therefore the properties of different elements within a period are largely due to the different filling of the subshells. Because the number of principal levels stay the same across the period and adding each new electron (plus a proton in the nucleus) increases the attractive force between the nucleus and the electrons, the atomic radius decreases going from left to right across a period. This means the outer electrons are more tightly bound to the nucleus, which results in an increase in ionization energy going from left to right across a given period. Recall that the ionization energy is the energy required to remove the outermost electron, or even more precisely the weakest bound electron. The ionization trend also influences the conductivity of the elements to the extent that for an element to be a good electrical conductor, it needs electrons that are only weakly bound to the nuclei of their atoms. This shows how different filings of subshells influence or determine the properties of the elements.

Note: The atomic radius of a chemical element is usually defined as the mean or typical distance from the nucleus of its atoms to the boundary of the surrounding cloud of electrons. Strictly speaking, the radius is originally a classical concept because neither atom is entirely spherical nor does it have a fixed radius. Practically, the radius of an atom is usually found by measuring the distance between the nuclei of two touching atoms in a bond, and then halving that distance.

Figure 8.5 demonstrates the separate grouping of elements with different groups corresponding to different properties due to different subshells. On the extreme left, are two groups of elements (1A and 2A) called alkali metals and alkaline earth metals, respectively, in which the valence s orbitals are being filled. On the extreme right, is a block of six groups (columns), in which the valence p orbitals are being filled. This p block on the right and the s block (with two columns) on the left are called representative or main-group elements.

In the middle of Figure 8.5 is a block of 10 groups (columns), in which the valence d orbitals are being filled. This d block consists of transition metals. Below the main part of the table, there is a block of two rows and 14 columns at the bottom, in which the valence f orbitals are being filled. These two rows are also called the Lanthanide and Actanide series, respectively, and the

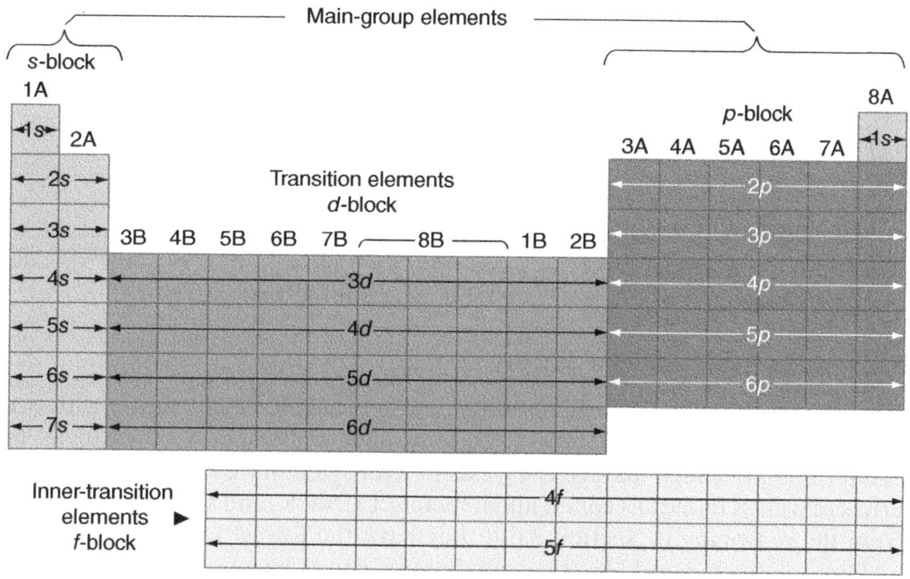

Figure 8.5. Block structure of the periodic table: The elements arranged in a column (group) have the same type of valence electron configuration.

f-block metals collectively. In Figure 8.5, note that 2 columns in the s block, 6 columns in the p block, 10 columns in the d block, and 14 columns in the f block correspond to precisely the number of electrons that can fill the s, p, d, and f subshells, respectively.

Caution: The general trends discussed here are true for a good number of representative elements. Exceptions happen due to complex interactions among electrons in many-electron atoms.

Going down a group, the number of principal levels increases; that is, the outermost electrons occupy higher energy levels farther from the nucleus. Therefore the atomic radius (and hence atomic size) increases from the top to the bottom of a given group. Accordingly, the ionization energy decreases going down a group. Nevertheless, the elements in the same group tend to have similar physical and chemical properties. This is because there is another very important factor involved: The elements in the same group have the same valence electron structure. This means they have the same number of electrons in an l orbital, and therefore are able to make similar chemical bonds. For example, Li, Na, and K all have one electron in the $l = 0$ (s) orbital of the outermost shell.

Note: The vertical columns in the periodic table are called groups and the horizontal rows are called periods. Elements in a group have similar properties, whereas elements in a period have different properties. This is largely due to the similarity and differences in the valence electron structure of the atoms. Valence electrons are the electrons in the outermost energy level (principal level) of an atom.

Now that you understand the physical reasoning behind some general tendencies in the periodic table, let us explore the physical basis of some specific groups of elements in the periodic table with similar properties.

8.6.2 Alkalis and Alkaline Earths

For each period n, the ns orbital in the outermost shell n is filled first with two electrons. Each element in the 1A group have only one electron in its ns orbital, and are called alkalis. They, except hydrogen, share the common properties of a class of metals called alkali metals: Li, Na, K, and so on. As expected from the reasoning in Section 8.6.1, they have the lowest ionization energies and are relatively good electrical conductors. This is because the lone valence electron is available to move around easily from one atom to another. All alkali metals are soft, have low melting points, and are chemically very active.

Note: Hydrogen, in spite of its top position in the alkali metal column in the periodic table, does not behave as an alkali metal under standard conditions. However, it does behave like a metal after it undergoes a phase change to the solid phase under extremely high pressure. The solid phase of hydrogen consists of a crystal lattice of atomic nuclei (basically, protons), with a spacing comparable with a typical electron wavelength, the de Broglie wavelength. The electrons under these conditions are pretty much unbound and behave like the conduction electrons in a metal.

Alkaline earth metals (Be, Mg, Ca, etc.) have two electrons in the ns orbital of their outermost shell n, and therefore fall into Group 2 of the periodic table. Although the ns subshell is filled, the properties of alkaline earths, also called alkaline earth metals, are understandably more similar to alkalis than to elements with higher groups: the ionization energies are low and electrical conductivity is high. Due to the two valence electrons, alkaline earth metals are chemically active.

Following the completion of the 4s orbital (in Ca), the next set of orbitals ready to be filled is 3d. Electrons are added to the five 3d orbitals beginning with scandium (Sc) and ending with zinc (Zn). These elements are called transition metals.

8.6.3 Transition Metals

The elements in the periodic table from Groups 3 to 12 in periods 4–7 are called transition metals. Their similar properties are primarily determined by the electrons in the outermost s subshell while the inner d subshell is being filled. However, for elements in the middle of the period, the filing of the 3d subshell leads to some significant characteristics. For example, electrons in the d subshell orbitals of Fe, Co, and Ni are unpaired, that is, they have unpaired spins because they are distributed according to Hund's rule. As a result, the unpaired spins of neighboring atoms in a crystal lattice align themselves to produce large magnetic moments, which give rise to certain properties called ferromagnetic properties. Again, alignment of unpaired spins is yet another attempt to gain stability. As the d subshell is filled, the electron spins in the orbitals are paired off, the magnetic moment and the tendency to align with the neighboring atoms diminishes.

Molecules with one or more unpaired electrons are drawn into a magnetic field. The greater the number of unpaired electrons, the stronger is this force of attraction. This property of a substance is called paramagnetism. Substances made of molecules with no unpaired electrons are slightly repelled from a magnetic field and are called diamagnetic.

Note: Strictly speaking, and also according to the definition of IUPAC (International Union of Pure and Applied Chemistry), a transition metal is an element whose atom has an incomplete d subshell, or that can give rise to cations with an incomplete d subshell. zinc, cadmium, and mercury do not meet this requirement because they have *their* d *subshell filled:* d^{10} *configuration*. They do, however, form ions with a +2 oxidation state, but retain the d^{10} configuration.

Due to the two valence electrons, alkaline earth metals are chemically active.

On the contrary, inert gases are chemically inactive.

8.6.4 Inert Gases

A group of elements called inert gases are in Group 8A (also called Group 18), farthest away from alkalis in Group 1. These two groups are equally different in their properties as well due to their different valence electron structure. All the inert elements have the outermost subshell (s subshell in He and p subshell in all others) completely filled. Because they have no valence electron available, they are chemically very inactive and are therefore called inert gases. They do not easily form bonds, have high ionization energy, or poor electrical conductivity. Due to their nonbonding tendency and as a result their weak interactions among atoms, their boiling points are quite low; they are monatomic gases at room temperature.

8.6.5 Halogens

Elements in Group 7A (F, Cl, Br, etc., also called Group 17), immediately to the left of Group 8A inert gases, are one electron short of having the outermost subshell filled and therefore become stable like the inert gases (recall the stability principle). As a result, halogens are chemically very active in a quest to gain stability. Recall that alkalis also are chemically very active due to a similar, but complimentary reason: They have one valence electron, and therefore to remain stable, they will be happy to lose it. As a result, alkalis and halogens form bonds with each other. They are strong bonds called ionic bonds. For example, the alkali atom Na loses an electron to become the sodium ion, Na^+, and the chlorine atom, Cl, gains the electron to become the chlorine ion, Cl^-. These two ions are attracted by the electrostatic force and make a strong bond that becomes the compound sodium chloride, NaCl.

8.6.6 Lanthanides and Actinides

The elements with atomic numbers from 58 to 71 are called lanthanides, or rare earths, have similar properties because they all have the outer 6s subshell filled and the inner 4f subshell filled. Just like the 3d orbitals of the transition metals, the 4f orbitals of lanthanides often have unpaired electrons. Consequently, the unpaired spins of neighboring atoms align and give rise to high magnetic moments and properties, such as ferromagnetism. Due to a larger number of orbitals than the d subshells, the f subshells may have more unpaired electrons and a larger angular momentum giving rise to an even greater magnetic moment and ferromagnetic effect. For example, the element holmium (Ho) can have an extremely large internal magnetic field at low temperatures, even muh larger than in iron (Fe).

Elements with atomic numbers from 90 to 103, called actinides, are very similar to lanthanides because they have their 7s subshell filled (instead of the 6s), and the inner subshells are in the process of being filled. A commonality among all of the actinides is that all of them are radioactive.

In the beginning of this chapter, we explained that the quantum mechanics developed so far could not explain the periodic table and Zeeman effect. We have already seen how by introducing the spin angular momentum quantized by the spin quantum number lays the foundations for the physical basis of the periodic table. The concept of spin can also be used to explain the Zeeman effect. First, however, we need to understand how the spin angular momentum contributes to the total angular momentum.

8.7 COMPLETING THE STORY OF ANGULAR MOMENTUM

By now, we have introduced two kinds of angular momentum: orbital angular momentum L quantized by orbital quantum number l, and spin angular momentum S quantized by spin quantum number s. Each electron in an atom has a certain amount of L and a certain amount of S. Both combine to give the total angular momentum J of the electron.

For example, consider alkalis with one valence electron in their outermost shell outside the closed inner shells. For such atoms, the magnetic moment and

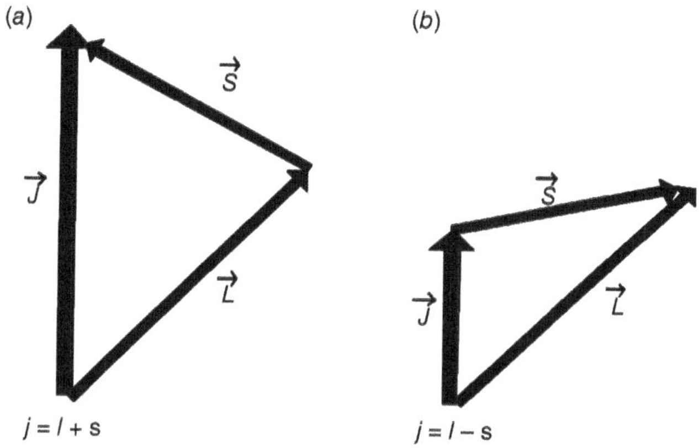

Figure 8.6. An example of the two ways in which the vectors $\vec{L}$ and $\vec{S}$ can add to form $\vec{J}$ given $l = 1$, $s = 1/2$.

the total angular momentum is given by this single valence electron. The total angular momentum J is the vector sum of L and S:

$$\vec{J} = \vec{L} + \vec{S} \tag{8.6}$$

where $\vec{J}$, $\vec{L}$, and $\vec{S}$ are vectors with J, L, and S as corresponding magnitudes, as shown in Figure 8.6.

STUDY CHECKPOINT 8.6

Will the total angular momentum of the ions He^+, Ba^+, and Mg^+ be determined by a single electron just like the alkali metals?

Solution:

The answer is yes, because these ions have only one valence electron because they have lost the second electron.

Note: According to the exclusion principle, the total angular momentum and magnetic moment of a closed shell is zero; another important outcome of the exclusion principle.

As shown in Figure 8.6, in the case of one electron, there are only two relative orientations:

$$\begin{aligned} j &= l + s \\ j &= l - s \end{aligned} \tag{8.7}$$

where j is the quantum number that quantizes the magnitude of the total angular momentum J to

$$J = \sqrt{j(j+1)}\hbar \tag{8.8}$$

where j is given by

$$j = l + s = l \pm \frac{1}{2} \tag{8.9}$$

To rule out the third possibility, $j = s - l$ is left as an exercise for you: Problem 8.9. The concept behind Eq. 8.9 is illustrated in Figure 8.6. This means that when only one electron is involved, there are only two relative orientations of L and S: alignment corresponding to $l + s$ in which case J > L, and antialignment corresponding to $l - s$ in which case J < L. For example, for the $l = 1$ state the alignment will result in the quantum number $j = 3/2$, and antialignment will result in the quantum number $j = 1/2$.

STUDY CHECKPOINT 8.7

Calculate the values of the quantum number j for an electron in the s subshell.

Solution:

The s subshell corresponds to $l = 0$.
Therefore, from Eq. 8.9:

$$j = 1/2, -1/2$$

But −1/2 is not a physical solution.
Therefore $j = 1/2$ is the only possibility.

Equation 8.8 represents the quantization of total angular momentum, J. Furthermore, just like L, J is also quantized in direction, which may be expressed in the z direction by the following equation:

$$j_z = m_j\hbar \tag{8.10}$$

where m_j is given by

$$m_j = -j, -j+1, \ldots, j-1, j$$

PROBLEM 8.5

An electron in a p subshell has its orbital angular momentum and spin angular momentum aligned. What are the possible orientations of the total angular momentum J, that is, the values of m_j?

Solution:

p subshell, which implies $l = 1$
Alignment, which implies $j = 1 + 1/2 = 3/2$

$$m_j = -j, -j+1, \ldots, j-1, j$$

which implies

$$m_j = -\frac{3}{2}, -\frac{1}{2}, \frac{1}{2}, \frac{3}{2}$$

Next, we will see how coupling between L and S helps explain the Zeeman effect.

8.8 UNDERSTANDING THE ZEEMAN EFFECT

The Zeeman effect refers to the splitting of an expected spectral line into multiple lines, such as three lines in the presence of an external magnetic field. As illustrated in Figure 8.7, there exist several possible electron configurations

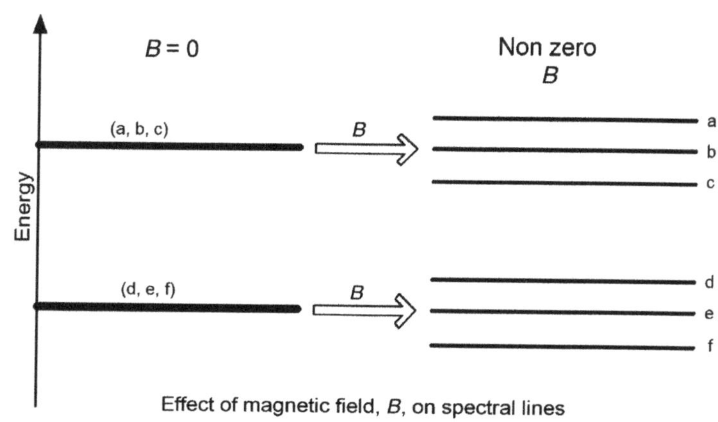

Figure 8.7. Illustration of the Zeeman effect.

in an atom with the same energy, called degenerate states. Therefore, the transitions between this set of configurations and another correspond to a single spectral line. The presence of a magnetic field breaks this degeneracy, because the magnetic field interacts differently with electrons with different quantum numbers, and thereby slightly modifies their energies. The result of this interaction follows: Where there were several configurations that previously had the same energy, they now have different energies; one spectral line in the absence of a magnetic field has been split into many close spectral lines in the presence of a magnetic field.

Now, let us see how the electron spin causes the Zeeman effect. You know from Chapter 1 that any motion of charge generates the magnetic field, and hence the magnetic moment. An electron in an atom has an angular momentum due to its circular-like motion according to the right-hand rule (rhr) explained in Section 1.14.

As shown in Problem 8.8 and demonstrated in Figure 8.8, the magnetic moment of electron μ in an atom is given by

$$\vec{\mu} = -\frac{e}{2m}\vec{L} \tag{8.11}$$

When the electron is in an external magnetic field, its magnetic moment interacts with the external field B to cause the potential energy V_B:

$$V_B = -\vec{\mu}.\vec{B} \tag{8.12}$$

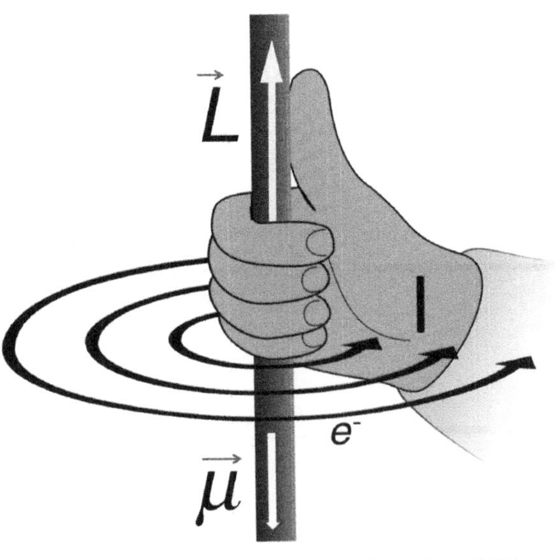

Figure 8.8. Magnetic moment produced by the orbiting electron.

where the expression $\vec{\mu}.\vec{B}$ represents the dot product between the two vector quantities μ and B, and is equal to the product of their magnitudes times the cosine of the angle their directions make with each other. Now, the key point is that the magnetic moment μ due to the total angular momentum is the vector sum of two magnetic moments from L and S. Therfore, Eq. 8.11 is replaced with:

$$\vec{\mu} = -\frac{e}{2m}\vec{J} = -\frac{e}{2m}(\vec{L}+\vec{S}) \qquad (8.13)$$

From Eqs. 8.12 and 8.13 you can see that the potential energy V_B will have multiple values possible, and hence multiple spectral lines. In the absence of spin, Eq. 8.13 will reduce to Eq. 8.11, and the potential energy will have a single value. Therefore, the Zeeman effect is the direct result of the intrinsic spin of the electron.

8.9 LIVING IN THE QUANTUM WORLD

We discussed in this chapter how quantum mechanics successfully explains the Zeeman effect; that is, splitting a spectral line into several components in the presence of a magnetic field. The Zeeman effect exists due to the quantum number s based on the electron spin that can have one of two possible values: +1/2, −1/2. Owing to its understanding by quantum mechanics, the Zeeman effect is successfully used in applications, such as nuclear magnetic resonance spectroscopy (NMR), electron spin resonance (ESR) spectroscopy, and magnetic resonance imaging (MRI). Magnetic resonance imaging is used to get images of internal parts, such as bones, muscles, and organs, of an organism, such as a human, for the purpose of medical diagnosis. Until recently X-rays were used for this purpose. However, using X-rays for imaging has some drawbacks. First, due to limited resolution, X-rays do not yield sufficiently resolved images of overlapping structures, and the images of diseased tissues and healthy tissues look alike. As a result, X-ray images often fail to detect injury or disease. Furthermore, X-rays are high-energy radiation, and therefore can cause physiological damage. The MRI is free of these drawbacks because it works in the environment of low-energy radiation and high resolution.

As we explained in this chapter, protons and neutrons have a spin of ½ just like electron, and therefore can cause the Zeeman effect as well. Magnetic resonance imaging is based on the spin quantum number of the spin of the atomic nuclei, which are made up of protons and neutrons. The hydrogen nucleus (containing one proton) is the most common nucleus studied using MRI because hydrogen is a major constituent of an aqueous body fluid and fatty tissues: Think of molecules of water and fatty acids that contain hydrogen.

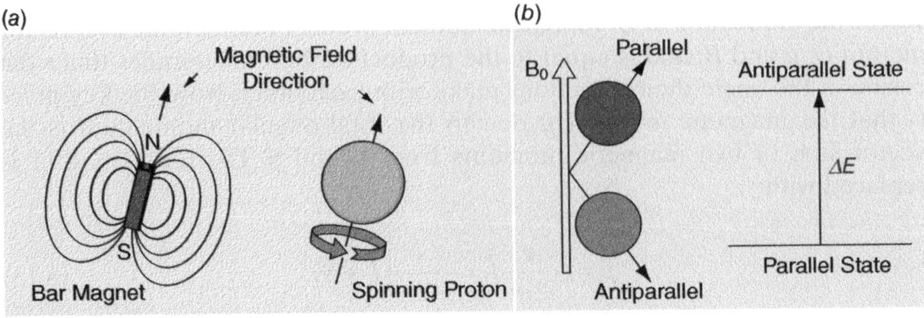

Figure 8.9. The protons align themselves either parallel or antiparallel to an external magnetic field, B_0 with the parallel state slightly lower in energy than the antiparallel state. Based on an image by Kieran Maher.

According to quantum mechanics the nucleus of hydrogen, a proton, can have only two spin states (orientations): ½ and −½. Due to their intrinsic spin, these protons contain a magnetic moment and the two states corresponding to each spin value have the same energy. However, when an external magnetic field is applied, some of these protons align parallel to the magnetic field, and others antiparallel to the field depending on their spin state, as shown in Figure 8.9. These two orientations correspond to slightly different energies.

If we shine radio waves on these protons, and the radio waves have just the right frequency, the lower energy protons can absorb photons of radiation and flip over, for example, flipping from parallel to antiparallel alignment. Recall from Chapter 2 that according to quantum mechanics the photon's energy is proportional to its frequency. The detection of this flipping between two spin states gives rise to the NMR spectrum. In other words, the details of energy absorption required for flipping gives information about the environment of the protons (if the protons are in water or in a fat, etc). Putting this information together, we get a structural image of a slice of the body, such as an organ. Because the difference between the energies of the two spin states is very small, it takes little energy for the flip over, and hence the radiation used in the MRI technique is in the radio frequency range, which is far less energetic than X-rays, and therefore harmless.

The key here is the interaction of the nuclei (a proton in the case of hydrogen) with the magnetic field. By varying the magnetic field strength with location enables us to code different parts of the body with different frequencies. Through measurement of frequencies with enough precision, an MRI expert can reconstruct a two- (2D) or three-dimensional (3D) image that accurately depicts the interior part of a body. In a nutshell, by varying the magnetic field and using these sophisticated techniques, body parts can be imaged to different depths at great resolution, and the diseased tissues can be differentiated from the healthy tissues. As an example, Figure 8.10 presents the MRI images of a human head (*a*) and nerve fibers (*b*).

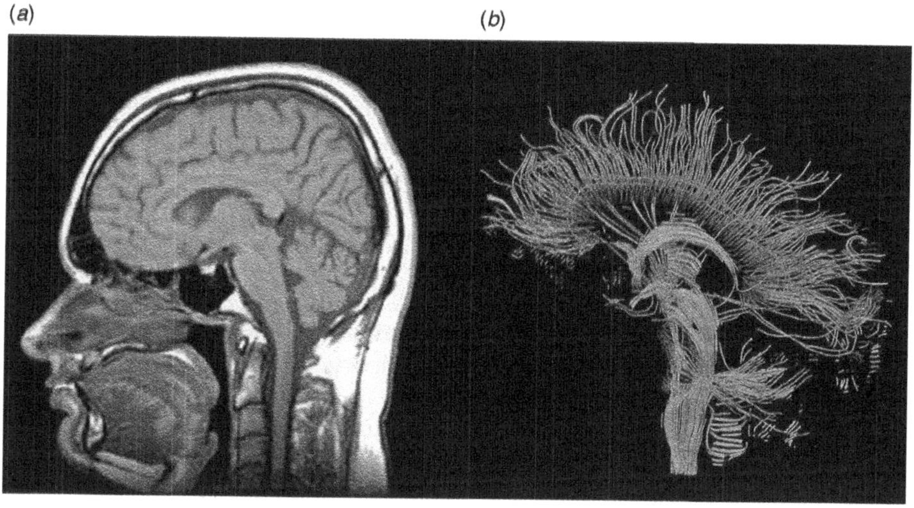

Figure 8.10. (*a*) The MRI image of a human head and (*b*) reconstruction of nerve fibers.

Due to the incredible success of MRI techniques and its potential in the practice of medicine, a physicist Peter Mansfield and a chemist Paul Lauterbur were awarded the 2003 Nobel Prize in physiology or medicine for their discoveries concerning MRI.

8.10 SUMMARY

When you consider the atomic number of elements, you find that similar physical and chemical properties appear in the elements at regular intervals. This is the basis of the periodic table. In other words, a periodic table is an arrangement of the elements in increasing atomic number in a series of rows so that the elements in the vertical columns have similar properties. Several properties of atoms are periodic in the sense that the trends regarding those properties repeat themselves from row to row. This periodicity and trends provide the periodic table with its predictive value. As a result, to fill the gaps between the known elements, several elements were predicted to exist before they were actually discovered. Elements in the same row (period) have different properties. Atomic radius and ionization energy are examples of the periodic properties of elements. Atomic radius tends to decrease, and hence the ionization energy tends to increase from the left to right inside a row (period). Going from top to bottom in a group, the radius increases and therefore the ionization energy tends to decrease.

These similarities and differences can be explained with the help of quantum number n corresponding to the principal energy level, l and m_l corresponding to orbital angular momentum, and m_s corresponding to spin angular momentum.

A set of values for n, l, m_l, m_s uniquely identifies a quantum state. According to the exclusion principle, no two electrons can occupy the same quantum state. This is the physical basis for the distribution of electrons in an atom among different orbitals (corresponding to different spin) in different subshells (corresponding to different l values) at each principal energy level or shell (corresponding to different values of n). The distribution is made in accordance with the stability principle, which dictates that the shells and subshells in a given shell be filled in increasing order of energy. This gives rise to a different valence electron structure along a period and similar valence electron structure along a group.

This provides the physical basis for the similarity in properties of elements in the same column group and different properties of elements in the same period. The coupling of orbital angular momentum and spin angular momentum to give rise to the total angular momentum explains the split of spectral lines into multiple spectral lines in an atom in an external magnetic field.

As mentioned several times earlier in this book, classical physics is an approximation of quantum physics, which is more general theory that can be applied to any structural levels. In this and previous chapters, we have applied quantum physics to atomic structure. In Chapter 9, we will explore how quantum physics can be applied to molecules or molecular structures.

8.11 ADDITIONAL PROBLEMS

Problem 8.6 Consider a proton as a charged sphere of diameter 2×10^{-14} m spinning around its axis. Determine the equatorial velocity that it will need to generate an angular momentum of $(\sqrt{3}/2)\hbar$.

Problem 8.7 Write down the electron configuration of the following atoms:

A. Na^{11}

B. Ar^{18}

C. Ca^{20}

D. Co^{27}

Problem 8.8 Consider Bohr's model of an electron in an atom revolving around the nucleus in a circle with radius r. Calculate the magnetic moment μ of this electron.

Problem 8.9 In the body of this chapter, you have learned

$$j = l \pm s$$

Argue against the possibility of

$$j = s - l$$

Problem 8.10

A. Which of the quantum numbers govern the shape of an orbital: n, l, m_l, or m_s?

B. Which of the quantum numbers govern the energy of an orbital: n, l, m_l, or m_s?

C. Which of the quantum numbers govern the spatial orientations of an orbital: n, l, m_l, or m_s?

D. Which of the quantum numbers govern the spin properties of an electron: n, l, m_l, or m_s?

Problem 8.11

A. How do the sizes of atoms change as we move from left to right across a row in the periodic table? Increase, decrease, stay constant?

B. How do the sizes of atoms change as we move from top to bottom in a group in the periodic table? Increase, decrease, stay constant?

C. Arrange the following atoms in order of increasing atomic radius:

As, Bi, Cl, Hg, Se

QUANTUM MECHANICS OF MOLECULES 9

All life involves movement.

Aristotle

Figure 9.0. Francis Crick (standing) and James D. Watson (sitting) (*c*) discovered the structure of the DNA molecules based on X-ray diffraction pattern (*b*) taken by Rosalind Franklin (*a*). Courtesy of the National Institute of Health.

Francis Crick, James D. Watson, and Maurice Wilkins (a colleague of Rosalind Franklin) were jointly awarded the 1962 Nobel Prize for Physiology or Medicine for their discoveries on deoxyribonucleic acid (DNA). Franklin did not share the Nobel Prize, as she had died in 1958 at the age of 37, and the rules of the Nobel Prizes state that each

Quantum Physics for Scientists and Technologists, by Paul Sanghera

recipient must be alive to receive the prize. Physics made a grand entrance into biology when according to legend, Francis Crick, originally a physicist and then a biologist, as well, and James Watson, a biologist, walked into the Eagle pub in Cambridge, England, on February 28, 1953, and Crick announced, "We have found the secret of Life". What they had really figured out was the now famous double helix structure of the DNA molecule. This discovery of the DNA structure catalyzed a revolution in biology by activating the already existing fields of molecular biology and biotechnology. Nevertheless, there was still a gigantic gap between determining the structure of DNA and understanding the exact role of DNA. It took the next quarter century of Crick's life along with other scientists to develop the ideas that now constitute the basics of molecular biology: the genetic code, messenger ribonucleic acid (mRNA), and the translation of mRNA into proteins. After more than one-half of a century since that event in the Eagle pub, the DNA molecule is still the biggest celebrity in biology.

In This Chapter

9.1 INTRODUCTION

In Chapter 8 we explored how quantum mechanics explains the behavior of atoms and elements. Atoms bond together to make crystalline and amorphous (noncrystalline) matter. We see different kinds of material (or matter as a physicist would say) around us. A particular kind of matter that has the same composition and properties is called a pure substance, or a substance for short. A substance is one kind of matter that has its own unique set of physical and chemical properties, not exactly the same as any other substance. Substances are broadly categorized into two groups: elements and compounds. Elements of the periodic table, discussed in Chapter 8, are examples of simple pure substances. The smallest unit of an element that retains the characteristics of that element is called an atom. A compound is a substance that consists of two or more elements always in the same proportion. The atoms of these substances are held together by bonds, such as ionic or covalent bonds, to make the building blocks for the compound. The building block, that is, the smallest unit that retains the properties of the compound, of a compound made of covalent bonds, is called a molecule. For example, water (H_2O), is a compound that in its pure form always contains hydrogen and oxygen atoms in a 2:1 ratio. The smallest unit of water is a molecule of water, H_2O, held together through covalent bonds between the hydrogen and oxygen atoms. Similarly, table salt, NaCl, is a compound that contains sodium and chlorine atoms in a 1:1 ratio. In table salt, sodium and chlorine atoms are held together through ionic bonds. Although atoms and ions play a great role in maintaining life in an organism, the true story of life begins at the molecular level.

The smallest functioning unit of life is a cell, which is largely composed of four types of molecules, commonly known as molecules of life: carbohydrates, proteins, nucleic acids, and lipids. At the current stage of developments in biology, it is very important for biologists to realize a universal physical principle: Each physical system left alone tends to acquire the most stable state, which is usually the state of minimum energy. Here are some examples of this principle expressing itself: water flowing from higher places to lower places, a pendulum when disturbed from its most stable state vibrating around it, chemical reactions happening spontaneously by losing energy called Gibbs free energy, and chemical reactions reaching a state of equilibrium. This principle certainly has its hand in the evolution of life at the molecular level: A quest for stability brings the atoms together to make molecules and crystals, and unquestionably this stability principle must have played an important role in bringing the molecules together to make the first cell and the first unicellular organism. It also contributed to how those organisms evolved into multicellular and more complex organisms. The stability principle may have made its mark on the existence (or lack) of symmetry in the shape of organisms: Why they are developed into one shape and not into another.

How do we study matter at the atomic and molecular levels? One of several methods is spectroscopy, the study of electromagnetic (EM) radiation (energy) emitted or absorbed by substances. It has been a great tool to probe the structure of matter to the molecular and atomic levels including cells, tissues, and body fluid. As a matter of fact, a substantial body of knowledge about atomic and molecular structure comes from spectroscopy.

As mentioned earlier, spectroscopy is based on studying the energy emitted or absorbed by the system. We can learn a great deal about the physical behavior of a system (a molecule is also a system) by examining its motion. That makes energy an important physical variable to study systems of all size scales ranging from planets to molecules, to atoms, and to subatomic particles. The internal energy of a system is the sum of all kinetic and potential energies of all of its components. For example, the energy of a diatomic molecule, such as H_2, includes the translational motion of the molecule through space, rotation of the molecule, vibration of atoms in the molecule, and energies of the electrons in the atoms of the molecule. Also, keep in mind that molecules, like atoms, are bound systems and as we learned in the previous chapters, the energy of bound systems turns out to be quantized, that is, it only has certain discrete values.

So, the central theme of this chapter is to explain the behavior of molecules using quantum mechanics and spectroscopy. To this end, we will explore three features of molecular behavior: rotational, vibrational, and electronic states.

9.2 A SYSTEM OF MOLECULES IN MOTION

In our everyday life, we deal with a collection (or a system) of molecules rather than individual molecules. In solids and liquids the molecules have very little or no freedom to move around unlike what is found in a gas. A gas in a container is composed of atoms or molecules, which are in rapid motion, bouncing off each other and off the walls of the container. Their motion is called translational motion, that is, the motion of an object as a whole. As a result of work by many scientists including Robert Boyle, Jacques Charles, and Joseph Louis Gay-Lussac over a long time spanning from the seventeenth to the nineteenth century, it was realized that at low pressure, the product of pressure and volume is directly proportional to the absolute temperature:

$$PV = CT$$

where C is the proportionality constant for a given gaseous system, that is, it is proportional to the number of molecules N in the system. Therefore, C can be written as:

$$C = \kappa N$$

where κ is the proportionality constant called the Boltzmann constant. It has been confirmed experimentally that the Boltzmann constant has the same value for any kind or amount of gas:

$$\kappa = 1.381 \times 10^{-23} \text{ J/K}$$

Therefore, equation $PV = CT$ can be rewritten as:

$$PV = \kappa NT \tag{9.1}$$

It is convenient to refer to the amount of gas in terms of the number of moles; a mole of molecules contains a constant number of molecules called Avogadro's number N_A:

$$N_A = 6.022 \times 10^{23} \text{ mol}^{-1}$$

Chemistry students know that 1 mol of gas molecules will fill a volume of 22.4 L at STP (Standard Temperature and Pressure), which is 1-atm pressure and 0°C. Assuming the molecular system has n mol, which amount to N molecules.

This means,

$$N = nN_A$$

Therefore, Eq. 9.1 can be rewritten as:

$$PV = \kappa\, nN_A T = nRT \tag{9.2}$$

where R is given by

$$R = \kappa N_A \tag{9.3}$$

Equation 9.2 is a famous equation in chemistry known as the ideal gas equation.

STUDY CHECKPOINT 9.1

How many molecules does a 22.4-L container filled with a mixture of nitrogen and oxygen gases contain at STP?

Hint: No calculations needed.

Solution:

A 22.4 L sample of any gas is equivalent to 1 mol of gas, which contains the number of molecules equal to Avogadro's number, that is, 6.022×10^{23}.

An ideal gas is defined as the gaseous system that obeys Eq. 9.2 for all pressures and the temperature T is known as the absolute temperature, given in units of kelvin (K). The standard practice for the kelvin scale is to refer to the temperature as kelvin and not degree kelvin, and to use the symbol K not k. The kelvin is related to units of Celsius °C, by the following equation:

$$\mathrm{K} = {}^\circ\mathrm{C} + 273.15$$

Therefore the absolute zero temperature is equal to −273.15°C. Note from the ideal gas equation that the volume of gas will be zero at absolute zero temperature; however, most of the gases turn into a liquid or a solid before that temperature is reached.

The ideal gas equation specifies how gases behave, but is not based on the reason they behave as they do. In 1857, Rudolf Clausius put the scattered work performed over a century into a theory called the kinetic molecular theory. The core points of this theory are summarized below:

1. Gases consist of a large number of molecules, which are continuously in random motion. The word molecule refers to the smallest unit of a given gas: an atom or a molecule. The random motion provides the kinetic energy.
2. The attractive and repulsive forces between gas molecules are negligible.
3. The volume of the gas molecules is negligible compared to the volume of the gas container.
4. The molecular collisions do not change the average kinetic energy of the molecules; the molecular collisions are elastic.
5. The kinetic energy of the molecules is due to and directly proportional to the temperature of the gas. Therefore the average kinetic energy of molecules of different gases at the same temperature is the same.

So, in an ideal gas, molecules have energy due to the random motion in three-dimensional (3D) space, that is, the translational motion. Considering that molecules also exist in liquid and solid states, this is only part of the story because each molecule in the system in general will also have other motions, such as rotation and vibration, which will add to the total energy of the molecule, and hence the system. To understand those motions, we need to understand the bonds between atoms in the molecules.

9.3 BOND: THE ATOMIC BOND

The EM force, also called the electrostatic or Coulomb force, which is discussed in Chapter 1, is the dominant force that governs the dynamics and behavior of matter at the atomic and molecular levels. Recall from Chapter 1 that due to the EM force, particles of opposite charge are attracted to each other and particles of like charges repel each other. This behavior governs the structure of atoms and molecules that tend to have a net zero charge, that is, an equal number of electrons and protons. As discussed in Chapter 1, the EM force between entities of atomic and molecular sizes is much greater than the gravitational force between these entities; thus gravity does not play a role. Because life starts from a cell, which is largely composed of four types of molecules, EM force is also the dominant force that keeps organisms, including humans, together and functioning.

Through the EM force of attraction, atoms bond with each other to make molecules. These bonds are called atomic or molecular bonds. Although an atom overall is neutral, its neutrality does not come from having a zero-point charge; it comes from the balance of opposite charges, as it has both negative charge on its electrons and positive charge on protons in the nucleus. Therefore, when two atoms get close enough, they can experience each other's charge and get affected by it. One of these affects is the bond, the atomic bond. These EM affects determined by the electron configuration of the involved atoms, discussed in Chapter 8, give rise to two types of bonds: ionic and covalent bonds.

STUDY CHECKPOINT 9.2

Why it is the EM and not the nuclear or gravitational force that is responsible for binding atoms into molecules?

Solution:

As shown in Section 1.19.4, the EM force is the only force that has enough strength and the range to be effective at the distance involved in binding atoms, that is, of the order of 10^{-10} m.

Gravitational force is too weak because atoms and electrons have such a small mass.

The range of the nuclear force is at a scale $\leq 10^{-15}$ m, which is much smaller than the scale of atoms and molecules.

An ionic bond is formed when the two atoms involved in the bond are easily ionized. Sodium chloride, NaCl, is an example, where Na^+ and Cl^- ions (and not Na and Cl atoms) bond with each other. This happens because Na, with an electronic configuration $1s^2 2s^2 2p^6 3s^1$, readily loses its outermost electron $3s^1$ to gain more stability and becomes Na^+, whereas the Cl atom with an electronic configuration $1s^2 2s^2 2p^6 3s^2 3p^5$ readily accepts this electron to attain a more stable structure and becomes Cl^-. Note that both Na^+ and Cl^- have completely filled shells. However, now they have opposite electric charges, and therefore get attracted to each other to form a bond called an ionic bond.

Note: An ionic substance, such as NaCl, exists as 3D crystal arrays. Such a substance in its aqueous form breaks down into its ions. For example, NaCl exists in a 3D structure, which breaks down into Na^+ and Cl^- ions when dissolved in water. A molecular substance, such as water and methane (CH_4), exists in units of molecules.

When the two bonding atoms share one or more pairs of electrons rather than losing or gaining electrons, the bond between the two is called a covalent, rather than an ionic bond. The hydrogen molecule (H_2), and water (H_2O)

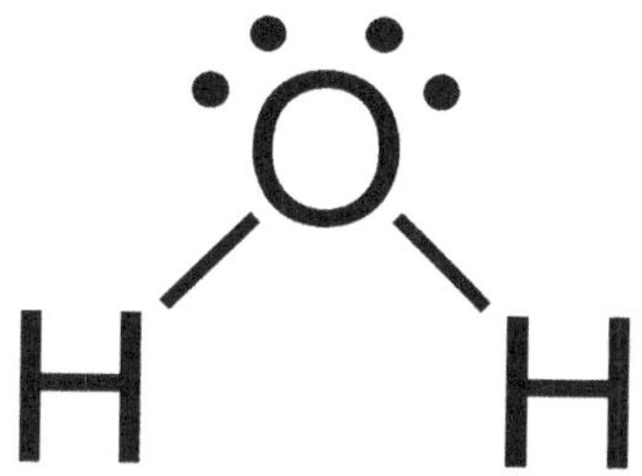

Figure 9.1. A water molecule formed from covalent bonds between one oxygen atom and two hydrogen atoms.

TABLE 9.1. Binding Energy of a Hydrogen Chloride Molecule, a Hydrogen Molecule, a Hydrogen Atom, and a Deuterium Nucleus

System	Binding Energy
HCl (intermolecular force/molecule)	0.16 eV
H_2 (hydrogen molecule)	4.5 eV
H (hydrogen atom)	13.6 eV
${}^2_1H^+$ (deuterium nucleus)	2.224 MeV

molecule, illustrated in Figure 9.1, are examples of covalent bonds. The shared electrons spend more time in between the two atoms. Although there will be a repulsive force between the two electrons in the pair, each of these electrons will be attracted to the protons in the nuclei. The natural tendency will be to acquire a balance between the attractive and repulsive forces in order to obtain stability. For example, the balance between these repulsive and attractive forces in a hydrogen molecule occurs when the two atoms are at a separation of $\sim 7.42 \times 10^{-11}$ m, where the molecule has a total negative energy of 4.5 eV. You can consider this negative energy as a potential well in which the molecule is constrained. This provides a stable equilibrium for the molecule.

If you want to break up the molecule, you need to raise the energy from a negative value to zero. The amount of energy required to separate a component from a system, such as a molecule from another molecule in a liquid, an atom from the molecule, an electron from an atom, and a proton from a nucleus, is called the binding energy. As Table 9.1 shows and Problem 9.8 demonstrates, it only takes 0.16 eV /molecule of energy to overcome the intermolecular forces of attraction among the HCl molecules in order to vaporize liquid HCl. This energy is a lot less than the energy required to break up the H_2 molecule. Also, it takes more energy to separate an electron from the hydrogen atom (break up the atom) than to break up the H_2 molecule; that is, separate one H atom bonded to another H atom. It takes even greater energy to break up the nucleus of deuterium (D_2, a heavy isotope of hydrogen with a proton and a neutron in its nucleus). From this, you can realize that it needs more and more energy to probe matter to smaller scales. For example,

Figure 9.2. Large hadron collider: The world's most powerful particle accelerator used to recreate the conditions that existed just after the Big Bang, by colliding two beams of protons accelerated to energy of 7TeV/proton. Courtesy of CERN.

you can break up the nucleus inside a lab the size of a room. However, to break up a proton, a constituent of the nuclei, you need a lab, such as hadron collider at CERN (Fig. 9.2). The lab includes a circular tunnel of 17 miles in circumference beneath the Franco–Swiss border near Geneva, Switzerland, used to provide protons an energy of 7TeV before they are smashed together head-on.

STUDY CHECKPOINT 9.3

What is the energy of the H_2 molecule in its most stable state?

How much energy is needed to break up the H_2 molecule?

Solution:

Because the attractive and repulsive forces balance out at an energy of −4.5 eV, this is the energy at which the H_2 molecule is most stable.

Also, this determines the amount of enegy (4.5 eV) needed to break up the H_2 molecule.

As mentioned earlier, atoms share valence electrons with one another to gain stability and in the process they make bonds. As a result of these bonds, molecules emerge. There are two electrons involved in a bond. As shown in Figure 9.1 for the water molecule, there could be nonbonding electron pairs, called lone pairs, as well. These nonbonding pairs and bonds around the central atom in a molecule are called electron domains. Electron domains are negatively charged, and therefore repel each other. The molecule takes the shape to minimize this repulsion energy, and therefore gain stability. The resulted molecular shape is defined by the bond lengths and angles made by the bonds with one another, called bond angles. In other words, the quest for stability gives rise to beautiful natural molecular shapes based on electron domains. Some examples of electron domain geometry are presented in Table 9.2.

As stated earlier, all substances are either elements or compounds. It is useful to categorize compounds into two categories: inorganic and organic. Organic compounds are the compounds that contain carbon and hydrogen, often in combination with other elements, such as oxygen and nitrogen.

TABLE 9.2. Geometry and Molecular Shapes for Molecules with a Different Number of Electron Domains around the Central Atom[a]

Number of Electron Domains	Geometry of the Molecule	Example
2	 Linear	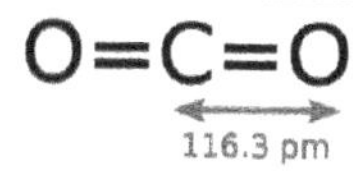 Carbon dioxide
3	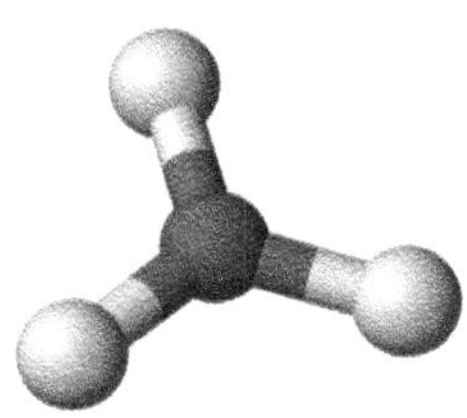 Trigonal planar	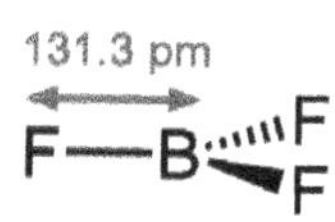 Boron trifluoride
4	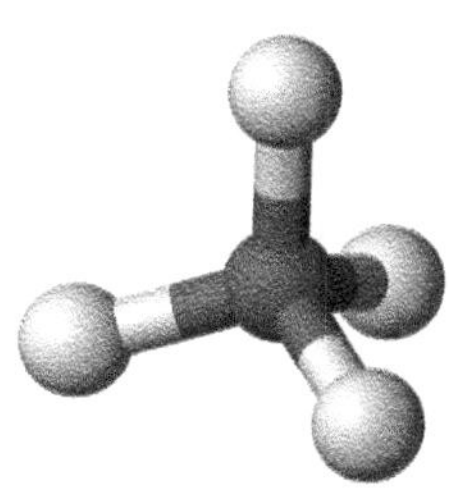 Tetrahedral	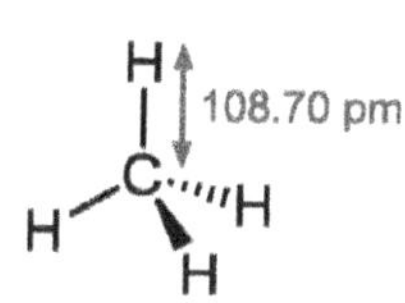 Methane

TABLE 9.2. ***Continued***

Number of Electron Domains	Geometry of the Molecule	Example
5	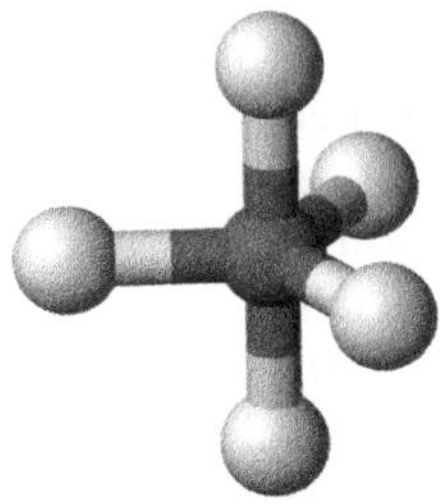 Trigonal bipyramidal	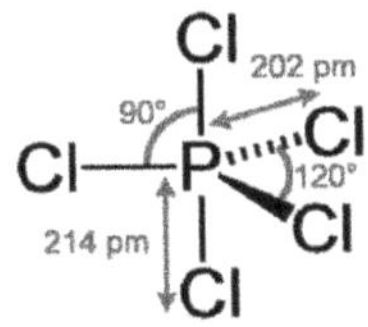 Phosphorus pentachloride
6	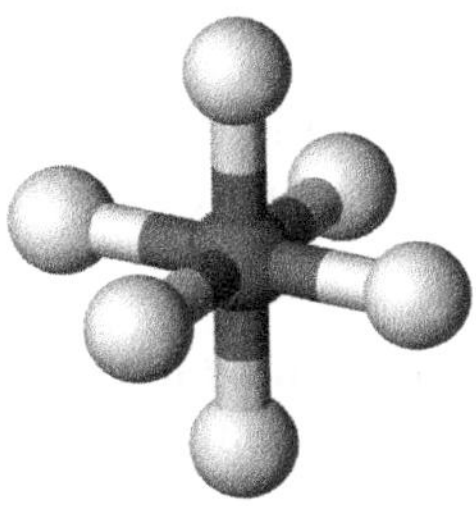 Octahedral	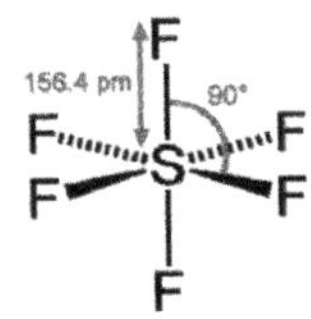 Sulfur hexafluoride

[a] Courtesy of Benjamin Mills.

Compounds that are not organic are called inorganic compounds. Inorganic compounds are mostly ionic, whereas organic compounds are mostly covalent. The smallest unit of a covalent compound that retains the properties of the compound is a molecule. In a nutshell, a molecule is an electrically neutral system of two or more atoms that are held together strong enough so they behave as a single particle. The simplest molecules are made of two atoms, called diatomic molecules.

9.4 DIATOMIC MOLECULES

As mentioned earlier, quantum mechanics can be applied at various levels of the material structure, such as electrons, atoms, and molecules. However, to truly study a molecule, we cannot ignore the fact that it is made of multiple atoms. It is often seen in physics, that we can simplify a problem by using the appropriate frame of reference. In this case, we treat a molecule in a reference frame called a center-of-mass (C.M.) system. You may be familiar with the C.M. reference frame from a course in introductory physics. For example, consider the C.M. of a system of two atoms. With respect to this C.M., the momenta p of the two nuclei will always be equal and opposite to each other. Therefore, the energy equation can be written as:

$$E = \frac{p^2}{2m_1} + \frac{p^2}{2m_2} + V = \frac{p^2}{2}\left(\frac{m_1 + m_2}{m_1 m_2}\right) + V = \frac{p^2}{2\mu} + V$$

where p is the magnitude of the momentum of each of the two nuclei, and m_1 and m_2 are the masses of the two atoms, which are the masses of the two nuclei. We have substituted

$$\mu = \frac{m_1 m_2}{m_1 + m_2} \tag{9.4}$$

So we can consider the diatomic molecule, the system of two atoms, as a single entity with mass μ called the reduced mass, for which the total energy equation can be rewritten as:

$$E = \frac{p^2}{2\mu} + V \tag{9.5}$$

In other words, a system of two particles with masses m_1 and m_2 is equivalent to a single particle with mass μ.

STUDY CHECKPOINT 9.4

Question:

Why will the momenta of the two nuclei in a molecule be equal and opposite in the C.M. frame?

Answer:

Because by definition in a C.M. reference frame, the total momentum of the system is zero.

A great deal can be learned about the molecules by studying how they absorb, emit, and scatter EM radiation (energy); a study called molecular spectroscopy. Energy comes in two forms: potential and kinetic. The kinetic energy comes from the motion of the molecules. In that context, if you look down on matter from the molecular level, you see the following three activities giving rise to three kinds of molecular motion and the resulting energy states:

1. *Rotation.* Rotation of a molecule as a whole gives rise to rotational states separated by relatively small energy intervals typically on the order of 10^{-3} eV. The spectra arising from the transitions among these states fall in the microwave region with wavelengths ranging from 1 cm to 0.1 mm.
2. *Vibration.* Vibrations of atoms in the molecule relative to one another give rise to vibrational states separated by energy intervals typically on the order of 0.1 eV, larger than that of rotational states. The spectra arising from the transitions among these states fall in the infrared (IR) region with wavelengths ranging from 0.1 mm to 1 μm.
3. *Electronic Configuration.* Changes of electron configuration inside atoms of the molecule, such as an electron, jumping from one principal level to another principal level, give rise to molecular electronic states. The typical separation between adjacent energy levels for these states is several electronvolts. The corresponding spectra fall in the visible and ultraviolet (UV) regions of EM spectrum. We have explored the absorption and emission spectra resulting from transition between electronic states.

All three kinds of activities mentioned above contribute to the energy of the molecule. Also, recall that emission and absorption spectra represent the release and absorption of energy, respectively. Therefore all three kinds of states will in general be represented by the absorption and emission spectra. We explored the electronic configuration in Chapter 8. Next, we discuss rotation of the molecule and the vibration of atoms that make the molecule.

9.5 ROTATIONAL STATES OF MOLECULES

Here is a question that you can ask at this point: Why are we interested in the rotation of a molecule? The rotation of a molecule produces absorption and emission spectra due to transitions among different rotational states. By studying these spectra, we can learn much about the properties of the molecules. This kind of study is generally referred to as molecular spectroscopy, very much like atomic spectroscopy based on emission and absorption by atoms.

Although in general the main principles discussed and conclusions drawn here apply to all molecules, we discuss diatomic molecules for the sake of simplicity. Consider a diatomic molecule, such as N_2 or CO, with two atoms of masses m_1 and m_2, as illustrated in Figure 9.3. You can model the bond between the two atoms as a massless rigid rod with length R from the center of one atom to the center of the other atom when the molecule is in an equilibrium (stable) state.

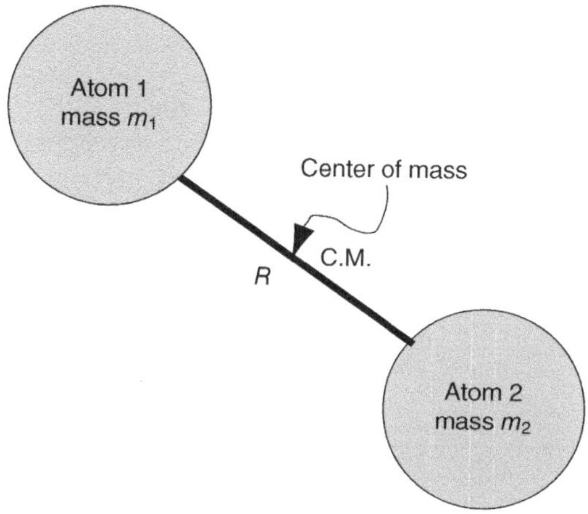

Figure 9.3. Illustration of a diatomic molecule as a rigid rotator with R as the distance between the center of the two atoms when the molecule is in the equilibrium (stable) state.

A polar molecule will rotate around its C.M.; that is, around the axis perpendicular to the line joining the two atoms and passing through the C.M.. The kinetic energy due to this rotation is given by

$$E_{\text{rot}} = \frac{L^2}{2I} \tag{9.6}$$

where I is the moment of inertia and L is the angular momentum quantized, as shown here:

$$L = \sqrt{l(l+1)}\,\hbar \qquad l = 0, 1, 2, 3, \ldots \tag{9.7}$$

Recall that l is the angular momentum quantum number. Therefore, the rotational energy can be written in terms of energy levels determined by l:

$$E_{\text{rot}} = \frac{l(l+1)\hbar^2}{2I} \tag{9.8}$$

Assume that atoms 1 and 2 are at distances r_1 and r_2, respectively, from the C.M., that is

$$r_1 + r_2 = R \tag{9.9}$$

The moment of inertia, I, of the molecule can be calculated as:

$$I = m_1 r_1^2 + m_2 r_2^2 = \frac{m_1 m_2}{m_1 + m_2}(r_1 + r_2)^2 = \frac{m_1 m_2}{m_1 + m_2} R^2 = \mu R^2 \tag{9.10}$$

Also,

$$E_{\text{rot}} = \frac{l(l+1)\hbar^2}{2I} = \frac{1}{2}I\omega^2$$

where ω is the angular velocity.

Therefore,

$$\omega = \frac{\sqrt{l(l+1)}\hbar}{I} = \sqrt{\frac{2E}{I}}$$

PROBLEM 9.1

Prove Eq. 9.10:

$$I = \mu R^2$$

Solution:

$$\begin{aligned}
I = m_1 r_1^2 + m_2 r_2^2 &= \frac{m_1 + m_2}{m_1 + m_2}\left(m_1 r_1^2 + m_2 r_2^2\right) \\
&= \frac{1}{m_1 + m_2}\left(m_1^2 r_1^2 + m_2^2 r_2^2 + m_1 m_2 r_1^2 + m_1 m_2 r_2^2\right) \\
&= \frac{1}{m_1 + m_2}\left(m_1 m_2 r_1 r_2 + m_1 m_2 r_1 r_2 + m_1 m_2 r_1^2 + m_1 m_2 r_2^2\right) \\
&= \frac{m_1 m_2}{m_1 + m_2}\left(r_1 r_2 + r_1 r_2 + r_1^2 + r_2^2\right) = \frac{m_1 m_2}{m_1 + m_2}(r_1 + r_2)^2 \\
&= \frac{m_1 m_2}{m_1 + m_2} R^2 = \mu R^2
\end{aligned}$$

We have used the fact that in the C.M. frame:

$$m_1 r_1 = m_2 r_2$$

The transitions between various rotational states are constrained by the quantum mechanical selection rule:

$$\Delta l = \pm 1 \tag{9.11}$$

This means that upon emission, the photon carries away its intrinsic angular momentum of one quantum unit, $\hbar$. Equations 9.8 and 9.11 mean that a line in the absorption spectrum corresponds to the event when a molecule in rotation state l absorbs a photon to move to the excited state $l + 1$. Then the excited molecule will return from the $l + 1$ state to the l state by emitting a photon. This will give rise to a line in the emission spectrum. The energy emitted by the molecule jumping from the $l + 1$ state to the l state is equal to the energy gap between these two states calculated from Eq. 9.8 as follows:

$$\Delta E = E_{\text{rot}}^{l+1} - E_{\text{rot}}^{l} = \frac{\hbar^2}{2I}[(l+1)(l+2) - l(l+1)] = \frac{\hbar^2}{2I}[l^2 + 3l + 2 - l^2 - l]$$
$$= \frac{\hbar^2}{I}[l+1] \tag{9.12}$$

STUDY CHECKPOINT 9.5

Assume a rotational constant R_c is defined by

$$R_c = \frac{h^2}{8\pi^2 I}$$

Show that transition energy ΔE between two consecutive rotational states l and $l + 1$ is given by

$$\Delta E = 2R_c(l+1)$$

Solution:

Equation 9.12 can also be written as:

$$\Delta E = \frac{\hbar^2}{I}[l+1] = \frac{h^2}{4\pi^2 I}(l+1) = \frac{2h^2}{8\pi^2 I}(l+1) = 2R_c(l+1)$$

Equation 9.12 combined with Eq. 9.11 predicts that ΔE for two consecutive rotational states of a diatomic molecule increase linearly with l producing a regularly spaced rotational spectrum. Experimental observations support this prediction. For example, Figure 9.4 illustrates, according to the quantum theory discussed here, the typical absorption IR spectrum (rotational + vibrational) of a diatomic molecule, such as $^{12}C^{16}O$, where you can see the regularly spaced spectral lines (peaks). The missing peak at the center corresponds to the forbidden $\Delta l = 0$ transition and is the prediction of the selection rule $\Delta l = \pm 1$. In Section 9.7, we will take a look at the spectral lines of a diatomic molecule (HCl) based on experimental data and see how they compare with quantum mechanical predictions just discussed.

Here is how you gain insight into the molecule from the spectral lines. The frequency (and hence energy) of the spectral lines can be measured and the

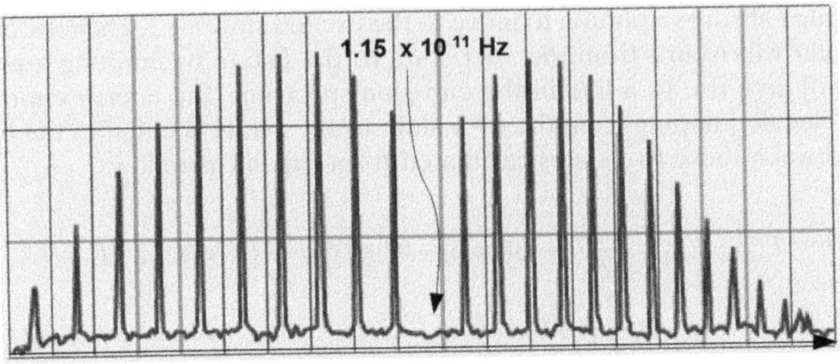

Figure 9.4. Illustration of an IR spectrum of CO molecules in the gas phase. Based on data from Dr. Tandy Grubbs, Professor, Department of Chemistry, Stetson University, DeLand, FL.

transition that it corresponds to (i.e., the value of l) can be determined from the sequence of the spectral lines. From these data, the moment of inertia can be calculated by using Eq. 9.12. From the moment of inertia, bond length R can be calculated by using Eq. 9.10. Problem 9.4 demonstrates this.

Freely moving molecules in a gas are always rotating regardless of whether they are vibrating or not because the rotational excitation energies are much smaller than the vibrational excitation energies. However, in liquids, the molecules are closer to each other and the interactions among them inhibit rotation. Therefore, molecules in liquids can exhibit pure vibrational spectra, discussed next, in Section 9.6.

9.6 VIBRATIONAL STATES OF MOLECULES

You learned about the harmonic oscillator in Section 6.6. Diatomic molecules are the simplest molecular systems with harmonic vibrations. They present a simple but effective test bed for the quantum theory of harmonic vibrations. So, it is time to apply what we learned about harmonic vibrations in Section 6.6.

As illustrated in Figure 9.5, the two constituent atoms in a diatomic molecule vibrate around the equilibrium point, which is reached when the nuclei of the two atoms are ~2-Å (0.2 nm) apart. Therefore, the restoring force for small displacements from equilibrium is modeled by a harmonic motion potential. Although a diatomic molecule may also have a rotational motion, which was treated in Section 9.5, here we only treat the vibrational motion.

There are two atoms involved here, but the Schrödinger equation is written for a single particle. So, this case is also another good example of how we model real-world problems to fit into the existing theoretical framework in

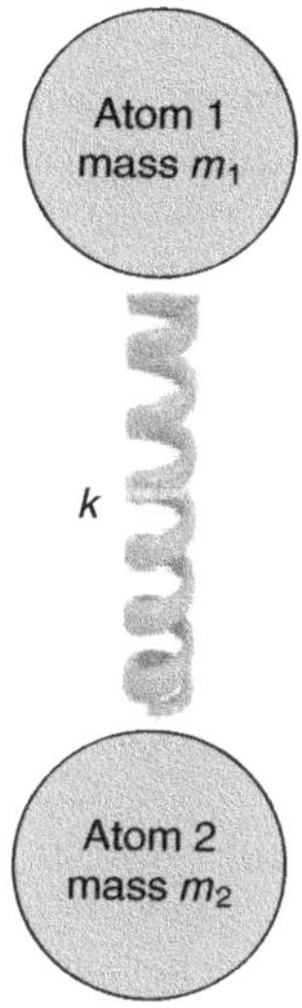

Figure 9.5. Illustration of a diatomic molecule with the two bonding atoms connected by a massless spring with force constant κ.

order to solve them. So, we will convert this two-atom problem into a one system problem in order to write the Schrödinger equation for it. Let us start by writing the total energy of the diatomic system:

$$E = \frac{1}{2}m_1 v_1^2 + \frac{1}{2}m_2 v_2^2 + V(x) \tag{9.13}$$

where m_1 and m_2 are the masses of the nuclei for the two atoms, v_1 and v_2 are their speeds, and $V(x)$ is the potential energy due to the restoring force acting at a point x distance away from the equilibrium point $x = 0$, that is, $V(x) = \frac{1}{2}\kappa x^2$, where κ is the force constant, also called the spring constant. Therefore, the energy equation can be rewritten as:

$$E = \frac{p^2}{2m_1} + \frac{p^2}{2m_2} + \frac{1}{2}\kappa x^2 = \frac{p^2}{2}\left(\frac{m_1 + m_2}{m_1 m_2}\right) + \frac{1}{2}\kappa x^2 \tag{9.14}$$

In a C.M. reference frame, we can consider this molecule, the system of two atoms, as a single vibrating entity with mass μ. So the total energy of the system now can be written as:

$$E = \frac{p^2}{2\mu} + \frac{1}{2}\kappa x^2 \tag{9.15}$$

This is equivalent to a single particle of mass μ in a potential $\frac{1}{2}\kappa x^2$. So far, we have not applied any quantum mechanics, which comes into the picture when we write the Schrödinger equation for such a particle:

$$\frac{\hbar^2}{2\mu}\left(\frac{d^2\Psi(x)}{dx^2}\right)+\left(E-\frac{1}{2}\kappa x^2\right)\Psi(x)=0 \tag{9.16}$$

As you learned in Section 6.6.2, Eq. 9.16 represents a harmonic oscillator of mass μ, and has the following solution for the energy E:

$$E_n=\left(n+\frac{1}{2}\right)\hbar\omega=\left(n+\frac{1}{2}\right)hf \qquad n=0,1,2,\ldots \tag{9.17}$$

where n is the vibrational quantum number, and

$$\omega=\sqrt{\frac{\kappa}{\mu}}=2\pi f$$

which implies

$$\kappa=\mu\omega^2=4\mu\pi^2 f^2$$

The quantum mechanical selection rule for transitions between two consecutive vibrational states is

$$\Delta n=\pm 1 \tag{9.18}$$

This means that an oscillating dipole whose frequency is f can only emit or absorb a photon of energy hf, which makes sense. Note that from Eq. 9.17 the vibrational energy states are spaced at equal intervals: ½ hf, 3/2 hf, 5/2 hf. This combined with Eq. 9.18 will give rise to equally spaced spectral lines with ΔE between any two consecutive states (neighbor lines) equal to hf.

You can ask How do you verify or compare these results with the experimental observations? There are various ways of doing it. Here is one of many possible methods for this comparison:

1. Experimentally observe the spectrum of radiation for a molecule. See if the spectral lines are equally spaced as predicted by the quantum number n in the equation for E_n.
2. Determine the frequency from the observed transition energy between two vibrational levels (energy states or spectral lines).
3. From this frequency, calculate the force constant κ for the molecule.
4. These results can be used to check the consistency of the theory among different experiments.

TABLE 9.3. Values of Force Constants for Various Diatomic Molecules Derived from an Experimentally Observed Vibrational Frequency at Which the Transition from $n = 0$ to 1 Occurs[a]

Diatomic Molecule	Molecular Mass μ (amu)	Observed Transition Frequency f ($\times 10^{13}$ Hz)	Force Constant $\kappa = \mu\omega^2 = 4\mu$ ($\times 10^{29}$ amu Hz^2)	κ (N/m)
H_2	0.504	12.5	3.11	517
CO	6.86	6.43	11.2	1860
Cl_2	17.7	1.67	1.95	324
O_2	8.00	4.67	6.89	1140
N_2	7.01	7.00	13.6	2260
HCl	0.982	8.66	2.91	482
HBr	0.997	7.68	2.32	386
HF	0.959	8.72	2.88	478
HI	1.00	6.69	1.77	293
NO	7.47	5.63	9.35	1552

[a] Observed frequency values are based on data presented in *Introduction to Molecular Spectroscopy* by Gordon M. Barrow, McGraw-Hill, New York, 1962.

The experimentally observed frequencies and the calculated force constant for this frequency for multiple molecules are presented in Table 9.3.

Equation 9.17 predicts that the ground-state energy (minimum radiation energy) of diatomic molecules is nonzero. This prediction was experimentally tested to be true by R.S. Mullikan in 1924; about the time when quantum theory was being developed.

PROBLEM 9.2

Calculate the value of force constant κ for Cl_2 listed in Table 9.3. Show the calculation for μ as well.

Solution:

Molecular mass of a Cl atom = 35.45 amu

Reduced mass for the Cl_2 molecule $= \mu = \dfrac{35.45\text{ amu} \times 35.45\text{ amu}}{35.45\text{ amu} + 35.45\text{ amu}} = 17.7\text{ amu}$

From the Cl_2 entry in Table 9.3:

$$f = 1.67 \times 10^{13}\text{ Hz}$$

Substituting these values in

$$\kappa = 4\mu\pi^2 f^2 = 1.95 \times 10^{29}\text{ amu Hz}^2$$

TABLE 9.4. Characteristic Frequencies of Some Groups in Molecules

Group	Characteristic Frequency ($\times 10^{13}$ Hz)
$-\overset{\vert}{\underset{\vert}{C}}-\overset{\vert}{\underset{\vert}{C}}-$	3.3
$\overset{\vert}{\underset{\vert}{C}}=\overset{\vert}{\underset{\vert}{C}}$	5.0
$-C\equiv C-$	6.7
$-NH_2$	10.
$-OH$	11.

A careful reader may be a little confused here: While discussing rotation, we assumed that the bonds were rigid, and now while discussing vibrations, we are saying they vibrate. However, notice that by definition vibration is around an equilibrium point. Therefore, assuming the bonds are rigid for treating rotation works well.

As mentioned earlier, although for simplicity we have only considered diatomic molecules, the main ideas can also be applied to more complex molecules. For example, a more complex molecule may have more modes of vibration (and rotation). Furthermore, some of these modes may apply to the whole molecule, and others only to specific parts of the molecule, such as the $-OH$ group or $-NH_2$ group. As shown in Table 9.4, these groups may have their own characteristics frequencies of vibrations. As listed in Table 9.4, the higher the number of bonds between the two carbon atoms, the greater is the frequency of vibrations. This is because the force constant increases, as expected, with an increase in the number of bonds between the two atoms.

As shown in Problem 9.7, it will take a temperature of ~4200 K to excite a molecule of HCl from the ground to the first excited vibrational state. This finding should give you a realization that the vibrational levels are not thermally excited at normal temperatures, such as room temperature. However, at an appropriate temperature, it is possible that a molecular system will have both the rotational and vibrational modes simultaneously.

9.7 COMBINATION OF ROTATIONS AND VIBRATIONS

In general, a molecule can absorb enough energy to get into excited rotational and excited vibrational states simultaneously. This general situation can be represented by the energy equation obtained by adding Eq. 9.8 and 9.17:

$$E = \frac{l(l+1)\hbar^2}{2I} + \left(n + \frac{1}{2}\right)\hbar\omega \tag{9.19}$$

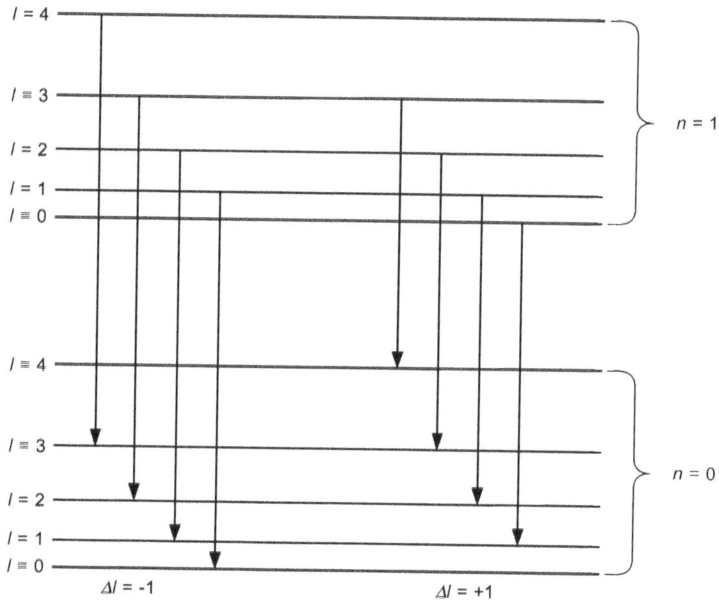

Figure 9.6. Illustration of emission spectra resulting from the transitions between the adjacent vibrational states from $n = 1$ to 0.

Once a molecule is stimulated to an excited state with the absorption of a suitable energy, it emits a photon to come back to a more stable state with less energy. As illustrated in Figure 9.6, in general, a wide spectrum of emitted photons corresponding to various rotational and vibrational transitions can be observed.

STUDY CHECKPOINT 9.6

If EM radiation of energy greater than the energy difference between the two consecutive levels (ΔE), falls on a molecule, will the molecule absorb the energy? If yes, how will the excessive energy be used by the molecule?

Solution:

The molecule will absorb the photon and move to an appropriate excited state. The left out energy will be emitted in terms of a scattered photon. This process is called Raman scattering.

As we saw in the previous sections, quantum mechanics predicts equally spaced spectral lines corresponding to both rotational and vibrational transitions. Following the same arguments, you can realize that Eq. 9.19 also predicts equally spaced spectral lines, as illustrated in Figure 9.7, which shows a more

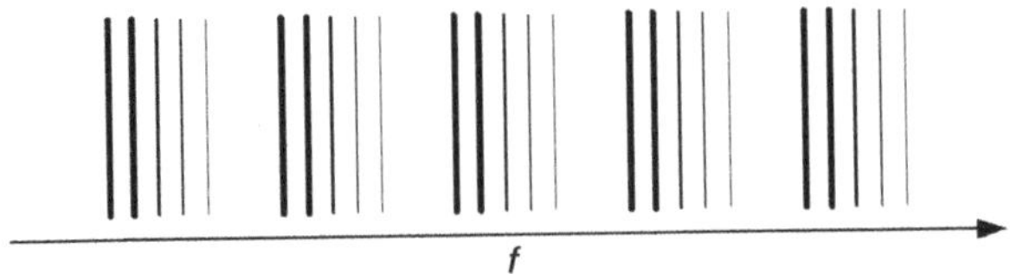

Figure 9.7. Illustration of the emission spectrum of a diatomic molecule as predicted by quantum mechanics.

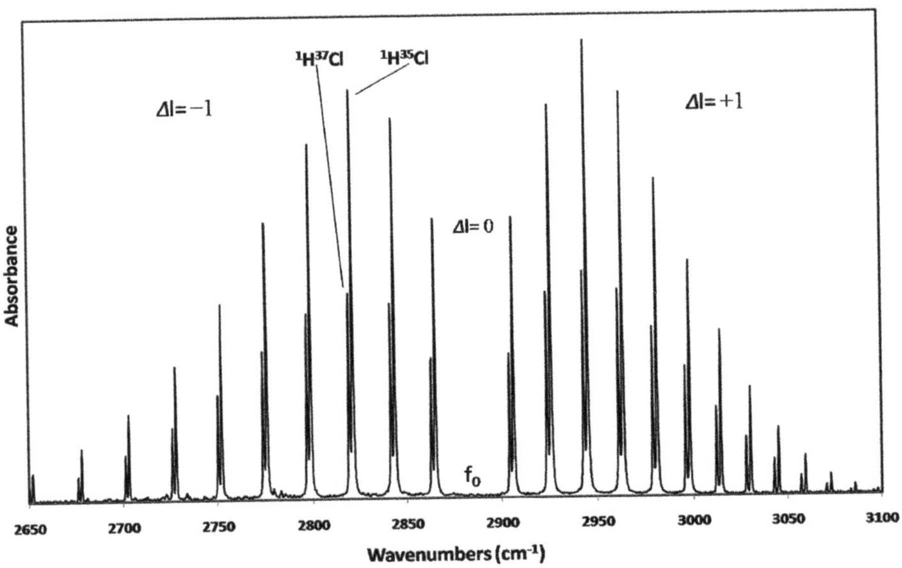

Figure 9.8. An Absorption vibration–rotation spectrum of gaseous HCl. Based on data from Dr. Tandy Grubbs, Professor, Department of Chemistry, Stetson University, DeLand, FL.

closely spaced rotational spectrum superimposed on each vibration spectral line. Different intensities of the spectral lines illustrated in Figure 9.7 are due to different transition probabilities. The greater the transition probability the greater is the intensity of the corresponding line. Spectral lines corresponding to some transitions will be missing according to the selection rules discussed in Section 9.6. These predictions are supported by the experimental observations, as demonstrated in Figure 9.8, which presents the classic rotation–vibration absorption spectrum of HCl.

The absorption lines shown in Figure 9.8 represent transitions from the ground to excited vibrational state of HCl, and also represent changes in the rotational state. To be specific, the rotational angular momentum changes by one during these transitions following the $\Delta l = \pm 1$ selection rule. Note that in Figure 9.8, the spectral lines are equally spaced as predicted by Eq. 9.19.

Let us take a closer look to see how Figure 9.8, which is based on experimental data, verifies the vibrational selection rule $\Delta n = \pm 1$ and the rotational selection rule $\Delta l = \pm 1$. The middle of the graph corresponding to frequency f_0 represents the ground vibrational state with vibrational energy $hf_0/2$. As discussed in the previous sections, each vibrational excitation from n to $n+1$ from here to either side involves an absorption of an energy hf_0. Also, from Study Checkpoint 9.5, each rotational excitation from l to $l+1$ absorbs an energy of $2R_c(l+1)$. This means that when starting with the ground vibrational state the transitions corresponding to $\Delta n = 1$ and $\Delta l = 1$ give the transition energies:

$\Delta E = hf_0 + 2R_c$ corresponding to transitions from $(n = 0, l = 0)$ to $(n = 1, l = 1)$

$\Delta E = hf_0 + 4R_c$ corresponding to transitions from $(n = 0, l = 1)$ to $(n = 1, l = 2)$

$\Delta E = hf_0 + 6R_c$ corresponding to transitions from $(n = 0, l = 2)$ to $(n = 1, l = 3)$

and so on.

The right half of the plot 9.8 ($\Delta l = +1$) represents the spectrum corresponding to these transition energies. In contrast, the left half of the plot ($\Delta l = -1$) represents a change in rotational quantum number of minus 1:

$\Delta E = hf_0 - 2R_c$ corresponding to transitions from $(n = 0, l = 1)$ to $(n = 1, l = 0)$

$\Delta E = hf_0 - 4R_c$ corresponding to transitions from $(n = 0, l = 2)$ to $(n = 1, l = 1)$

$\Delta E = hf_0 - 6R_c$ corresponding to transitions from $(n = 0, l = 3)$ to $(n = 1, l = 2)$

and so on.

Also, note in Figure 9.8 that each spectral line is split due to the mass difference between ^{35}Cl and ^{37}Cl, which gives rise to different rotational inertia for these two isotopes. The higher line belongs to ^{35}Cl and the lower line represents ^{37}Cl. The absence of a spectral line ~8.65×10^{13} Hz (in the middle of the graph) corresponds to the forbidden $\Delta l = 0$ transition, in accordance with the selection rule $\Delta l = \pm 1$ for HCl.

Caution: For diatomic molecules, such as NO, which have a nonzero electronic orbital angular momentum, you will see a spectral line in the middle of the graph, that is, corresponding to $\Delta n = 1$, $\Delta l = 0$.

STUDY CHECKPOINT 9.7

According to the selection rules, is it possible to a observe a spectral line corresponding to a pure vibrational transition in diatomic molecules that have zero electronic orbital angular momentum?

Solution:

The answer is no according to the selection rule:

$$\Delta l = \pm 1$$

because the pure vibrational transition will correspond to

$$\Delta l = 0$$

In other words, because the absorbed photon carries one unit of angular momentum, the pure vibrational transition is not possible.

However, note that although selection rules are helpful in understanding the spectra, they may be violated when the conditions are not ideal.

PROBLEM 9.3

An HCl molecule is excited to its first excited state. It emits a photon of frequency 1.2×10^{12} Hz to come back to the ground state. By using the rigid rotator model, calculate the bond length of the HCl molecule. Assume that the emitted energy is twice the rotational energy in the first excited state.

Solution:

According to the rigid rotator model:

From Eq. 9.8, the rotational energy in the first excited state is

$$E_{\text{rot}} = \frac{l(l+1)\hbar^2}{2I}$$

For $l = 1$,

$$\Delta E = 2E_{\text{rot}} = 2\frac{2\hbar^2}{2I} = 2\frac{\hbar^2}{I} \quad (1)$$

$$\text{Also, } \Delta E = hf = 1.2 \times 10^{12} \text{ Hz } h \quad (2)$$

From Eqs. 1 and 2

$$I = \frac{2\hbar^2}{h \times 1.2 \times 10^{12} \text{ Hz}} = \frac{2h}{4.8\pi^2 \times 10^{12} \text{ Hz}} \quad (3)$$

$$\mu = \frac{m_{\text{H}} m_{\text{Cl}}}{m_{\text{H}} + m_{\text{Cl}}} = \frac{1.008 \times 35.45}{1.008 + 35.45} \text{amu} = 0.9801 \text{ amu} \quad (4)$$

Also,

$$I = \mu R^2$$

which implies

$$R = \sqrt{\frac{I}{\mu}} \tag{5}$$

From Eqs. 3–5:

$$\begin{aligned} R^2 &= \frac{2h}{4.8\pi^2 \times 10^{12}\ \text{Hz} \times 0.9801\ \text{amu}} \\ &= \frac{2 \times 6.626 \times 10^{-34}\ \text{J s}^2}{4.8\pi^2 \times 10^{12} \times 0.9801\ \text{amu}} \times \frac{\text{amu}}{1.661 \times 10^{-27}\ \text{kg}} \\ &= 0.01718 \times 10^{-18}\ \text{m}^2 \end{aligned}$$

which implies

$$R = 0.1311\ \text{nm}$$

Therefore the bond length of the HCl molecule is ~1.3 Å (0.13 nm).

To understand the uses of IR spectroscopy, realize how it works: When radiation in the IR region strikes a molecule, it causes the molecule to stretch, bend, or vibrate, and thereby changes the polarity of the molecule. The energy needed to stretch, bend, or vibrate is quantized, and a given molecule can vibrate only at specific frequencies corresponding to specific energy levels. Infrared spectroscopy is widely used in both research and industry as a reliable technique for making measurements and controlling quality. For example, it is used in both criminal and civil forensic analysis to identify polymer degradation. It is important to understand that IR techniques exploit the fact that molecules vibrate (and rotate) at specific frequencies corresponding to discrete energy levels. This fact is used to identify organic compounds because each compound produces a characteristic pattern that can be used to identify the molecule. The region from 200 to 1400 cm^{-1} is especially useful for this purpose and is therefore called the *fingerprint region*. Furthermore, IR helps to confirm or rule out the existence of functional groups in a molecule under study. This is how it works. When an organic molcule is exposed to EM radiation, it absorbs the energy of a certain frequency (or wavelength) and transmits energy of some other frequency. When information is collected about energy released or absorbed in an experiment, information about functional groups can be deduced. This occurs because functional groups have characteristic IR absorptions that do not change from one compound to another.

Because most of the mass of an atom is concentrated in the nucleus, the rotational and vibrational energies of a molecule are mostly provided by the motion of the nuclei of the atoms in the molecule. The electrons in a molecule can also be excited to higher states, called higher energy levels, to give rise to electronic spectra.

9.8 ELECTRONIC STATES OF MOLECULES

In electronic transition spectra, each transition appears as a series of closely spaced spectral lines called a band, which demonstrates the fact that different rotational and vibrational states exist in each electronic state. All this means that in general, a molecule with a given electronic configuration can rotate and vibrate.

However, note that all molecules exhibit electronic spectra, but not all the molecules exhibit rotational and vibrational spectra. For example, H_2 and N_2 do not exhibit rotational and vibrational spectra because they do not have a permanent dipole moment; but they do exhibit electronic spectra.

> *Note*: Even though molecules with no electric dipole moment do not normally exhibit rotational spectra, the rotational states may be excited due to the collision of the molecules with each other or with the walls of the container.

Once a molecule in a ground state gets into an excited electronic state by absorbing some energy, it can lose that energy to come back to the original state by two possible ways:

1. The molecule emits a photon of the same frequency (energy) as that of the photon that was absorbed.
2. The molecule loses some of its energy in a collision with another molecule, and then loses the rest of the energy by transitioning from a lower to the original (ground) energy state. This process is called *fluorescence*.

Obviously, the fluorescent radiation has a lower frequency than that of the absorbed radiation. This is the key to fluorescence.

STUDY CHECKPOINT 9.8

As you know by now, electrons in atoms bind atoms into crystals and molecules. Electrons in molecules (or atoms of molecules) also play an important role in the anatomy and physiology of biomolecules called molecules of life, such as DNA and proteins. Now, consider two biomolecules and assume that electrons in each of these molecules are too tightly bound to leave the molecule, according to the laws of classical physics.

According to quantum mechanics, however, the electrons from one biomolecule can jump into the other molecule, even if they are forbidden by classical physics to do so. Which quantum mechanics principle allows this phenomenon?

Solution:

Quantum tunneling.

To summarize, changes of electron configuration inside atoms of the molecule, such as an electron jumping from one principal level to another principal level, give rise to molecular electronic states. These transitions are discussed in several chapters in different contexts, including Chapters 4 and 8. So, we will not give more coverage to this topic here. However, recall that we discussed earlier how the emission spectra of molecules can be used in the chemical analysis of substances. Similarly, looking from substance level, the emission spectrum of a chemical element (or chemical compound) represents the relative intensity of each frequency of EM radiation emitted by the atoms and molecules of the substance when they return to a ground state. Because the emission spectrum of each element is unique, spectroscopy can be used to identify the elements in matter of unknown composition.

9.9 LIVING IN THE QUANTUM WORLD

Based on its understanding through quantum mechanics, molecular spectroscopy has numerous applications in various fields, including health sciences. For example, fluorescence in molecules excited by UV light has several applications, including identifying minerals and biochemical compounds. Another example is a fluorescent lamp (see Fig. 9.9 for an example) that contains a mixture of mercury vapors and an inert gas, such as argon, inside a glass tube, which has obvious practical applications. When a current is passed through the tube, the mixture inside it gives off UV radiation. The inside of the tube is coated with a fluorescent material called a phosphor, which absorbs UV radiation and emits visible light. The fluorescent process is much more efficient than the process of heating a filament in an ordinary light bulb called an incandescence bulb.

Molecular Raman spectroscopy is an optical technique that provides a measure of the molecular composition of tissue by analyzing the way that tissue scatters laser light. It has long been considered a valuable tool for early detection and diagnosis of diseases. For example, it offers us the ability to provide specific biochemical information that may foreshadow the onset of cancer and other life-threatening illnesses. Here is how it works. Disease

Figure 9.9. Photograph of an illuminated incandescent replacement fluorescent bulb. Courtesy of J. Darwin.

symptoms are almost always preceded by changes in the molecular composition of tissue or body fluids. Raman spectrum provides structured and detailed information about cell and tissue composition without the need for using dyes, labels, or other contrast-enhancing agents. Peaks in the Raman spectrum correspond to known molecular signatures in tissues. The disease symptoms can be detected by comparing the Raman spectrum of the suspect tissue with that of healthy tissue. This way Raman spectroscopy offers us the ability to precisely diagnose a wide range of disease states. In short, the chemical structural change of cells and tissues detected by Raman spectroscopy is a good indicator of disease.

Devices have been developed that demonstrated the ability of Raman spectroscopy to successfully distinguish between normal and diseased tissues of, for example, the arteries, brain, breast, colon, cervix, and skin. Furthermore, these devices can be used to monitor and analyze various body fluids, and to detect abnormalities in the eyes.

Raman spectroscopy has also found its way into the dental field. Researchers have found that areas of decay with significant bacterial activity are highly fluorescent. This could facilitate locating hard-to-detect infections beneath the surface of a tooth. The ability of Raman spectroscopy to locate these underlying infections by distinguishing them from healthy tissue could provide powerful diagnostic tool for dentists.

Understanding molecular vibrations in terms of IR radiation also helps us to understand issues, such as global warming. Earth absorbs radiation from the sun. Some of this radiation absorbed by the Earth is re-emitted into the atmosphere as IR radiation. Some gases in the atmosphere, such as carbon

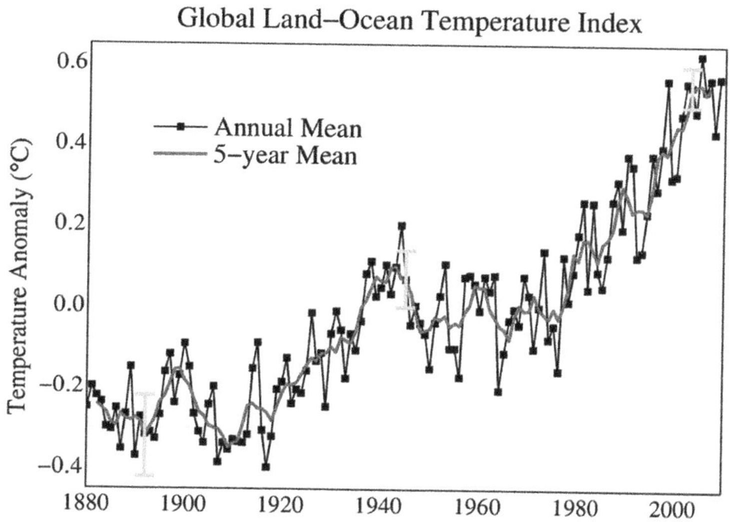

Figure 9.10. The global annual surface temperatures relative to the 1951–1980 mean temperatures. Taken from Goddard Institute for Space Studies. Courtesy of NASA.

dioxide, water vapors, nitrous oxide, and methane, have the property of absorbing IR radiation; and some other gases, such as oxygen and nitrogen (the major gases in the atmosphere) do not have this property. The gases that absorb IR radiations are called greenhouse gases because they behave much like the glass panes in a greenhouse, which absorb the heat of the sun, but keep it from escaping. When a greenhouse gas molecule absorbs IR radiation, it bends and vibrates, and therefore gains kinetic energy by this absorption of IR radiation. Some of this kinetic energy flows to other molecules, such as oxygen and nitrogen. The result is general heating of the atmosphere called global warming. While the greenhouse gases on Earth help maintaining life, problems arise when the atmospheric concentrations of these gases increase above certain levels.

For example, Figure 9.10 presents global annual surface temperatures relative to 1951–1980 mean temperatures from the data collected by the Goddard Institute for Space Studies, NASA. This analysis summarizes satellite measurements from 1982 onward, and ship-based analysis from years earlier. The estimated error reflected from the 95% confidence level accounts for the incomplete spatial coverage.

9.10 SUMMARY

The EM force, also called the Coulomb force, holds the atoms together through bonds in a molecule. Molecules, in addition to the electronic states of their atoms, have molecular, rotational, and vibrational states because they can rotate and vibrate. Quantum mechanics can be used to predict the rotational and vibrational energies of a diatomic molecule:

$$E_{\text{rot}} = \frac{l(l+1)\hbar^2}{2I}$$

$$E_{\text{vib}} = \left(n + \frac{1}{2}\right)\hbar\omega = \left(n + \frac{1}{2}\right)hf$$

Transitions between rotational states and between vibrational states give rise to absorption and emission spectra. Useful information about a molecule can be obtained from these spectra including bond angles, bond length, and force constants.

Because excitation energies for molecular rotation is much smaller, the molecules in a gas are almost always rotating. However, in the liquid phase, interactions between adjacent molecules inhibit rotation. Therefore, a pure vibrational spectrum can only be observed in the liquid phase. Molecules usually do not vibrate at normal temperatures, such as room temperature, because the size of excitation energies for vibration requires very high temperatures.

In our physical world, so many phenomena, even at the microscale level, involve systems composed of a great number of fundamental entities, such as atoms and molecules. Therefore, to understand matter at the molecular and atomic level, not only do we need to understand the behavior of individual atoms and molecules, but we also need to understand the collective behavior for a set of atoms and molecules. This is where statistical quantum mechanics enters into the picture, which we discuss in Chapter 10.

9.11 ADDITIONAL PROBLEMS

Problem 9.4 As shown in Figure 9.4, experiments show that in a system of carbon monoxide (CO) molecules in the gas phase, the absorption spectral line corresponding to transition from the $l = 0$ rotational state to the $l = 1$ rotational state appears at a frequency of 1.15×10^{11} Hz.

A. What is the moment of inertia of the CO molecule?

B. What is the reduced mass of the molecule.

C. What is the bond length of the CO molecule?

Problem 9.5 Consider a CO molecule in its lowest rotational state.

A. What is the rotational energy of the CO molecule.

B. What is the angular velocity of the rotating molecule?

Problem 9.6 Calculate the force constant by treating the vibrational transition in the HCl spectrum from its ground state to the first excited state, as a quantum harmonic oscillator. Consider the absorption spectra of HCl presented in this chapter. Assume that the midpoint frequency in Figure 9.8 is 8.66×10^{13} Hz, at which the first vibrational state is excited.

Problem 9.7 Consider the absorption spectra of the HCl presented in this chapter. Assume that the midpoint frequency at which the first vibrational state is excited is 8.66×10^{13} Hz.

A. Calculate the spacing between the two energy levels by considering the midpoint frequency .

B. What will be the temperature required to thermally excite the HCl molecule to make the transition mentioned in Problem 9.4.

Problem 9.8 About 16 kJ/mol of energy is required to overcome the intermolecular forces of attraction among the HCl molecules in order to vaporize liquid HCl. Show that this is equivalent to 0.16 eV/molecule.

STATISTICAL QUANTUM MECHANICS

10

God does not throw dice.

Albert Einstein, 1926

God not only plays dice, but He sometimes confuses us by throwing them where they cannot be seen.

Stephen Hawking, 1994

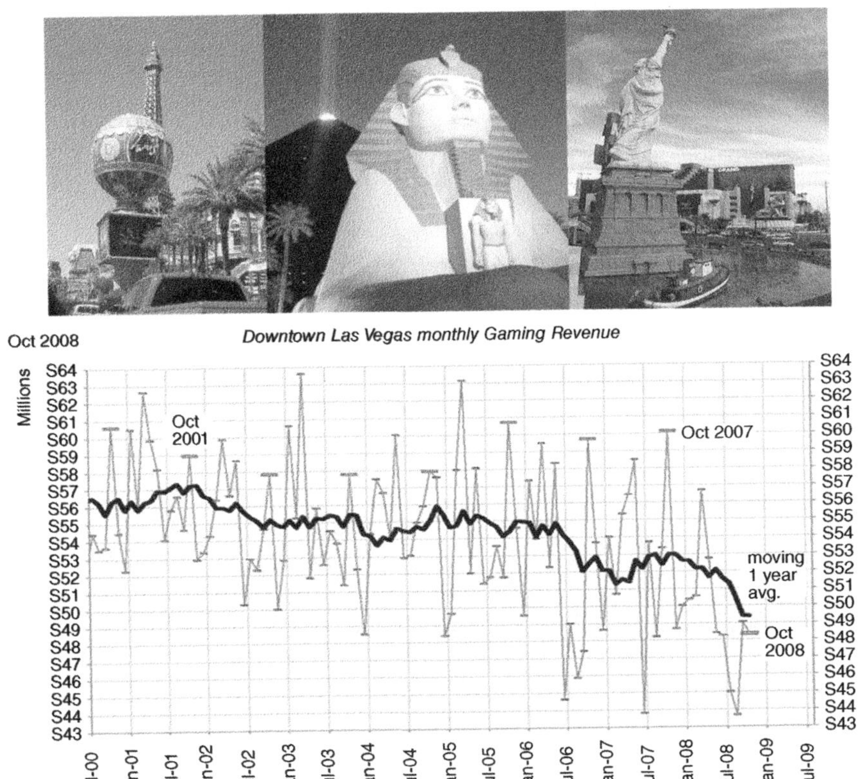

Figure 10.0. Gaming revenue for downtown Las Vegas. Data drawn from Nevada Gaming Commission. One year moving average 10 December 2008. Courtesy of Hajor for Luxor Hotel, Nail Li for New York Hotel, and Frank Martin for the graph.

Quantum Physics for Scientists and Technologists, by Paul Sanghera

Mathematics is the language of science. Statistics, which is the science of collecting and making effective use of numerical data relating to groups of individual entities including people, atoms, and molecules, is an important part of mathematics. The data may come from surveys or from experiments. Statistics deals with different aspects of data including planning for data collection, data collection, analysis, and interpretation. Planning for data collection includes designing surveys or experiments. The beauty of statistics for a physicist lays in the fact that its basic principles apply equally to both classical and quantum physics.

In This Chapter

10.1 INTRODUCTION

Matter is composed of fundamental entities, such as atoms and molecules, which we loosely refer to in this book as particles. Even at the microscale level, many phenomena in our physical world involve systems of a large number of particles. For example, the cell, the basic unit of life, is composed of largely four kinds of molecules (carbohydrates, proteins, nucleic acids such as deoxyribonucleic acid [DNA], and lipids) with a great number of them in each cell. For example, based on the chemical instructions written on the DNA molecules, a cell makes thousands of types of protein molecules to support our human body. Each cell of our body has 46 DNA molecules, and each molecule is packed in a structure called a chromosome; we inherit these molecules from our parents. Keep in mind that there are a multi-trillion cells in an average human body. We will discuss all four of these types of molecules further in Chapter 11.

To understand the behavior of a collection of molecules, it is apparent that we need to understand the overall behavior of a system of a great many particles and how it is related to the properties of individual constituent particles. This is exactly the issue that statistical mechanics addresses. However, recall that quantum mechanics brought statistics into the picture at the individual particle level. You learned very early on in this book that the quantum mechanical wave function, Ψ, which contains all the information about a particle it represents, is based on probability: a fundamental concept in statistics. For example, the position of an electron in an atom is described in terms of probability distribution represented by $|\Psi|^2$.

The three most important things to realize about statistical mechanics follow:

1. Statistical mechanics does not directly deal with the motion and interactions of individual particles. It deals instead with the whole system, and with the question of what is most likely to happen.
2. Statistical mechanics cannot be used to track an individual particle in the system over time, it does help us to determine the probability that the particle has a certain amount of energy or is at a certain position at a given moment.
3. The basic general concepts of statistics apply equally to classical and quantum mechanics.

To keep things simple, we are considering systems of identical particles that may or may not be distinguishable (physically) from one another. Therefore, the central theme of this chapter is to explain how statistical quantum mechanics can be used to study systems of particles, such as electrons, photons, atoms, and molecules. To this end, the three underlying streams of thought in this chapter are the framework for statistical distributions, statistical distributions for distinguishable identical particles, and statistical distributions for identical indistinguishable particles.

10.2 STATISTICAL DISTRIBUTIONS

You already know that the motion or dynamics of a particle is described by using some variables, such as speed and energy. However, in a system of particles, in general, all the particles will not have the same value for a physical quantity, such as speed and energy, at a given time. In statistics, you will also be dealing with the same basic variables, such as speed and energy, but with two changes: We will be concerned with the system of particles, not just individual particles, and with the probability of these variables, for example, how many particles have this much speed, and so on. Variables with probabilities attached to them are called probability or statistical distributions. In this book, you should have realized by now that the concept of energy and the corresponding energy variable is common to an individual particle, and a system of particles, in classical and quantum mechanics.

Thermodynamics, a field that originally belonged to physics, but is very relevant to many other disciplines (e.g., chemistry, mechanical engineering, and health sciences) is based on the concept of energy. A system of N particles, such as a collection of molecules or atoms, has an overall energy E as a result of the energy of the individual particles. Furthermore, there will be a disturbance as a result of the energy, but the natural tendency of the system will be to settle into a state of stability, that is, a state of thermal equilibrium. A relevant question to ask is How can we determine the most probable way in which the total amount of energy E is distributed among the N particles in the system that is in thermal equilibrium at absolute temperature T? We can use statistical mechanics to determine that way. In other words, by using statistical mechanics we can determine how many particles have energy E_1, how many particles have energy E_2, and so on.

A system is after all made up of particles. So, to start with, note the following three features of particles in the system:

1. At a given moment, each particle is in a certain state called the particle state, and in general more than one particle state may correspond to the same specific energy.
2. Particles in the system are not strongly correlated. Within this limit, however, they can interact with each other and with the walls of the container to establish equilibrium.
3. If the particles in the system are not subject to the exclusion principle, discussed earlier in this book, more than one particle can exist in the same state.

Given this, realize that there is more than one possible way in which the total energy of the system can be distributed among the N particles of the system, each way corresponding to one energy distribution. Furthermore, realize that there are in general multiple ways in which the particles can be arranged among the available states to yield a specific energy distribution. Assume that each state of a particular energy has an equal chance of being occupied by a particle. This scenario leads to a basic principle of statistical mechanics:

The greater the number of different ways (N_W) in which particles can be arranged among different available states to achieve a specific energy distribution, the more probable is that energy distribution.

> *Note*: To some of us, the assumption that *each state of a certain energy has an equal chance of being occupied* is statistical common sense. However, this assumption is also supported by experiments.

The core underlying issue throughout this chapter is to determine the different fractions of molecules in different states, or the population probability distribution. In general, you can find several ways in which the particles can be arranged among available states, and the most probable way will be for which N_W is maximum. Each way will involve determining the number of particles with energy E_i by using the following formula:

$$n(E_i) = g(E_i) f(E_i) \tag{10.1}$$

where,

- The term $n(E_i)$ is the number of particles with energy E_i in a discreet distribution, and $n(E_i)dE_i$ is the number of particles that contain energy between E_i and $E_i + dE_i$ in a continuous distribution.
- The term $g(E_i)$ is the number of available states of energy E_i in a discreet distribution, and $g(E_i)dE_i$ is the number of available states with energy in the range of E_i and $E_i + dE_i$ in a continuous distribution. This can also be interpreted as the statistical weight or the density of states, for example, states per unit energy.
- The term $f(E_i)$ is the probability that each state of energy E_i will be occupied, which is equal to the average number of particles in each state of energy E_i. It is also called the distribution function.

Make sure that you feel comfortable with Eq. 10.1, as it simply states the fact that the number of particles in a given energy state is equal to the number of available states with that specific energy multiplied by the probability that each of these states will be occupied by a particle. Therefore, Eq. 10.1 is a very general equation that applies to classical as well as quantum situations. Most of the statistical efforts go into figuring out the distribution function $f(E_i)$, which depends on the nature of the particles in the system.

Similarly, we can write distributions for other variables. For example, we can rewrite Eq. 10.1 for velocity as:

$$n(v_i) = g(v_i) f(v_i) \tag{10.2}$$

In our treatment here, the particles in the system under study are assumed to be identical. However, the question is Can those particles be physically distinguished from each other? Identical particles with an effective quantum spin cannot be distinguished from each other because their wave functions will overlap. However, the identical particles with no spin can be distinguished from each other assuming they are sufficiently apart from each other. This is the case we consider next, in Section 10.3.

10.3 MAXWELL–BOLTZMANN DISTRIBUTION

James Clerk Maxwell, the same scientist who put together Maxwell's equations discussed in Section 1.16, was among the first scientists to apply mathematical theories of probability and statistics to physical problems in the field of thermodynamics. He derived equations for the distribution of speeds and velocities of molecules in an ideal gas and demonstrated that these distributions can be used to explain the phenomena observed at the macroscopic level. According to the distribution originally derived by Maxwell in 1859, the number of molecules with a speed between v and $v + dv$ is given by

$$n(v)\,dv = 4\pi N\left(\frac{m}{2\pi\kappa T}\right)^{3/2} v^2 e^{-\left(mv^2/2\kappa T\right)}dv \tag{10.3}$$

where we are using v instead of v_i for simplicity of notation, m is the mass of the molecule, $\boldsymbol{\kappa}$ is the Boltzmann constant, and T is the absolute temperature of the system of N molecules in kelvins.

> *Note*:
> In Eq. 10.3, dv is three dimensional (3D):
>
> $$dv = dv_x dv_y dv_z$$
>
> and
>
> $$v^2 = v_x^2 + v_y^2 + v_z^2$$

Figure 10.1 illustrates this distribution for oxygen at three different temperatures: 90.20 K, the boiling point of oxygen; 298.2 K, room temperature; and 873.2 K the temperature between the melting point of zinc (Zn) and plutonium (Pt). Note that the distribution spreads out and the peak shifts to the right with an increase in temperature.

Let us try to understand Eq. 10.3 in terms of Eq. 10.2. By comparing these two equations, the speed (or velocity) distribution function can be written as:

$$f_{\mathrm{MB}}(v) = Ce^{-\left(mv^2/2\kappa T\right)} \tag{10.4}$$

where MB is the Maxwell–Boltzmann distribution and C is the proportionality factor given by

$$C = \left(\frac{m}{2\pi\kappa T}\right)^{3/2}$$

and the number of available states $g(v_i)$ is

$$g(v) = 4\pi N v^2$$

By realizing that

$$E = \frac{1}{2}mv^2$$

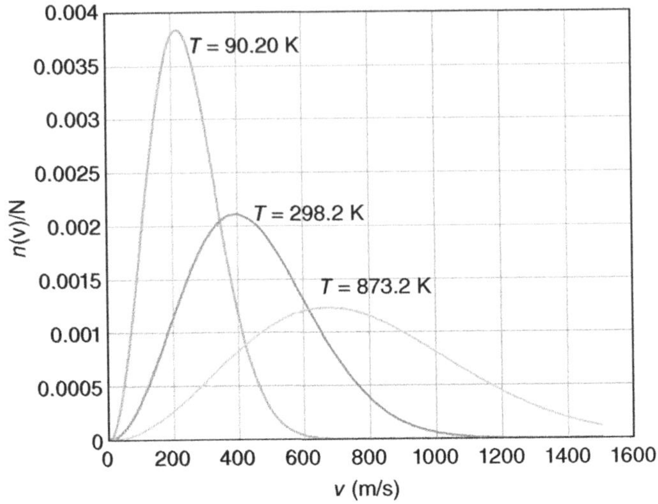

Figure 10.1. A fraction of the molecules of oxygen as a function of their speeds at three different temperatures.

the Maxwell–Boltzmann speed distributions in Eq. 10.4 in terms of speed can be rewritten as an energy distribution:

$$f_{MB}(E_i) = Ae^{-(E_i/\kappa T)} \tag{10.5}$$

Here A is the proportionality constant, κ is the Boltzmann constant, and T is the absolute temperature of the system. The function $f_{MB}(E_i)$ is called the Maxwell–Boltzmann energy distribution function, which represents the probability that the state (or states) corresponding to energy E_i will be occupied. The constant A in Eq. 10.5 called the normalization constant, depends on the number of particles in the system.

Note: The readers with a chemistry background should be familiar with a distribution mathematically equivalent to the Maxwell–Boltzmann distribution presented in Eq. 10.5. The distribution embodies what is called the Arrhenius equation given by

$$k = Ae^{-E_a/RT}$$

where k is the reaction rate constant, E_a is the activation energy for the chemical reaction, R is the ideal gas constant, and T is the temperature in kelvins. The constant A, called the frequency factor, depends on the frequency of molecular collisions and on how much the molecules are properly oriented so that the reaction will happen. The factor $e^{-E_a/RT}$ represents the fraction of molecules with energy equal to or greater than the activation energy of the reaction, that is, the minimum energy required for the reaction to happen.

James Clerk Maxwell (1831–1879)

James Clerk Maxwell, one of the greatest physicists of the nineteenth century, was born on June 13, 1831 in Edinburgh, Scotland. Maxwell studied physics at the University of Cambridge, UK, taught at a college in Scotland, and later at King's college, London. Maxwell's greatest contribution to physics is the unification of apparently three different phenomena called electricity, magnetism, and light into one phenomenon called electromagnetism (EM) in terms of four equations, famously known as Maxwell's equations.

Maxwell also developed a successful theory of kinetic gases. His work *Theory of Heat*, in 1871, laid the foundations for what is now called statistical mechanics. Maxwell, the greatest theoretical physicist of the nineteenth century, lost his life to cancer at the young age of 48 in 1879, the same year in which Albert Einstein was born. Einstein became the greatest theoretical physicist of the twentieth century.

Ludwig Boltzmann (1844–1906)

Ludwig Boltzmann was born on February 20, 1844 in Vienna, Austria. He studied physics at the University of Vienna. He taught physics and mathematics at the University of Graz and later at the University of Vienna. Boltzmann is mostly remembered for his famous work on the statistical nature of entropy. Both Maxwell and Boltzmann worked independently to develop laws that govern the statistical behavior of a system of classical particles. Boltzmann was a champion of the atomic theory of matter that was still controversial during the late nineteenth century, and he had a tremendous admiration for Darwin's theory of evolution.

Boltzmann spent the last years of his life mostly defending his theories and battling with nonbelieving scientists. In the process, he became victim to health problems, such as asthma, headaches, and increasingly poor eyesight, which threw him into depression. He committed suicide on September 5, 1906 at the age of 62.

Let us use this probability distribution to find the distribution of energies among the N molecules in the system. Assume that N is large enough so that we could consider the distributions as continuous. Therefore, by using Eqs. 10.1 and 10.5, the number of molecules $n(E_i)dE_i$ that contain energies in the range of E_i and $E_i + dE_i$ can be represented by the equation:

$$n(E_i)dE_i = Ag(E_i)e^{-(E_i/\kappa T)}dE_i \tag{10.6}$$

Accordingly, the number of molecules $n(E_i)$ that contain energy E_i can be represented by the equation:

$$n(E_i) = Ag(E_i)e^{-(E_i/\kappa T)} \tag{10.7}$$

PROBLEM 10.1

Recall the rigid rotator model of diatomic molecules discussed in Section 9.5. Consider such a diatomic molecule with angular momentum L from rotation with l as the corresponding quantum number.

A. Calculate the rotational energy corresponding to $l = 0$.
B. Determine the number of available rotational states, $g(E_L)$.
C. Calculate the value of the normalization constant A in the Maxwell–Boltzmann distribution:

$$n(E_i) = Ag(E_i)e^{-(E_i/\kappa T)}$$

D. Prove that the Maxwell–Boltzmann distribution for the rotational states can be written as:

$$n(E_L) = n_0(2l+1)e^{-l(l+1)(\hbar^2/2I\kappa T)}$$

where n_0 is the number of molecules in the state $l = 0$.

Solution:

A.
As shown in Section 9.5, the rotational energy of the molecule is given by

$$E_L = \frac{l(l+1)\hbar^2}{2I} \tag{1}$$

Therefore, for $l = 0$:

$$E_{l=0} = 0 \tag{2}$$

B.
There are $2l + 1$ states corresponding to a specific l.
Therefore,

$$g(E_L) = 2l + 1 \tag{3}$$

This means there is only one state at $l = 0$

$$g(E_{l=0}) = 1 \tag{4}$$

C.
From Eq. 10.7, by using Eqs. 2 and 4:

$$n(0) = n_0 \tag{5}$$

D.
By using Eqs. 3 and 5, Eq. 10.7 becomes:

$$n(E_L) = Ag(E_L)e^{-(E_L/\kappa T)} = n_0(2l+1)e^{-[l(l+1)\hbar^2/2I\kappa T]}$$

Hence, the proof.

Our next task is to determine the expression for $g(E_i)dE_i$, the number of available states in the energy range of E_i and $E_i + dE_i$. For each energy E_i, there is a magnitude of momentum P_i, which has three degrees of freedom represented by its three components: P_{ix}, P_{iy}, and P_{iz}. Each set of values for P_{ix}, P_{iy}, and P_{iz} represent a different momentum state. Because there are three degrees of freedom, you can think of the momentum space as a sphere with radius P_i. The number of states in momentum space therefore is proportional to the volume of this sphere: The underlying assumption here is that you are dealing with point particles, that is, structureless particles: no rotational or vibrational evnrgy.

The number of momentum states in the spherical shell of thickness dP_i is proportional to the volume of this spherical shell, which is equal to $4\pi P_i^2 dP_i$. In other words, the number of momentum states, $g(P_i)dP_i$ in the momentum range of P_i to $P_i + dP_i$ are given by

$$g(P_i)dP_i = b4\pi P_i^2 dP_i \tag{10.8}$$

where b is the proportionality constant.

Each value for the magnitude of momentum gives rise to a unique value of energy. Therefore the number of energy states $g(E_i)dE_i$ are equal to the number of momentum states $g(P_i)dP_i$. This means, by using Eq. 10.8, we obtain

$$g(E_i)dE_i = g(P_i)dP_i = b4\pi P_i^2 dP_i \tag{10.9}$$

From

$$E_i = \frac{P_i^2}{2m}$$

we can write

$$dP_i = \frac{m}{P_i}dE_i = \frac{m}{\sqrt{2mE_i}}dE_i \tag{10.10}$$

By substituting this into Eq. 10.9, we obtain

$$g(E_i)dE_i = b4\pi P_i^2 dP_i = b4\pi(2mE_i)\frac{m}{\sqrt{2mE_i}}dE_i = B\sqrt{E_i}dE_i \tag{10.11}$$

Therefore, Eq. 10.6 for the number of molecules with energy in the range of E_i to $E_i + dE_i$ can be rewritten as:

$$n(E_i)dE_i = Ag(E_i)e^{-(E_i/\kappa T)}dE_i = AB\sqrt{E_i}e^{-(E_i/\kappa T)}dE_i = C\sqrt{E_i}e^{-(E_i/\kappa T)}dE_i \tag{10.12}$$

The constant C can be calculated by using the normalization condition that the total number of molecules in the system is N:

$$N = \int_0^\infty n(E_i)dE_i = C\int_0^\infty \sqrt{E_i}e^{-(E_i/\kappa T)}dE_i \tag{10.13}$$

The value of the definite integral in Eq. 10.13 is given by

$$\int_0^\infty \sqrt{x}e^{-ax}dx = \frac{1}{2a}\sqrt{\frac{\pi}{a}}$$

This results in

$$N = \frac{C}{2}\sqrt{\pi}\,(\kappa T)^{3/2} \tag{10.14}$$

Solving for the constant C:

$$C = \frac{2\pi N}{(\pi\kappa T)^{3/2}} \tag{10.15}$$

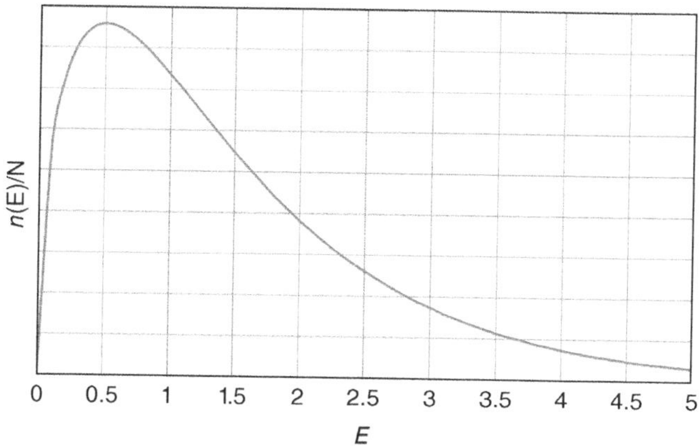

Figure 10.2. The Maxwell–Boltzmann probability distribution for energy plotted in the range of 0–5 κT. Note that the average molecular energy is ~3/2 κT.

Substituting this back into Eq. 10.12:

$$n(E_i)dE_i = \frac{2\pi N}{(\pi\kappa T)^{3/2}}\sqrt{E_i}e^{-(E_i/\kappa T)}dE_i \tag{10.16}$$

Equation 10.16 gives the distribution for the number of molecules with energy between E_i and $E_i + dE_i$ in a system of N molecules assumed to be an ideal gas with absolute temperature T. The probability distribution $n(E_i)/N$ from Eq. 10.16 is plotted in Figure 10.2 as a function of E_i, with E_i shown along the x-axis in units of κT. Note that the distribution curve is not symmetrical around the most probable energy. This is because there is a lower limit on energy to be 0, whereas there is no upper limit in principle.

STUDY CHECKPOINT 10.1

What does $n(E_i)/N$ represent the probability of in Figure 10.2.

Solution:

Probability that a given molecule will have energy E_i.

As Problem 10.13 demonstrates, you can derive the particle (or molecular) velocity distribution $n(v)$ given in Eq. 10.3 from the particle energy distribution $n(E_i)$ given by Eq. 10.16. Also, from Eq. 10.16, we can calculate the average energy of a molecule in this sample using the expectation value. To this end, first let us calculate the total internal energy of the system from Eq. 10.16, as shown below:

$$E = \int_0^\infty E_i n(E_i) dE_i = \frac{2\pi N}{(\pi \kappa T)^{3/2}} \int_0^\infty E_i^{3/2} e^{-(E_i/\kappa T)} dE_i$$

Substituting for the definite integral:

$$\int_0^\infty x^{3/2} e^{-ax} dx = \frac{3}{4a^2}\sqrt{\frac{\pi}{a}}$$

we obtain

$$E = \left[\frac{2\pi N}{(\pi \kappa T)^{3/2}}\right]\left[\frac{3}{4}(\kappa T)^2 \sqrt{\pi \kappa T}\right] = \frac{3}{2} N \kappa T \tag{10.17}$$

This is the total energy of N molecules. Therefore, the average energy per molecule is

$$\bar{E} = \frac{E}{N} = \frac{3}{2}\kappa T \tag{10.18}$$

Note: As Eqs. 10.17 and 10.18 demonstrate, the energy of the system of molecules and the average energy of a molecule in the system are independent of the molecular mass.

As mentioned earlier, a molecule in an ideal gas system has three degrees of freedom corresponding to motion in three independent directions: x, y, and z. Because the average energy of a molecule is $\frac{3}{2}\kappa T$, we can assign an energy of $\frac{1}{2}\kappa T$ to each degree of freedom. This is called equipartition of energy.

Equipartition Theorem: For a system of particles in thermal equilibrium at absolute temperature T, the average energy per degree of freedom of a particle is $\frac{1}{2}\kappa T$.

PROBLEM 10.2

Consider a rigid diatomic molecule in a system of molecules that acts as an ideal gas system in thermal equilibrium at temperature T. The molecule rotates around an axis perpendicular to the bond axis between the two atoms. The molecule also has translational motion.

A. How many degrees of freedom does the molecule have?
B. What is the average energy of the molecule as a function of T?

Solution:

A. A rigid diatomic molecule will have three degrees of freedom corresponding to translational motion, and two degrees of freedom corresponding to rotation. This is because translational motion can happen in 3D space, that is, a particle can move in the x, y, or z dimension, and rotational motion only needs two dimensions. Therefore the molecule has five degrees of freedom.
B. The average energy per degree of freedom is $\frac{1}{2}\kappa T$. Therefore energy for five degrees of freedom $= \frac{5}{2}\kappa T$.

The Maxwell–Boltzmann distribution function is generally used to handle classical systems and works remarkably well in explaining the behavior of gases. However, as Problem 10.1 demonstrates, it can also be used in semi-classical (or semiquantum) treatment in which the number and energy of states are determined from quantum mechanics, a situation discussed next in Section 10.4.

10.4 MOLECULAR SYSTEMS WITH QUANTUM STATES

According to Eq. 10.5, the probability that the state corresponding to energy E_i will be occupied is directly proportional to $e^{-(E_i/\kappa T)}$. In quantum mechanics, the energy E_i of a state corresponds to one of a set of discrete energy states as opposed to a continuum of states or energies in classical mechanics. This means, the probability that the molecular system occupies the energy state E_i is proportional to $e^{-(E_i/\kappa T)}$. In other words, the number of molecules N_i with energy E_i is given by

$$N_i = Ce^{-(E_i/\kappa T)} \qquad i = 1, 2, 3, \ldots \tag{10.19}$$

where C is the proportionality constant.

STUDY CHECKPOINT 10.2

True or False: The population number N_i that occupies a certain energy state E_i decreases with an increase in the value of E_i.

Solution:

True
Use Eq. 10.16 to realize this answer.

In quantum mechanics, there is a discrete set of energy states with each state occupied by a certain number of molecules at a given temperature. A valid question to ask is What is the probability of finding a given molecule in the quantum state i with energy E_i? This is the same question as How many molecules, or what fraction of the molecules, are occupying energy state E_i? To find the fraction, we need to bring the total number of molecules in the system into the picture. Because we have a set of discrete energy states with N_1 molecules in energy state E_1, N_2 molecules in energy state E_2, and so on; we can sum over all population numbers corresponding to different states to obtain the total number of molecules N in the system:

$$N = \sum_{i=1}^{\infty} N_i = C \sum_{i=1}^{\infty} e^{-(E_i/\kappa T)} \tag{10.20}$$

Therefore, from Eqs. 10.19 and 10.20 we can determine the probability P_i of finding a given molecule in state i with energy E_i as:

$$P_i = \frac{N_i}{N} = \frac{e^{-(E_i/\kappa T)}}{\sum_{i=1}^{\infty} e^{-(E_i/\kappa T)}} \tag{10.21}$$

The quantity Z given by

$$Z = \sum_{i=1}^{\infty} e^{-(E_i/\kappa T)} \tag{10.22}$$

is called the partition function. From Eqs. 10.21 and 10.22, we can write

$$P_i = \frac{1}{Z} e^{-(E_i/\kappa T)}$$

Note: The probability that a given molecule will occupy a certain state i with energy E_i is given by the fraction of molecules occupying that state.

The letter Z for the partition function stands for the German word *Zustandssumme*, which means "sum over states". It is called the partition function because it encodes how the probabilities are split (or partitioned) among the different quantum states, based on their individual energies. As demonstrated in Problem 10.3, the partition function also plays the role of a normalization constant.

PROBLEM 10.3

Calculate $\sum_{i=1}^{\infty} \frac{1}{Z} e^{-(E_i/\kappa T)}$

Solution:

$$\sum_{i=1}^{\infty} \frac{1}{Z} e^{-(E_i/\kappa T)} = \frac{1}{Z} \sum_{i=1}^{\infty} e^{-(E_i/\kappa T)} = \frac{Z}{Z} = 1$$

Let us now enquire how a partition function enters into the expression for a number of discrete energy states.

From Eqs. 10.20, and 10.22:

$$C = \frac{N}{Z}$$

Therefore, we can rewrite Eq. 10.19 as:

$$N_i = \frac{N}{Z} e^{-(E_i/\kappa T)} \qquad i = 0, 1, 2, 3, \ldots \tag{10.23}$$

To sum up, the equations in this section allow us to determine the distribution of N molecules of a system, which is in thermal equilibrium at temperature T, over all given quantum states corresponding to discrete energies E_i. This treatment may sound very theoretical. To make the situation more practical, we apply these equations to vibrating and rotating molecules in the forthcoming sections. Recall from Chapter 9 that molecules may have vibrational, rotational, and translational energies corresponding to vibrational, rotational, and translational motion.

10.5 DISTRIBUTION OF VIBRATIONAL ENERGIES

In Section 9.6, we explored vibrations of a molecule by using quantum mechanics. Now, we are dealing with a system of molecules. Here, we are interested in finding out the following two distributions for the vibrating molecules:

1. The total vibrational energy of a system of N molecules as a function of temperature T.
2. The number of molecules N_n, and hence the fraction of molecules in a vibrational state n, where n is the vibrational quantum number.

10.5.1 Vibrational Energy

The vibrational mode of a molecule can be modeled to a harmonic oscillator. To arrive at the distributions for the vibrating molecules, we begin with one vibrating diatomic molecule. As we know from Section 9.6, the vibrational energy E_n of a diatomic molecule with atomic masses m_1 and m_2 is

$$E_n = \left(n+\frac{1}{2}\right)\hbar\omega = \left(n+\frac{1}{2}\right)hf \qquad n = 0, 1, 2, 3, \ldots \tag{10.24}$$

where n is the vibrational quantum number, and

$$\omega = \sqrt{\frac{\kappa}{\mu}} = 2\pi f \tag{10.25}$$

with κ as the force constant and μ as the reduced mass given by

$$\mu = \frac{m_1 m_2}{m_1 + m_2} \tag{10.26}$$

PROBLEM 10.4

What is the ground-state vibrational energy for a CO molecule with force constant $\kappa = 1860\,\text{N/m}$.

Solution:

$$E_n = \left(n+\frac{1}{2}\right)\hbar\omega = \left(n+\frac{1}{2}\right)hf$$

$$E_0 = \frac{1}{2}hf = \frac{1}{2}h\frac{1}{2\pi}\sqrt{\frac{\kappa}{\mu}} = \frac{1}{4\pi}6.626\times10^{-34}\,\text{Js}\sqrt{\frac{1860\,\text{N/m}}{11.39\times10^{-27}\,\text{kg}}} = 2.13\times10^{-20}\,\text{J}$$

$$= 2.13\times10^{-20}\,\text{J}\times\frac{6.242\times10^{18}\,\text{eV}}{1\,\text{J}} = 0.133\,\text{eV}$$

Now the question is What is the probability that a given molecule will occupy the energy state E_n corresponding to a quantum number n. The same question can be phrased as What is the population probability density; that is, the fraction of molecules occupying the energy state E_n. Let us enquire in the next section.

10.5.2 Population Probability of Vibrational States

Population probability P_n is given by

$$P_n = \frac{N_n}{N}$$

where N_n is the number of molecules in the energy state E_n and N is the total number of molecules in the system. We rewrite Eq. 10.23 for N_n in the following form:

$$N_n = \frac{N}{Z_v} e^{-(E_n/\kappa T)} \qquad n = 0, 1, 2, 3, \ldots \tag{10.27}$$

As shown in Problem 10.5, the value of Z_v for the vibrating molecules can be written as:

$$Z_v = \frac{e^{-(hf/2\kappa T)}}{1 - e^{-(hf/\kappa T)}} \tag{10.28}$$

The function is plotted in Figure 10.3, where energy is plotted along the x-axis in units of κT.

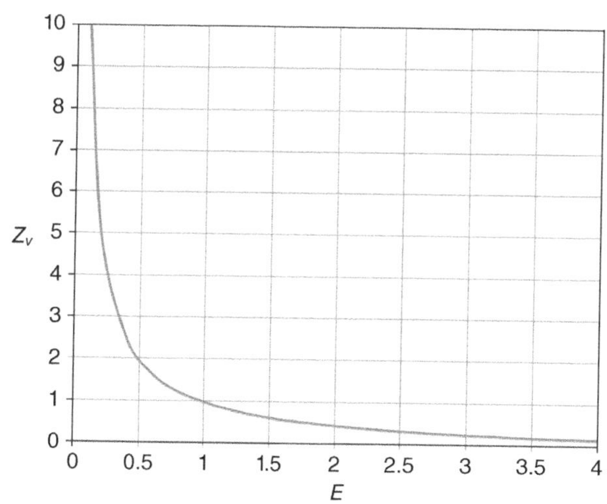

Figure 10.3. Partition function (Z_v) as a function of energy (E) in units of κT.

PROBLEM 10.5

Prove that the partition function for vibrating molecules is given by

$$Z_v = \frac{e^{-(hf/2\kappa T)}}{1-e^{-(hf/\kappa T)}}$$

Solution:

For vibrating molecules in quantum state n:

$$E_n = \left(n+\frac{1}{2}\right)hf$$

Therefore, from Eq. 10.22:

$$\begin{aligned} Z_v &= \sum_{n=0}^{\infty} e^{-(E_n/\kappa T)} = \sum_{n=0}^{\infty} e^{-\left[\left(n+\frac{1}{2}\right)hf/\kappa T\right]} = \sum_{n=0}^{\infty} e^{-\left(n+\frac{1}{2}\right)x} \\ &= e^{-x/2}\sum_{n=0}^{\infty}\left(e^{-x}\right)^n = e^{-x/2}\sum_{n=0}^{\infty} y^n \end{aligned} \qquad (1)$$

where we have substituted

$$x = \frac{hf}{\kappa T} \quad \text{and} \quad y = e^{-x}$$

Given that the number y is <1, we can write

$$\sum_{n=0}^{\infty} y^n = 1 + y + y^2 + y^3 + \cdots = \frac{1}{1-y} = \frac{1}{1-e^{-x}} = \frac{1}{1-e^{-(hf/\kappa T)}}$$

Substituting this in Eq. 1, we obtain

$$Z_v = \frac{e^{-(hf/2\kappa T)}}{1-e^{-(hf/\kappa T)}}$$

At sufficiently low temperature, almost all the molecules will be in the ground state. With an increase in temperature more molecules start occupying excited states.

The other quantity of interest is the total vibrational energy, E_v, of the system of N molecules:

$$E_v = \sum_{n=0}^{\infty} N_n E_n = \frac{N}{Z_v}\sum_{n=0}^{\infty} E_n . e^{-(E_n/\kappa T)} \qquad (10.29)$$

As shown in Problem 10.6, the expression on the right-hand side (rhs) of the equation can be solved to obtain

$$E_v = N\left(\frac{1}{e^{(hf/\kappa T)}-1}+\frac{1}{2}\right)hf \tag{10.30}$$

where f is the frequency of vibration.

PROBLEM 10.6

Prove from Eq. 10.28 that:

$$E_v = N\left(\frac{1}{e^{(hf/\kappa T)}-1}+\frac{1}{2}\right)hf$$

Solution:

Starting from Eq. 10.28 and using Problem 10.9:

$$E = N\kappa T^2 \frac{d}{dT}(\ln Z) \quad \text{(from problem 10.9)} \tag{1}$$

$$Z_v = \frac{e^{-(hf/2\kappa T)}}{1-e^{-(hf/\kappa T)}} = \frac{e^{-x/2}}{1-e^{-x}} \quad \text{(from eq 10.28)} \tag{2}$$

where,

$$x = \frac{hf}{\kappa T} \tag{3}$$

Therefore,

$$\frac{dx}{dT} = -\frac{hf}{\kappa T^2} \tag{4}$$

From Eq. 2,

$$\ln Z_v = -\frac{x}{2} - \ln(1-e^{-x})$$

$$\frac{d}{dT}(\ln Z_v) = -\frac{1}{2}\frac{dx}{dT} - \frac{e^{-x}}{1-e^{-x}}\frac{dx}{dT}$$

Using Eqs. 3 and 4:

$$\frac{d}{dT}(\ln Z_v) = \frac{hf}{\kappa T^2}\left(\frac{1}{e^{(hf/\kappa T)}-1}+\frac{1}{2}\right)$$

Substituting it into Eq. 1:

$$E_v = N\left(\frac{1}{e^{(hf/\kappa T)}-1}+\frac{1}{2}\right)hf$$

As demonstrated in Problem 10.14, starting with Eq. 10.27 and using Eqs. 10.24 and 10.28, we obtain the following equation.

$$N_n = \frac{N}{Z_v}e^{-(E_n/\kappa T)} = N\left(1-e^{-(hf/\kappa T)}\right)e^{-(nhf/\kappa T)}$$

Therefore,

$$P_n = \frac{N_n}{N} = \left(1-e^{-(hf/\kappa T)}\right)e^{-(nhf/\kappa T)} \tag{10.30a}$$

This is the probability that a molecule will be in energy state E_n. Figure 10.4 illustrates this probability by plotting it at $T = 300\,\text{K}$ for two types of molecules: $^{19}F^{35}Cl$ and $^{127}I^{35}Cl$. Note that the term hf in Eq. 10.30a for P_n represents the energy gap ΔE between different vibrational states. This gap is known to be 15.6×10^{-21} J for $^{19}F^{35}Cl$ and 7.63×10^{-21} J for $^{127}I^{35}Cl$ at 300 K, which is the approximate room temperature.

As is clear from Figure 10.4, at room temperature most of the molecules for FCl and ICl are in the ground ($n = 0$) state. A small proportion is distributed over higher states. As the temperature increases, more and more molecules transistion into higher vibrational states.

10.5.3 Correspondence with Classical Mechanics

We can prove that at very high temperature the total vibrational energy of the molecular system is approximately equal to that predicted by classical statistics.

To simplify, substitute:

$$x = \frac{hf}{\kappa T}$$

Therefore,

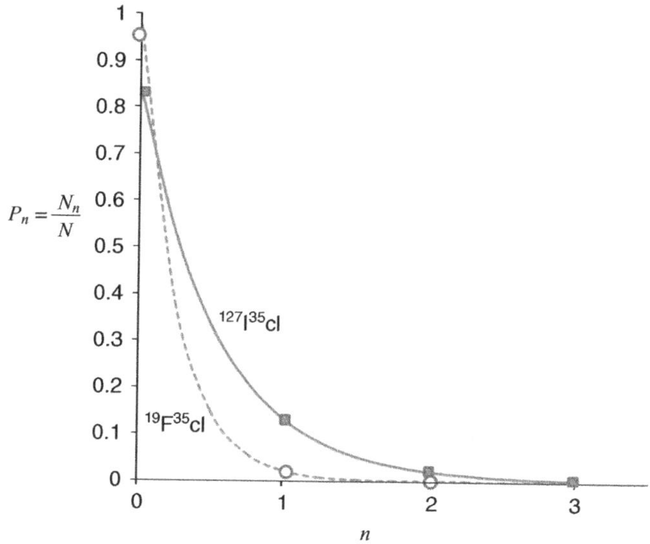

Figure 10.4. Population probability (N_n/N) plotted as a function of quantum number n at $T = 300\,K$ for two types of molecules: $^{19}F^{35}Cl$ ($\Delta E = hf = 15.6 \times 10^{-21}$ J) and $^{127}I^{35}Cl$ ($\Delta E = hf = 7.63 \times 10^{-21}$ J).

$$e^{(hf/\kappa T)} = e^x = 1 + x + \frac{x^2}{2!} + \frac{x^3}{3!} + \cdots$$

For very high temperature, and therefore a very small value of x:

$$e^{(hf/\kappa T)} - 1 = e^x - 1 = 1 + x + \frac{x^2}{2} - 1 = 1 + \frac{hf}{\kappa T} + \frac{1}{2}\left(\frac{hf}{\kappa T}\right)^2 - 1 = \frac{hf}{\kappa T} + \frac{1}{2}\left(\frac{hf}{\kappa T}\right)^2$$

Substituting this into Eq. 10.30:

$$E_v = Nhf\left(\frac{1}{2} + \frac{1}{x + x^2/2}\right) = Nhf\left(\frac{1}{2} + \frac{1}{x}\frac{1}{1 + x/2}\right) = Nhf\left(\frac{1}{2} + \frac{1}{x}\left\{1 - \frac{1}{2}x\right\}\right)$$
$$= Nhf\left(\frac{1}{2} + \frac{1}{x} - \frac{1}{2}\right) = \frac{Nhf}{x} = N\kappa T$$

This is in accordance with the equipartition theorem derived from the classical treatment seen earlier in this chapter. Because a vibrating molecule has two degrees of freedom, its energy will be κT, ½κT for each degree of freedom according to the equipartition theorem, and the energy of N molecules will be $N\kappa T$.

Therefore, for large enough T:

$$E_v = N\kappa T \tag{10.31}$$

In a nutshell, for high enough temperature, the system behaves as a classical system if we use the Maxwell–Boltzmann distribution function, even if we use quantum mechanics to determine the number of states. Molecules in the system can also have rotational energies.

10.6 DISTRIBUTION OF ROTATIONAL ENERGIES

In Section 9.5, we explored rotations of a molecule by using quantum mechanics. Now, we are dealing with a system of molecules. We are interested in finding out the following two distributions for the rotating molecules:

1. The total rotational energy of the system of N molecules as a function of temperature T.
2. The number of molecules N_l, and hence the fraction of molecules in a vibrational state l, where l is the rotational (or angular momentum) quantum number.

10.6.1 Rotational Energy

To arrive at the energy distribution for the rotating molecules, we begin with one rotating diatomic molecule. As we know from Section 9.5, the rotational energy E_l of a diatomic molecule with reduced mass μ in terms of atomic masses m_1 and m_2 is

$$E_l = \frac{l(l+1)\hbar^2}{2I} = \frac{l(l+1)\hbar^2}{2\mu R^2} = \frac{h^2}{8\pi^2\mu R^2} l(l+1) = R_c l(l+1) \qquad (10.32)$$

$$l = 0, 1, 2, \ldots$$

where I is the moment of inertia, R is the bond length, and we have defined the constant R_c as:

$$R_c = \frac{h^2}{8\pi^2\mu R^2}$$

Note that there is only one vibrational energy state corresponding to the vibrational quantum number n of a molecule, whereas there are $2l + 1$ rotational energy states corresponding to the rotational quantum number l of the molecule.

10.6.2 Population Probability of Rotational States

Population probability P_l is given by

$$P_l = \frac{N_l}{N}$$

where N_l is the number of molecules in the energy state E_l and N is the total number of molecules in the system. Because there are $2l + 1$ states corresponding to each l, the number N_l of molecules in energy state l, according to Eqs. 10.4 and 10.23 becomes:

$$N_l = (2l+1)\frac{N}{Z_r}e^{-(E_l/\kappa T)} \qquad l = 0, 1, 2, 3, \dots \tag{10.33}$$

Because corresponding to each quantum number l, there are $2l + 1$ states, the partition function will be given by

$$Z_r = \sum_{l=0}^{\infty}(2l+1)e^{-(E_l/\kappa T)} = \sum_{l=0}^{\infty}(2l+1)e^{-[R_c l(l+1)/\kappa T]} \tag{10.34}$$

As demonstrated in Problem 10.15, at high temperatures, higher than the one corresponding to the rotational energy spacing, there are enough occupied excited energy states that this expression approximates to

$$Z_r = \frac{\kappa T}{R_c}$$

Substituting this back into Eq. 10.33, we obtain the population probability P_l for rotational states:

$$P_l = \frac{N_l}{N} = (2l+1)\frac{R_c}{\kappa T}e^{-(E_l/\kappa T)} = \frac{R_c}{\kappa T}(2l+1)e^{-[R_c l(l+1)/\kappa T]}$$

Therefore,

$$P_l = \frac{N_l}{N} = \frac{R_c}{\kappa T}(2l+1)e^{-[R_c l(l+1)/\kappa T]} \tag{10.35}$$

In Eq. 10.35, P_l is the probability that a molecule will be in energy state E_l. It is also the fraction of molecules in energy state E_l.

Figure 10.5 illustrates this probability by plotting it at $T = 300\,K$, which is approximately the room temperature, for two types of molecules: $^{19}F^{35}Cl$ and $^{127}I^{79}Br$. The rotational constant R_c for FCl is taken to be 1.02×10^{-23} J corresponding to a bond length $a = 163$ pm, and the rotational constant R_c for IBr is taken to be 0.110×10^{-23} J corresponding to a bond length $a = 247$ pm.

As Figure 10.5 demonstrates, in contrast to vibrational states (Fig. 10.4), the most probable state of a molecule at room temperature is not the ground state, but rather an excited state. Furthermore, given that IBr has a smaller energy level spacing than FCl, Figure 10.5 demonstrates that as the energy level

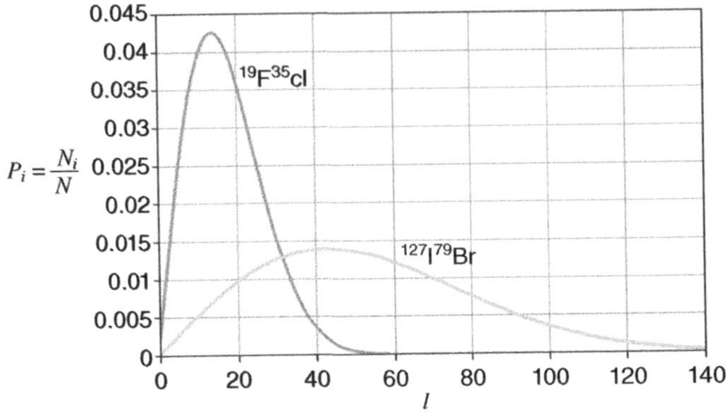

Figure 10.5. Population probability (N_l/N) plotted as a function of quantum number l at $T = 300\,\text{K}$ for two types of molecules: $^{19}F^{35}Cl$ ($\Delta E = hf = 15.6 \times 10^{-21}\,J$) and $^{127}I^{79}Br$.

spacing decreases, the rotational population probability distribution gets broader and its peak shifts toward the higher excited states.

Let us turn our discussion back to the rotational energy of the molecular system, to see if we could see the correspondence principle in action in the rotational energy as well.

10.6.3 Correspondence with Classical Mechanics

From Eq. 10.32, the rotational energy E_l of a diatomic molecule with reduced mass μ is

$$E_l = R_c l(l+1), \qquad l = 0, 1, 2, \ldots$$

Therefore, from Eq. 10.33, the energy for a system of N molecules will be given by

$$E_r = \sum_{l=0}^{\infty} N_l E_l = \frac{N}{Z_l} \sum_{l=0}^{\infty} (2l+1) E_l e^{-(E_l/\kappa T)} \tag{10.36}$$

Note that at a temperature $T = 0$, $E_r = 0$, almost all the molecules will be in their ground state (the lowest energy state). However, as the temperature increases, more and more molecules occupy the higher energy states. At high enough temperature, it can be shown that Eq. 10.36 will result in

$$E_r = N\kappa T \tag{10.37}$$

This is the same result as predicted by the equipartition theorem. Hence, in the high-temperature limit, the limit where a large number of quantum states are activated, quantum mechanics predicts the same total rotational energy of the system as classical physics; reappearance of the correspondence principle.

STUDY CHECKPOINT 10.3

At very low temperature, that is, $T = 0$:

A. What will be the value of total energy E_r?
B. What will be the value of the quantum number l?

Do not use Eq. 10.37.

Solution:

A.
For $T = 0$, the term $e^{-(E_l/\kappa T)}$ reaches 0. Therefore, from Eq. 10.36:

$$E_r = 0$$

B.
When

$$E_r = 0$$

each E_l will be zero as well. Therefore, from Eq. 10.32
$l = 0$; the ground state.

In addition to rotating and vibrating, a molecule also moves as a whole giving rise to translational energy.

10.7 DISTRIBUTION OF TRANSLATIONAL ENERGIES

Consider a system of molecules in a 3D box with the length of each side equal to a. From Chapter 6, the energy of a molecule in this box, that is, the translational energy, is given by

$$E_n = \frac{\hbar^2\pi^2}{2ma^2}\left(n_x^2 + n_y^2 + n_z^2\right) = \frac{h^2}{8ma^2}\left(n_x^2 + n_y^2 + n_z^2\right), \quad n_x, n_y, n_z = 1, 2, 3, \ldots \tag{10.38}$$

where m is the mass of the molecule.

The number of molecules N_n in the state (n_x, n_y, n_z) can be written by using Eq. 10.23:

$$N_n = \frac{N}{Z_t} e^{-(E_n/\kappa T)} \tag{10.39}$$

where the partition function is given by

$$Z_t = \sum_{n_x,n_y,n_z=1}^{\infty} e^{-(E_n/\kappa T)} = \sum_{n_x,n_y,n_z=1}^{\infty} e^{-\left[1h^2/\kappa T 8ma^2\right]\left(n_x^2+n_y^2+n_z^2\right)} \tag{10.40}$$

This can be evaluated to result in the following expression:

$$Z_t = a^3\left(\frac{2\pi m\kappa}{h^2}\right)^{3/2} T^{3/2}$$

Substituting this back into Eq. 10.39, we obtain the population probability P_n for translational states:

$$P_n = \frac{N_n}{N} = \frac{1}{a^3T^{3/2}}\left(\frac{h^2}{2\pi m\kappa}\right)^{3/2} e^{-\left(1h^2/\kappa T 8ma^2\right)\left(n_x^2+n_y^2+n_z^2\right)} \tag{10.41}$$

where P_n is the probability of finding a molecule in a state characterized by the quantum numbers (n_x, n_y, n_z).

Figure 10.6 presents the occupation probability of translational states determined by setting

$$\frac{h^2}{\kappa ma^2} = 1$$

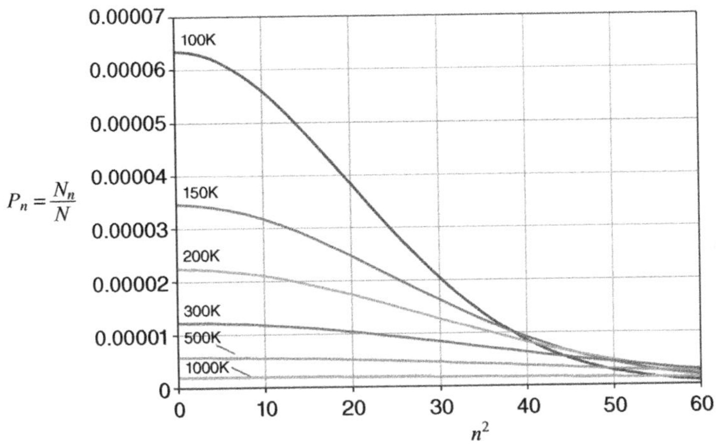

Figure 10.6. Population probability (N_n/N) of translational states plotted as a function of the quantum number $n^2 = n_x^2 + n_y^2 + n_z^2$ at various temperatures: 100, 150, 200, 300, 500, and 1000 K.

The graphs are shown at temperatures 100, 150, 200, 300, 500, and 1000 K. The figure demonstrates that the maximum probability at low temperatures is at $n = 0$. This means that a given molecule at low temperature is most likely to be found in its ground translational state ($n = 0$), or the large number of molecules at low temperature are in their ground states ($n = 0$). As the temperature increases, more and more molecules transition to excited states. Also, as the temperature increases, the graph is getting flat; that is, the population probability is becoming independent of the quantum numbers, and hence the predictions of quantum mechanics are getting closer to those of classical mechanics.

From Eq. 10.39, the total translational energy of the system can be written as:

$$E_t = \sum_{n=1}^{\infty} N_n E_n = \frac{N}{Z_n} \sum_{n_x,n_y,n_z=1}^{\infty} E_n e^{-(E_n/\kappa T)} = \frac{N}{Z_n} \sum_{n_x=1}^{\infty} \sum_{n_y=1}^{\infty} \sum_{n_z,=1}^{\infty} E_n e^{-(E_n/\kappa T)} \tag{10.42}$$

STUDY CHECKPOINT 10.4

What is the probability P_n that a given molecule is in the state

$$n = (n_x, n_y, n_z)?$$

Solution:

$$P_n = \frac{N_n}{N}$$

By using Eq. 10.39,

$$P_n = \frac{N_n}{N} = \frac{1}{Z_n} e^{-(E_n/\kappa T)}$$

To approximate the quantum consideration into the classical one, we assume that there are many quantum states in the system and their corresponding energies are very closely spaced. In other words, the discrete distribution in Eq. 10.42 can be replaced with continuous distribution represented by integrals. In this case, as Problem 10.10 demonstrates:

$$E_t = \frac{3}{2} N\kappa T \tag{10.43}$$

which is in accordance with the partition theorem given that the translational motion has three degrees of freedom: the correspondence principle again. Note that E_t in Eq. 10.43 is independent of the mass of the molecule.

In the last few sections, we covered a substantial amount of material to describe the vibrational, rotational, and translational states of molecular systems or assemblies of identical, but distinguishable, molecules (or particles). For some of you, the mathematics we have gone through may be rather overwhelming. So, it is worth the effort to summarize the core points to build a big picture of the quantum statistics of distinguishable particles.

10.8 QUANTUM STATISTICS OF DISTINGUISHABLE PARTICLES: PUTTING IT ALL TOGETHER

For vibrational, rotational, and translational motion of N molecules in a system of molecules, we explored the answers to three fundamental questions:

1. What is the quantum mechanical expression for the energy E_n of a molecule due to each of these three kinds of motion: vibrational, rotational, and translational?
2. What is the total energy of the molecular system for each of these three kinds of motion?
3. What fraction of molecules N_n/N are in a given quantum mechanical state corresponding to a specific energy state E_n, that is, a specific quantum number n? This fraction defines the population probability $P = N_n/N$ distribution for the quantum states of the system of particles, that is, how different available quantum states are occupied by the particles in the system.

The quantum mechanical expressions for the energy of individual molecules for these three kinds of motion were derived in Chapter 9. The basic expression for the fraction of molecules occupying a specific state (the population probability) is

$$P_i = \frac{N_i}{N} = \frac{1}{Z} e^{-(E_i/\kappa T)} \tag{10.44}$$

where Z is the partition function given by

$$Z = \sum_{i=1}^{\infty} e^{-(E_i/\kappa T)} \tag{10.45}$$

Because the expressions for E_i is different for each of three kinds of motion (vibrational, rotational, and translational) the specific expressions for Z and P_i are also different.

The specific expressions for these three quantities are presented in Table 10.1, where the rotational constant R_c in this table is given by

TABLE 10.1. Energy and Population Probability Distributions of Molecules for Vibrational, Rotational, and Translational Motions

Mode of Motion	Partition Function	Energy	Population Probability
Vibrational	$Z_v = \dfrac{e^{-(hf/2\kappa T)}}{1-e^{-(hf/\kappa T)}}$	$E_n = \left(n+\dfrac{1}{2}\right)hf$	$P_n = (1 - e^{-hf/\kappa T})e^{-(nhf/\kappa T)}$
Rotational	$Z_r = \dfrac{\kappa T}{R_c}$	$E_l = R_c l(l+1)$	$P_l = \dfrac{R_c}{\kappa T}(2l+1)e^{-[R_c l(l+1)/\kappa T]}$
Translational	$Z_t = a^3\left(\dfrac{2\pi m\kappa}{h^2}\right)T^{3/2}$	$E_n = \dfrac{\hbar^2\pi^2}{2ma^2}(n_x^2+n_y^2+n_z^2)$	$P_n = \dfrac{1}{a^3T^{3/2}}\left(\dfrac{h^2}{2\pi m\kappa}\right)^{3/2} e^{-\frac{1}{\kappa T}\frac{h^2}{8ma^2}(n_x^2+n_y^2+n_z^2)}$

$$R_c = \frac{h^2}{8\pi^2 \mu R^2} \tag{10.46}$$

For each of these three kinds of motion, you learned the following three important lessons:

1. At low temperature, most of the molecules are in their ground states, that is, minimum energy states.
2. As the temperature increases, more and more molecules transition to the excited states, that is, states with higher energies.
3. At high enough temperature, a large number of quantum states are occupied by the molecules and the quantum mechanical results begin to agree with the classical mechanics results. This is an expression of the correspondence principle.

So far, we have been treating the particles in the system under the assumption that although they are identical, they are distinguishable from one another. This means that the particles are physically separated from each other, which in quantum mechanics means that their wave functions either do not overlap or the overlap is negligibly small. This also means that particles in the system are not strongly correlated. Particles of an ideal gas obeying the kinetic molecular theory discussed in Section 9.2 are an example. Although within this limit, they can interact with each other and with the walls of the container to establish equilibrium, as mentioned in Section 9.2. Because their wave functions cannot overlap significantly, you can argue that we are in a classical regime or pseudo-quantum regime; after all we did quantize the energy.

What if the particles cannot be distinguished from each other? We deal with this situation next, in Section 10.9.

10.9 QUANTUM STATISTICS OF INDISTINGUISHABLE PARTICLES

The Maxwell–Boltzmann distribution function works well for a system of identical particles that can be distinguished from one another, that is, their wave functions do not overlap significantly. Molecules in the gas phase can be modeled to meet this requirement and therefore the system they form can be described by the Maxwell–Boltzmann distribution function. So far, this is how we have treated the molecules in this chapter. Now, let us consider the situation when the wave functions of the particles in the system overlap significantly so that they cannot be distinguished from each other.

> *Note*: The assumption here is that although the particles in the system cannot be distinguished from each other, they can still be counted.

From the standpoint of quantum mechanics, the particles whose wave functions overlap so much that they cannot be distinguished from one another, fall into two categories:

1. *Fermions.* These are the particles that have odd half-integer spins and obey Pauli's exclusion principle. This means that no two fermions can occupy the same quantum state simultaneously. Electrons with spin ½ are examples of fermions.
2. *Bosons.* These are the particles that have zero or integral spins and do not obey Pauli's exclusion principle. This means that more than one boson can occupy the same quantum state simultaneously. Photons, the quanta of light, are an example of bosons.

STUDY CHECKPOINT 10.5

Write down the first few fermion spins, that is, the first few odd half-integers.

Solution:

$$\frac{1}{2}, \frac{3}{2}, \frac{5}{2}, \frac{7}{2}, \ldots$$

For example, $\frac{4}{2}$ is not a fermion spin.

Recall that the number of particles $n(E_i)$ with energy E_i is given by

$$n(E_i) = g(E_i) f(E_i) \tag{10.47}$$

where $g(E_i)$ is the number of available states with energy E_i, and $f(E_i)$ is the probability that each state of energy E_i will be occupied. The function $f(E_i)$ is also called the distribution function. The distribution function that we have been using so far in this chapter is called the Maxwell–Boltzmann distribution function and is given by

$$f_{\mathrm{MB}}(E_i) = Ae^{-\beta E_i} \tag{10.48}$$

where A is the normalization constant and β is given by

$$\beta = \frac{1}{\kappa T} \tag{10.49}$$

For a system of fermions, the distribution function is called the Fermi–Dirac (FD) distribution and is given by

$$f_{\mathrm{FD}}(E_i) = \frac{1}{Fe^{\beta E_i} + 1} \tag{10.50}$$

For a system of bosons, the distribution function is called the Bose–Einstein (BE) distribution and is given by

$$f_{\mathrm{BE}}(E_i) = \frac{1}{Be^{\beta E_i} - 1} \tag{10.51}$$

where F and B are normalization constants that depend on the physical systems being represented by the distributions. Note from Eqs. 10.50 and 10.51 that for a given energy, the value for $f_{\mathrm{BE}}(E_i)$ will be higher than the value for $f_{\mathrm{FD}}(E_i)$. This accounts for the fact that multiple bosons can occupy the same energy state. Figure 10.7 presents the comparison of these three distributions as functions of energy over the same range shown along the x-axis in terms of κT. The normalization constants A, B, and F are set to 1. As the figure demonstrates, the three distributions behave the same way at higher energy and a higher number of degrees of freedom (represented by κT).

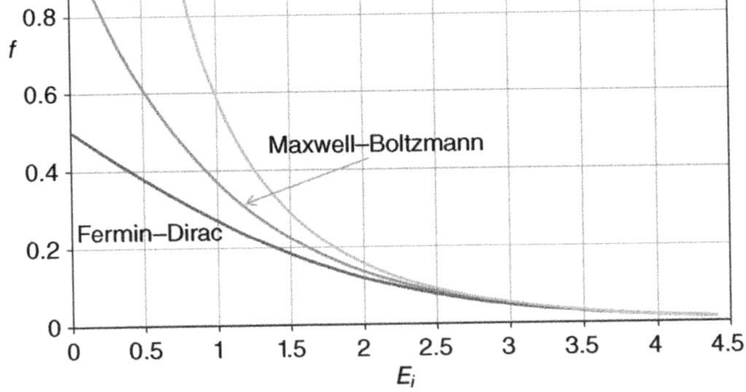

Figure 10.7. Comparison of Fermi–Dirac, Bose–Einstein, and Maxwell–Boltzmann distributions as a function of energy plotted in units of κT; the normalization constants A, B, and F are set to 1.

TABLE 10.2. Characteristics of the Three Statistical Distributions

Distributions	MB	FD	BE
Distribution Function	$f_{\mathrm{MB}}(E_i) = Ae^{-\beta E_i}$	$f_{\mathrm{FD}}(E_i) = \dfrac{1}{Fe^{\beta E_i} + 1}$	$f_{\mathrm{BE}}(E_i) = \dfrac{1}{Be^{\beta E_i} - 1}$
Applies to systems of	Identical particles that can be distinguished from one another	Identical, indistinguishable particles that obey Pauli's exclusion principle	Identical, indistinguishable particles that do not obey Pauli's exclusion principle
Examples of the systems	Molecules of a gas	Electron gas, free electrons in a metal, moving electrons in the cells of an organism	Photons in a cavity, liquid helium
Spin of particles in the system	Independent of spin or no spin	$\frac{1}{2}, \frac{3}{2}, \frac{5}{2}, \frac{7}{2}, \ldots$	$0, 1, 2, 3, \ldots$
Particles per state	No limit	Not > 1	No limit

Table 10.2 compares the three distribution functions with one another. From these three distribution functions, note the following three points:

1. Both Fermi–Dirac and Bose–Einstein distributions are significantly different from the Maxwell–Boltzmann distribution function.
2. The Fermi–Dirac and Bose–Einstein distributions are very similar; they only differ by their normalization constants and by the sign of the number 1 in the denominator. However, this sign difference gives rise to significant differences at certain limits.
3. At very high energies, all three distributions give the same results, that is, Fermi–Dirac and Bose–Einstein distributions look like the Maxwell–Boltzmann distribution.

STUDY CHECKPOINT 10.6

Prove that Maxwell–Boltzmann statistics can be used for both fermions and bosons in the high-energy limit.

Solution:

In the limit of very high energy, the factor $e^{\beta E_i}$ will be >>1. Therefore, from Eq. 10.50:

$$f_{\mathrm{FD}}(E_i) = \frac{1}{Fe^{\beta E_i}+1} = \frac{1}{Fe^{\beta E_i}} = \frac{1}{F}e^{-\beta E_i} = Ae^{-\beta E_i}$$

where $A = 1/F$

Also, from Eq. 10.51:

$$f_{\mathrm{BE}}(E_i) = \frac{1}{Be^{\beta E_i}} = \frac{1}{B}e^{-\beta E_i} = Ae^{-\beta E_i}$$

where $A = 1/B$

Hence, both the Fermi–Dirac and Bose–Einstein distribution functions behave like the Maxwell–Boltzmann distribution functions in the limit of high energies. So, in this limit, Maxwell–Boltzmann statistics can be used for both fermions and bosons.

Enrico Fermi (1901–1954)
Courtesy of the U.S. Department of Energy.

Enrico Fermi, one of the great physicists of the twentieth century, was born on September 29, 1901 in Rome, Italy. He studied at Scuola Normale Superiore in Pisa and worked and taught at various institutes including the University of Rome, Columbia University, and the University of Chicago. He did his work on quantum statistics in 1926 while he was in Rome. He was exiled from Italy in 1938. He then went to the United States where he led a team in Chicago that built the first self-sustaining nuclear reactor in 1942. Fermi also worked on the atomic bomb project (The Manhattan Project) during the Second World War. After the war, he became a faculty member at University of Chicago, where he worked for the rest of his life.

Fermi received the Nobel Prize in physics in 1938 for his "demonstrations of the existence of new radioactive elements produced by neutron irradiation, and for his related discovery of nuclear reactions brought about by slow neutrons". He died on November 28, 1954 at the age of 53 in Chicago. Fermium, a synthetic element created in 1952 at the Fermi National Accelerator Lab near Chicago, and a type of particles called fermions are named after him.

Satyendra Nath Bose
(1894–1974)

Satyendra Nath Bose was born on January 1, 1894 in Kolkata (Calcutta), India. He studied at the Hindu College and Presidency College in Kolkata, and taught and worked at the University of Calcutta and the University of Dhaka (now in Bangladesh).

In 1924, Bose derived the quantum radiation law, but his paper was rejected by a leading British research journal. After failing to get it published, he sent the paper to Einstein in Germany. Einstein, understanding the importance of Bose's work, translated the paper into German and sent it for Bose to the prestigious German research journal *Zeitschrift für Physik,* where it was published. Because Einstein extended Bose's work to apply it to material particles, both names are attached to the equation called the Bose–Einstein distribution.

The recognition that followed this publication, allowed Bose to leave India for the first time and he spent 2 years in Europe working with noted scientists of his time including Louis de Broglie, Marie Curie, and Albert Einstein.

Although he never worked to earn a Ph.D., Bose had diverse interests in science and studied biology, chemistry, geology, zoology, anthropology, and engineering. He died on February 4, 1974 in Kolkata, India at the age of 80.

So, atoms and molecules in the excited state transition to the lower state by emitting energy in the form of radiation. We realized in Chapter 2 that radiation is a collection of photons. Also, in Section 2.3 we presented Planck's radiation formula. Now that we know how to handle a collection of particles, let us see if we can derive Planck's radiation formula in the context of what we have discussed in this chapter.

10.10 PLANCK'S RADIATION FORMULA

We revisit the problem of black-body radiation discussed in Section 2.3 because radiation is just a collection of photons, that is, bosons. If we model the cavity as a cube with each edge equal to a, we can model the photon in the cavity as a particle in the 3D box. From Section 6.5, we know that the energy of a particle in a 3D box with each side equal to a is given by

$$E = \frac{p^2}{2m} = \frac{\hbar^2\pi^2}{2ma^2}\left(n_1^2 + n_2^2 + n_3^2\right) \tag{10.52}$$

Therefore,

$$p^2 = \frac{\hbar^2\pi^2}{a^2}\left(n_1^2 + n_2^2 + n_3^2\right) \tag{10.53}$$

and the energy of a photon can be written as:

$$E_p = pc = \frac{hc}{2a}\sqrt{\left(n_1^2 + n_2^2 + n_3^2\right)} \tag{10.54}$$

We are doing all this manipulation for one purpose: to calculate the density of quantum states. Recall that n_1, n_2, n_3 are the quantum numbers that quantize energy in 3Ds and therefore represent the quantum states. As illustrated in Figure 10.8, The number of states can be represented by the volume of the sphere in the number space with radius r, which is given by

$$r^2 = \left(n_1^2 + n_2^2 + n_3^2\right) \tag{10.55}$$

To be precise, the number of states corresponding to n_1, n_2, n_3 will be

$$N_n = 2 \times \frac{1}{8} \times \frac{4}{3}\pi r^3 \tag{10.56}$$

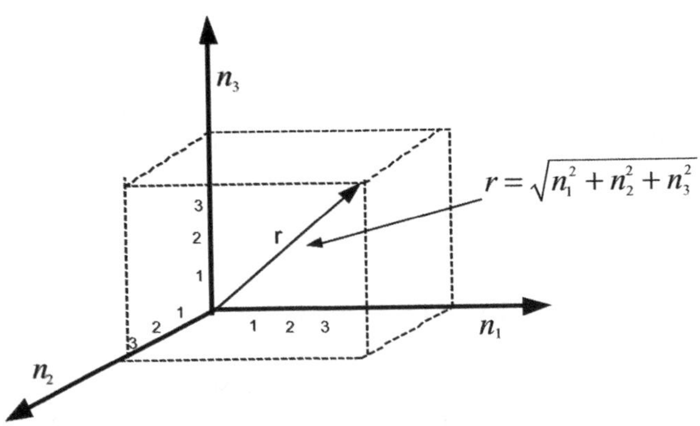

Figure 10.8. Counting allowed states in the 3D number space of states.

where the factor 2 accounts for the fact that there are two possibilities for the photon polarization, and the factor 8 appears in the denominator because we are only interested in that part of the sphere that corresponds to the positive values of n_1, n_2, n_3.

From Eqs. 10.54 and 10.55 we can write

$$E_p = \frac{hc}{2a} r \tag{10.57}$$

From Eqs. 10.56 and 10.57, we write

$$N_n = \frac{8\pi a^3}{3h^3c^3} E_p^3 \tag{10.58}$$

Therefore, the density of states can be calculated as:

$$g(E_p) = \frac{dN_n}{dE_p} = \frac{8\pi a^3}{h^3c^3} E_p^2 \tag{10.59}$$

Because photons are bosons, their distribution function can be written from Eq. 10.51 as:

$$f_{\mathrm{BE}}(E_p) = \frac{1}{e^{\beta E_p} - 1} \tag{10.60}$$

where the normalization constant B has been set to 1. This is because the total radiant energy in the cavity remains constant even though the number of photons that compose that energy may change, as photons are absorbed and emitted by the walls of the cavity. From Eqs. 10.59 and 10.60, we are now able to write the distribution for the number of photons carrying energy E_p:

$$n(E_p) = g(E_p) f_{\mathrm{BE}}(E_p) = \frac{8\pi a^3}{h^3c^3} E_p^2 \frac{1}{e^{\beta E_p} - 1} \tag{10.61}$$

From this distribution, we can write the distribution for the energy density, that is, the energy per unit volume of the cavity:

$$u(E_p) = \frac{n(E_p) E_p}{a^3} = \frac{8\pi a^3}{h^3c^3} \frac{E_p^3}{e^{\beta E_p} - 1}$$

For photons in the range of E_p to $E_p + dE_p$:

$$u(E_p) dE_p = \frac{n(E_p) E_p}{a^3} = \frac{8\pi}{h^3c^3} \frac{E_p^3}{e^{\beta E_p} - 1} dE_p \tag{10.62}$$

At frequency f corresponding to wavelength λ:

$$E_p = hf = \frac{hc}{\lambda} \tag{10.63}$$

This means,

$$dE_p = hdf = \frac{hc}{\lambda^2}d\lambda \tag{10.64}$$

We dropped the minus sign because we are only interested in the magnitude of energy. By using Eqs. 10.63 and 10.64, we obtain from Eq. 10.62:

$$u(f,T)df = \frac{8\pi f^2}{c^3}\frac{hf}{e^{hf/\kappa T}-1}df \tag{10.65}$$

and

$$u(\lambda,T)d\lambda = \frac{8\pi hc}{\lambda^5}\frac{1}{e^{hc/\lambda\kappa T}-1}d\lambda \tag{10.66}$$

where we have also substituted

$$\beta = \frac{1}{\kappa T}$$

Note that Eqs. 10.65 and 10.66 are the same as those introduced in Section 2.3 for Planck's radiation formulas. As shown in Problem 10.16, if we used the Maxwell–Boltzmann distribution instead of the Bose–Einstein distribution and followed the same procedure, we would have arrived at an energy density expression indicating that the energy density keeps increasing and approaching infinity as the wavelength keeps decreasing. Hence the frequency keeps increasing, which is consistent with the classical result from Section 2.3.1 called the classical catastrophe.

The quantum mechanical understanding of radiation emission by atoms and molecules led to the invention of lasers.

10.11 ABSORPTION, EMISSION, AND LASERS

The term laser is an acronym for *light amplification by stimulated emission of radiation.* A laser is a device that produces a nearly monochromatic and coherent beam of light (radiation). Monochromatic means that the light beam has a single frequency or a very narrow range of frequencies or wavelengths. A

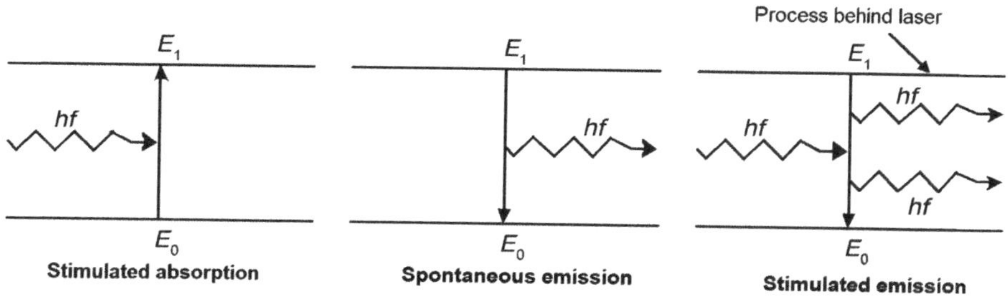

Figure 10.9. Three ways of transitioning between two energy levels in an atom or a molecule.

coherent beam means that the waves in the beam are in phase with each other or have a constant phase difference over time. The laser phenomenon was discovered with the help of quantum physics and obviously can be explained with quantum mechanics.

In Chapters 2 and 9, we explored how atoms and molecules in excited states can transition into a lower energy state by emitting radiation, that is, photons. How do these atoms and molecules transition into excited states? The answer is, of course: By absorbing photons. As shown in Figure 10.9, there are three ways in which these transitions between a lower energy level E_0 and higher energy level E_1 can occur:

1. A molecule (or atom) in the energy state E_0 transitions to the state E_1 by absorbing a photon of energy $hf = E_1 - E_0$. This process is called stimulated absorption.
2. If the molecule (or atom) is initially in the energy state E_1, it can transition to E_0 by emitting a photon of energy $hf = E_1 - E_0$. This process is called spontaneous emission.
3. A photon of energy hf incident on the molecule (or atom) in state E_1 stimulates the transition of the molecule (or atom) to the state E_0. This process is called stimulated emission, as opposed to spontaneous emission, which does not involve any external help.

The process of stimulated emission was first suggested by Einstein in 1917. Einstein argued that the probability for the stimulated emission is the same as that of the stimulated absorption. In other words, the likelihood that a photon of energy hf incident on a molecule in state E_0 will be absorbed is the same as the likelihood that this photon will cause the emission of another photon if it was incident on the molecule in state E_1.

In the case of stimulated emission, the emitted photon is in phase with the incident photon. The key idea behind the laser is to stimulate emission before spontaneous emission occurs. This is facilitated by the fact that many molecules (or atoms) consist of excited states whose lifetime is larger than usual. Such states are called *metastable* (almost stable) states. As an example of a simple laser, consider a system of molecules (or atoms) with the following three characteristics:

1. The molecules have a metastable state at energy *hf* greater than the energy of the ground state.
2. The molecules also have a higher excited state from which they can transition into the metastable state.
3. There are more molecules in the metastable state than in the ground state.

If light with frequency *f* is incident on this system, there will obviously be more stimulated photon emission from molecules in the metastable state than the stimulated absorption of photons by molecules in the ground state because there are more molecules in the metastable state than in the ground state. All the photons emitted through stimulated emission are in phase with the incident photons and have the same frequency as that of the incident photons. Therefore this results in the amplification of the original light in the form of a coherent and monochromatic light beam. This is the basic concept behind the operation of a laser.

A laser is one of those inventions that are used in the field of applications, as well as in the field of research as tools. It is not rare that sometimes the theoretical predictions, the necessary characteristic of a scientific theory, are so ahead of their times that they have to wait for the development of the right technology before they can be tested. Such was the case with one prediction of quantum theory called the Bose–Einstein condensation.

10.12 BOSE–EINSTEIN CONDENSATION

You already know three states (or phases) of matter: solid, liquid, and gas. Bose–Einstein condensation refers to the process that gives rise to a new state of matter called the Bose–Einstein condensate. This state was predicted by Bose and Einstein based on their quantum mechanical description of particles with integral spins, the Bose–Einstein statistics described earlier in this chapter.

Here is the account of the prediction. Consider a gaseous system of particles with each particle having an integer spin. Due to the integer spin of the particles, this system will obey Bose–Einstein statistics discussed earlier in this chapter. Because the system is in the gaseous sate, the volume of the container is large compared to the size of each particle. The quantum mechanics wave packet (or de Broglie wave) associated with each particle is so small in size compared to the spacing among the particles that the particles can move around fairly independently and be physically distinguished from each other. This is one extreme where wave packets are not so important, so let us move to the other extreme by cooling the gas. As shown in Study Checkpoint 10.7, as a particle is cooled, the size of its wave packet increases. You can arrive at the same result by using this form of the uncertainty principle:

$$\Delta x.\Delta p \geq \frac{\hbar}{2}$$

As the particles are cooling and their wave packets are growing, throw in the fact that the Bose–Einstein particles do not obey Pauli's exclusion principle, that is, multiple particles can occupy the same quantum state. This has a very profound implication for our cooling of the Bose–Einstein system of particles. As the particles get cold enough, their individual wave packets get larger than the spacing among the particles and therefore overlap. Eventually, at minimum temperature ($T = 0\,\mathrm{K}$), all the overlapping wave packets will merge into one wave packet because all the atoms will occupy the same quantum mechanical state (the lowest possible energy or ground state) and therefore all the atoms will come to a halt and act as one particle, called a superparticle or superatom if the gas is made of atoms. This state (or phase) of matter is called the Bose–Einstein condensate.

STUDY CHECKPOINT 10.7

Prove that according to quantum mechanics the wave packet size increases with the cooling of a particle.

Solution:

The wavelength of a particle is given by

$$\lambda = \frac{h}{p}$$

When the particle is cooled, the momentum p decreases, and therefore the wavelength increases. This is equivalent to saying that the wave packet size increases.

These Bose–Einstein condensates predicted in 1924–1925 by Bose and Einstein were experimentally observed 70 years later in 1995 and were reported on July 14, 1995, in an issue of Science magazine by a group led by Eric Cornell and Carl Wieman. These researchers were working at the University of Colorado at Boulder NIST-JILA lab. They used the rubidium (^{87}Rb) gas atoms, which they cooled by using lasers and a magnetic field. Lasers were used to hit the atoms precisely from the opposite direction with photons to slow them down. In this first phase, the atoms were cooled down to ~1 mK. In the second

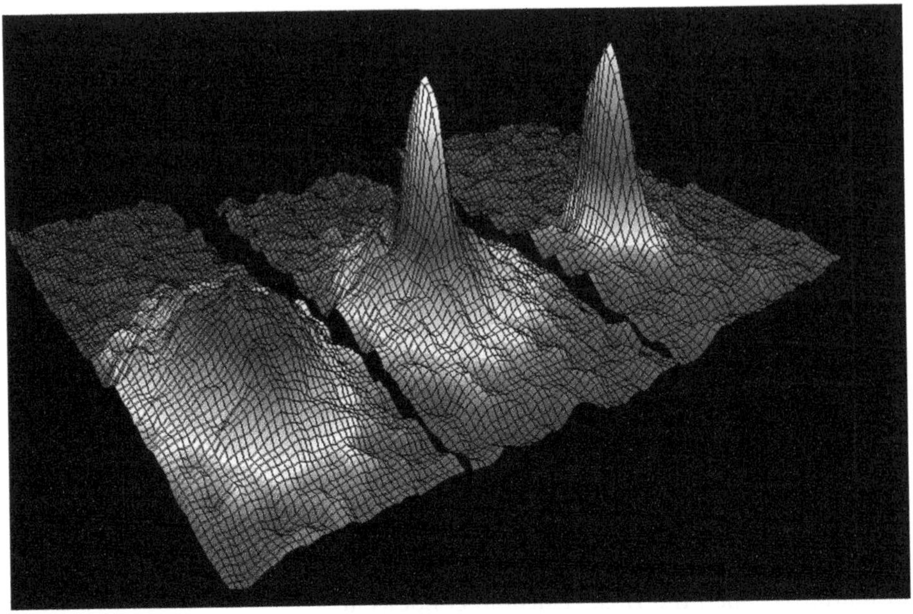

Figure 10.10. Three-dimensional illustration of the gradual formation of the Bose–Einstein condensate as the temperature decreases. The two horizontal axes represent the velocity components and the vertical axis represents the number of atoms corresponding to those velocities. Courtesy of the National Institute of Standards and Technology.

cooling phase, the already slowed atoms were trapped in a magnetic field and were cooled to ~20 nK: extremely close to 0 K. At ~170 nK, the researchers observed the transition of the atoms from a gas with a normal broad velocity distribution to a very narrow velocity distribution, indicating that all atoms had about the same velocity. As shown in Figure 10.10, the fraction of atoms in the Bose–Einstein condensate state increased as the temperature decreased.

Figure 10.10 is a 3D illustration of the gradual formation of the Bose–Einstein condensate as the temperature decreases. The two horizontal axes represent the velocity components and the vertical axis represents the number of atoms corresponding to those velocities. Three frames in this picture correspond to temperatures of ~200, 100, and 20 nK (from the bottom left to the top right). The higher and narrower peak growing at the center of these graphs represent the growing number of atoms getting into the singular quantum mechanical state with zero velocity.

These observations of the Bose–Einstein condensates were confirmed by other researchers including Wolfgang Ketterle at MIT, who shared the 2001 Nobel Prize in physics with Cornell and Wieman for their achievements on Bose–Einstein condensation. Since then, researchers have been exploring the applications of the Bose–Einstein condensates, one of which is creating a

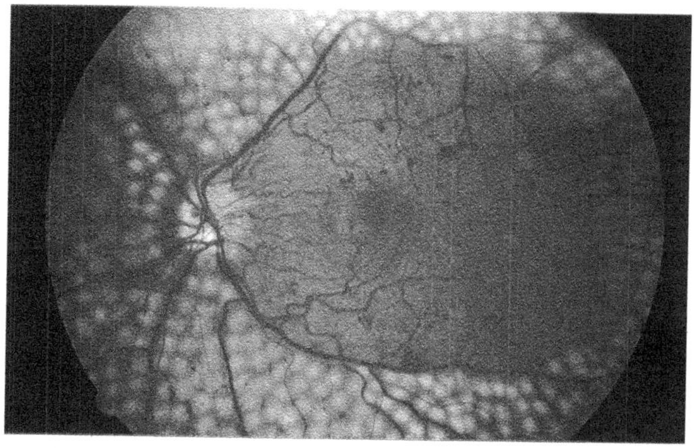

Figure 10.11. Fundus photo showing scatter laser surgery for diabetic retinopathy. Courtesy of National Eye Institute, National Institutes of Health.

coherent beam of atoms analogous to the coherent beam of photons produced by a laser.

10.13 LIVING IN THE QUANTUM WORLD

As mentioned earlier, the laser phenomenon was discovered with the help of quantum physics and obviously can be explained with quantum mechanics. In addition to their wide range of applications in science, lasers have a wide spectrum of uses in several other areas including surgery, welding and metal cutting, sound and video recording, and reproduction. One obvious scientific application of lasers is the precise determination of locations and precise measurements of distances by making use of the coherent and nondivergent nature of the laser beam.

Over the years, lasers have been used increasingly in medicine. Due to their better precision and ability to focus energy into a small spot than by conventional means, they are used in surgery to make precise incisions in different types of tissues. The use of a laser in surgery also reduces bleeding because the laser helps the blood coagulate during cutting. These days, lasers are frequently used in a number of eye operations, and are helping many preventing blindness. For example, Figure 10.11 presents the Fundus photo showing a scatter laser surgery for diabetic retinopathy. The fundus of the eye is the interior surface of the eye, opposite the lens, and includes the retina and optic disc.

The CD and DVD recording and reproduction are another common application of lasers. For example, laser light is directed toward the tracks (on CD or DVD) that contain the recorded (encoded) information. The light reflected from the tracks contains this information, which is then turned into electrical signals, which subsequently produce the output.

The huge scope of laser applications can be realized from the long list of laser types, which includes chemical, dye, free-electron, gas, metal–vapor, photonic crystal, and semiconductor lasers. In a nutshell, lasers, a gift from quantum mechanics, have enormous uses in our daily lives that range from a laser pointer to preventing blindness.

10.14 SUMMARY

A molecule has a definite energy at a given time, which includes electronic, rotational, vibrational, and translational energy. However, all the molecules in an assembly or collection of molecules do not have the same value of energy (or of other physical quantity, e.g., velocity) at a given time. Instead, the different values of energy of a collection of molecules are statistically distributed over molecules. The core issue is how to determine this distribution, in general, the fraction of molecules in each state, or the population probability distribution.

To describe a system of identical particles, the general formula to determine the number of particles with energy E_i is

$$n(E_i) = g(E_i) f(E_i)$$

where $n(E_i)$ is the number of particles with energy E_i, $g(E_i)$ is the number of available states of energy E_i, and $f(E_i)$ is the probability that each state of energy E_i will be occupied. The function $f(E_i)$ is also called the distribution function. The Maxwell–Boltzmann distribution function used for describing a system of identical, but distinguishable, particles is given by

$$f_{\rm MB}(E_i) = Ae^{-\beta E_i} = Ae^{-E_i/\kappa T}$$

where A is the proportionality constant. We can use this function with $g(E_i)$ determined by the availability of quantum mechanical states to describe the system of identical particles that can be distinguished from one another.

We can use the Maxwell–Boltzmann distribution function to treat the system of particles in quantum mechanics as well. For a system with many quantum states, the fraction of molecules occupying a specific state (the population probability) is

$$P_i = \frac{N_i}{N} = \frac{1}{Z} e^{-E_i/\kappa T}$$

where Z is the partition function given by

$$Z = \sum_{i=1}^{\infty} e^{-(E_i/\kappa T)}$$

Because the expressions for E_i is different for each of three kinds of motion (vibrational, rotational, and translational), the specific expressions for Z and P_i are also different. For each of these three kinds of motion, you learned the following three important lessons:

1. At low temperature, most of the molecules are in their ground states, that is, minimum energy states.
2. As the temperature increases, more and more molecules transition to the excited states, that is, states with higher energies.
3. At high enough temperature, a large number of quantum states are occupied by the molecules and the quantum mechanical results begin to agree with the classical mechanics results. This is an expression for the correspondence principle.

The Maxwell–Boltzmann distribution function does not distinguish between odd half-integer spin and integer spin. A system of particles with odd half-integer spins, called fermions, is described by the Fermi–Dirac distribution function:

$$f_{\text{FD}}(E_i) = \frac{1}{Fe^{\beta E_i} + 1}$$

A system of particles with zero or integer spins called bosons is described by the Bose–Einstein distribution function:

$$f_{\text{BE}}(E_i) = \frac{1}{Be^{\beta E_i} - 1}$$

The difference between the distribution functions for fermions and bosons stem from the fact that fermions obey the exclusion principle and bosons do not. Because a photon is a boson, the Bose–Einstein statistics can be used to derive Planck's radiation formula for black-body radiation:

$$u(\lambda, T)d\lambda = \frac{8\pi hc}{\lambda^5} \frac{1}{e^{hc/\lambda\kappa T} - 1} d\lambda$$

The quantum mechanical understanding of radiation emission by atoms and molecules led to the invention of lasers, which already have enormous applications in a multitude of areas including health, welding and metal cutting, sound and video, and recording and reproduction.

10.15 ADDITIONAL PROBLEMS

Problem 10.7 Cells are the fundamental building blocks of all forms of life. Biologists believe that the first cells on planet Earth evolved during the time when unstable crustal plates were colliding and forming the proto-continents. As a result, volcanoes were spewing lava into the superheated water and gases into the atmosphere. The first cells evolved when unstable systems of molecules were stabilizing by reaching thermal equilibrium. Consider a system of 3×10^{26} molecules.

A. Calculate the minimum volume of this system if molecules can be modeled as spheres of an average radius of 1 nm and if the molecules are free.
B. Find the number of molecules in their first excited state at 10,000 and at 100°C. For simplicity, model the molecule as a single atom whose states are represented by the effective principal quantum number n. Assume that the energy difference between the first excited state ($n = 2$) and the ground state ($n = 1$) of a molecule is 12 eV.

Problem 10.8 Consider a rigid diatomic molecule in a system of molecules that acts as an ideal gas system in thermal equilibrium at temperature T. The molecule rotates around an axis perpendicular to the bond axis between the two atoms. The molecule also has translational motion and vibrates like a one-dimensional harmonic oscillator.

A. How many degrees of freedom does the molecule have?
B. What is the average energy of the molecule as a function of T if you disregard the quantization?

Problem 10.9 Prove that the total energy of a system of N molecules (vibrational, rotational, or electronic) at temperature T is given by

$$E = N\kappa T^2 \frac{d}{dT}(\ln Z)$$

where Z is the partition function:

$$Z = \sum_{i=1}^{\infty} e^{-(E_i/\kappa T)}$$

Problem 10.10 Prove that for a collection of molecules N the total translational energy E_t in the limit of large quantum numbers is given by

$$E_t = \frac{3}{2} N\kappa T$$

Problem 10.11 Consider a system of diatomic molecules. Derive a formula for the temperature T_c of a molecule in terms of bond length and reduced mass at which the quantum description of its rotational excitation energy is equivalent to the classical one.

Problem 10.12 Consider the gaseous system of molecules in a box. The energy of a gas molecule in a one-dimensional box of length a is given by

$$E_n = \frac{\hbar^2 \pi^2}{2ma^2} n^2$$

Derive the formula for the number of states for this molecule.

Problem 10.13 Use the relationship

$$E = \frac{1}{2} m v^2$$

to derive the molecular velocity distribution

$$n(v)dv = 4\pi N \left(\frac{m}{2\pi \kappa T} \right)^{3/2} v^2 e^{-(mv^2/2\kappa T)} dv$$

from the molecular energy distribution

$$n(E_i)dE_i = \frac{2\pi N}{(\pi \kappa T)^{3/2}} \sqrt{E_i} e^{-(E_i/\kappa T)} dE_i$$

Problem 10.14 For molecular vibrations, use the following relationships

$$N_n = \frac{N}{Z_v} e^{-(E_n/\kappa T)}$$

$$Z_v = \frac{e^{-(hf/2\kappa T)}}{1 - e^{-(hf/\kappa T)}}$$

$$E_n = \left(n + \frac{1}{2} \right) hf$$

to prove

$$\frac{N_n}{N} = \left(1 - e^{-(hf/\kappa T)}\right) e^{-(nhf/\kappa T)}$$

Problem 10.15 The partition function for rotational motion in a system of particles is given by

$$Z_r = \sum_{l=0}^{\infty} (2l+1) e^{-(E_l/\kappa T)} \tag{1}$$

where

$$E_l = R_c l(l+1)$$

Assume that the temperature of the system is high enough compared to the spacing in consecutive energy states corresponding to quantum numbers l. Under this assumption:

(a) Show that Eq. 1 can be written in the form of the following integral:

$$Z_r = \int_0^{\infty} (2l+1) e^{-[R_c l(l+1)/\kappa T]} dl \tag{2}$$

(b) Show that Eq. 2 can be solved to

$$Z_r = \frac{\kappa T}{R_c}$$

Problem 10.16 Derive the energy distribution $u(\lambda, T)d\lambda$ by following the same procedure used to derive Eq. 10.66, but use the Maxwell–Boltzmann distribution instead of the Bose–Einstein distribution for the distribution function. Show from the derived expression that the energy density keeps increasing as the wavelength decreases, as predicted by classical physics in Section 3.3.1, Eq. 2.2.

Problem 10.17 Show that the energy density predicted by Planck's Law of radiation approaches zero for the following both cases:

A. In the limit of wavelength approaching infinity.
B. In the limit of wavelength approaching zero. (*Hint*: use l'Hospital's rule from calculus.)

Problem 10.18 Use the Taylor series to prove that Planck's law of quantum mechanics and the Rayleigh–Jeans Law of classical physics (discussed in Chapter 2) yields about the same value of energy density in the limit of large wavelength, that is, short frequencies.

QUANTUM MECHANICS: A THREAD RUNS THROUGH IT ALL

11

Those of us who learned quantum mechanics, we love this notion that quantum mechanics is going to save the world and rewire the grid.

Richard Smalley

Figure 11.0. Universe ATU from the Thoth Tarot. Courtesy of Leister Crowley and Frieda Harveys.

Quantum mechanics is the common thread linking many fields of science in the microworld into one field called nanoscience on which nanotechnology is based. This thread runs not only through different scientific disciplines in the microscale, but also runs through the whole scale from microscale (including nano) through macroscale. As we emphasized through the correspondence principle throughout this book, classical mechanics is just an approximation of quantum mechanics in the limit of large quantum numbers. In other words, classical mechanics is the quantum mechanics of macrosystems. In this chapter, we will explore this thread along different disciplines of science.

In This Chapter

11.1 INTRODUCTION

Because the macroworld we live in is made of building blocks from the microworld, nature does its most fundamental design work at the microlevel. Therefore, the everyday reality that we experience in the macroworld has its roots in the microworld. Quantum mechanics, being the underlying science of the microworld, is the common thread running through all scientific disciplines dealing with matter including biology, chemistry, computing, material science, and through the junction of it all nanoscience and nanotechnology. As is evident from the material covered in this book, outside of physics, chemists were the first to realize the importance of quantum mechanics. In fact, the areas of physical chemistry and atomic physics are often intertwined with each other. This is because chemistry for more than a century has been based on atoms and molecules, as opposed to biology, where molecular biology is still a relatively new field. Following chemists, the electrical engineers soon realized the power of quantum mechanics and solid-state electronic technology rapidly developed. The term electronic itself should give you a clue that electrical engineers were playing around with ways of managing electrons to produce useful technology. However, the time has come when all scientific disciplines involving matter, living or nonliving, cannot escape the uses of quantum mechanics. This is today's message and the messenger is nanotechnology, the application of science at the nanoscale. After their independent development, all the major scientific disciplines are meeting at the nanoscale juncture where the behavior of matter is governed by quantum mechanics. Therefore, quantum mechanics is gaining traction in all different fields ranging from computer science to biology.

In this chapter, we briefly explore the relevance, need, and importance of quantum mechanics in nanotechnology, computer science, and biology.

11.2 NANOSCIENCE AND NANOTECHNOLOGY

The history of science is a journey of the development of our understanding, starting from the macroworld and making our way to the micro- and nanoworld. Science at the scale of nanometers is called nanoscience, and its application is called nanotechnology. Nanoscience and nanotechnology are the natural progression (or evolution) of science and technology. It is the junction where several different disciplines of science intersect one another as they were destined to. The unifying theme of nanotechnology is the control of matter at the level of building blocks, such as molecules and atoms, normally on the scale of 1–100 nm, and the fabrication of devices with critical dimensions that lie within this size range. Of course, the control on the nanoscale level can also be used to build large structures and systems more effectively and with better features from bottom up. This implies nanoscience and nanotechnology will have a global impact of unparallel proportions on our world by influencing almost all existing fields in science and technology and by opening new doors to knowledge and opportunities. So, if nanotechnology comes from nanoscience, where does nanoscience come from?

11.2.1 Sciences behind Nanoscience

Nanoscience is the study of matter and material systems at the nanoscale level, that is, at least one dimension of the system under study is in the range of ~1–100 nm. All branches of science have developed at a different pace. Some of them, such as physics, ran into the nanoscale level much earlier than others. Nevertheless, nanoscience is emerging from the studies of different sciences on the nanoscale level. Let us not lose sight of the fact that the discipline of science that studies the fundamental building blocks of matter and the interactions between them is called physics. Because all things, living and nonliving, are formed of matter, basic physical laws should be at the bottom of any natural phenomenon in any discipline of science. So, undoubtedly, physics is the basis of all sciences and engineering. It should not come as a surprise that the idea of nanotechnology first came from a physicist, Richard Feynman,

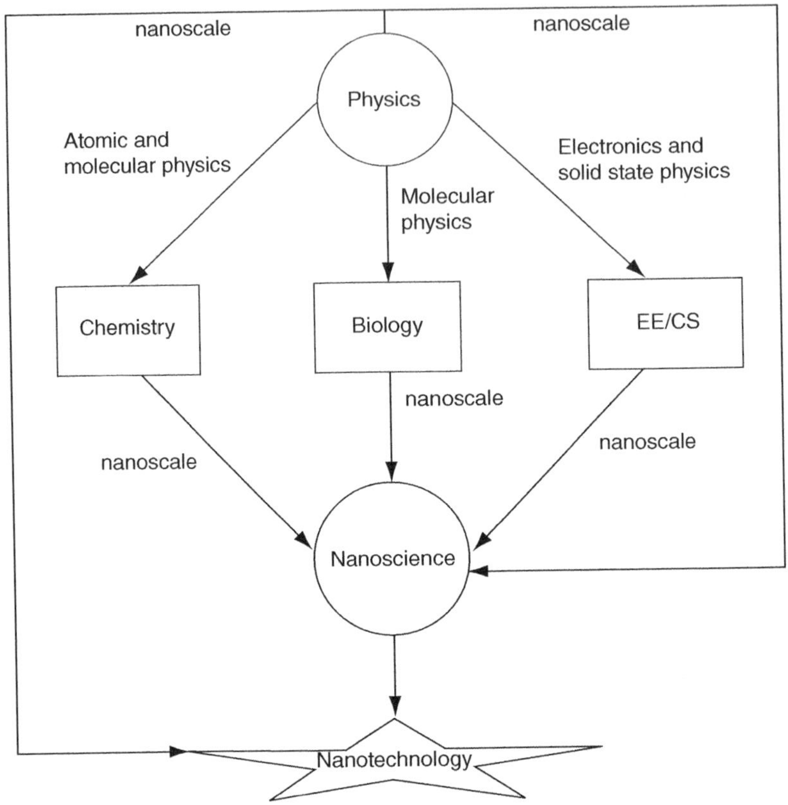

Figure 11.1. A schematic diagram illustrating the relationships between sciences behind nanoscience and nanotechnology. Electronic engeneering-EE and computer science-CS.

when on December 29, 1959, he gave a talk at the meeting of American Physical Society at California Institute of Technology titled: *There Is Plenty of Room at the Bottom.* This event is now considered as the beginning of nanotechnology.

The relationships between physics, some other sciences behind nanoscience, and nanoscience itself are illustrated in Figure 11.1.

The three common underlying characteristics that unify some aspects of the different sciences into nanoscience are the following:

1. They all follow the scientific method.
2. Studies in the different sciences that are encompassed by nanoscience are all on the nanoscale level.
3. Here is how quantum mechanics comes into the picture: Generally speaking, the laws of quantum physics apply to all these studies in different sciences.

Already popular in research labs, electron microscopes and atomic force microscopes, already discussed in this book, themselves are an example of the use of quantum mechanics in nanotechnology. These microscopes are helping us to see nanostructures.

Richard Feynman (1918–1988)
Courtesy of AIP

Richard Phillips Feynman, one of the outstanding theoretical physicists and great minds of twentieth century, was born on May 11, 1918 in Far Rockaway, a suburb of New York City. He studied at Massachusetts Institute of Technology and Princeton University, and worked at Cornell University and California Institute of Technology. After receiving his Ph.D. in physics, Feynman worked at the Manhattan Project to develop atomic bomb during the Second World War.

Feynman made significant contributions to theoretical particle physics, especially to quantum electrodynamics (QED), the quantum version of the electromagnetic (EM) theory, that describes the interaction between two charged particles, such as electrons and protons. He shared the 1965 Nobel Prize in physics with Julian Schwinger and Sin-ItiroTomonaga for his work in QED. This work had profound consequences for particle physics.

His lectures and books, such as *The Character of Physical Law*, has helped greatly to popularize physics and has also worked as a great source to learn about how physics thinking is done. His *Lectures on Physics*, in three volumes, has contributed greatly to physics teaching since their publication in 1963. Feynman was noted not only for his grasp of multiple subfields of physics, but also for his vision of the application of physics outside the discipline of physics. He is known to introduce the concept of nanotechnology.

Feynman developed two rare forms of cancer, Liposarcoma and Waldenström macroglobulinemia. He died in 1988 shortly after a final attempt at surgery for Liposarcoma. Feynman's last recorded words are noted as:

"I'd hate to die twice. It's so boring."

STUDY CHECKPOINTS 11.1

Explain how electron microscopes and atomic force microscopes, introduced in Section 3.8, themselves are an example of the use of quantum mechanics in nanotechnology.

Solution:

These microscopes generally use the quantum mechanical principles, such as wave particle duality and quantum tunneling, for their functioning.

11.2.2 You Need to See Them before You Could Control Them

In order to control anything on the nanoscale level, you must first be able to "see" them. However, as shown in Problem 11.1, the smallest objects you can see with even the most powerful light microscope are ~200 nm in size. In other words, optical microscopes using the visible light are not powerful enough to help you see the nanostructures, such as molecules. The reason is simple: wavelengths at the frequency of visible light are too large to resolve nanoscale structures.

PROBLEM 11.1

Estimate the smallest size of an object that you can see separately from its surrounding by using the most powerful optical microscope?

Solution:

The wavelength of visible (white) light puts the lowest limit on the size of the object you can see even with the use of an optical microscope. Visible light that we use to see things with naked eyes has a wavelength in the range of 380–750 nm, that is, the smallest wavelength of visible light is 380 nm. Therefore, the minimum size of two small objects that you can distinguish from each other is equal to 380/2 = 190 nm, that is, ~200 nm.

Therefore, to observe nanostructures, you need to use electron microscopes, which use a beam of higher energy particles, such as electrons, instead of a beam of visible light photons, to generate an image of the target specimen. Electron microscopes are based on the wave–particle duality of electrons. From the quantum physics discussed so far, you have learned that an electron can behave like a wave with wavelength given by the following formula:

$$\lambda = \frac{h}{mv} \tag{11.1}$$

As demonstrated in Problem 11.5, the wavelength of an electron going through the electric potential V is given by

$$\lambda = \frac{1.23}{\sqrt{V}} \tag{11.2}$$

For example, for $V = 60\,\text{kV}$

$$\lambda = \frac{1.23}{\sqrt{V}} = \frac{1.23}{\sqrt{60\,\text{kV}}} = 0.005\,\text{nm}$$

Compare this wavelength of 0.005–200 nm to that of photons, the maximum theoretical resolution of an optical microscope, and you will realize that electrons can probe much deeper and with greater resolution into the structure of matter. However, as Problem 11.6 suggests, the electrons speed is getting closer to the speed of light, so the relativistic quantum mechanical effects will begin to become significant. There are various types of electron microscopes: Those in which electrons pass through a material, and those in which electrons do not penetrate the material, but rather scatter off the surface. For example, a transmission electron microscope (TEM) is an electron microscope that uses the transmission of electrons through the target material to take the image of the target specimen. There are several ways an electron beam can be used to generate these images, which give rise to different kinds of electron microscopes. The basic idea is that the electrons transmitting through the specimen will be scattered by the constituent particles of the specimen, and the scattering will expose the internal pattern (or nanostructure) of the specimen, which will be captured in an image. The imaging process used in a TEM is described in the following:

1. A source, such as an electron gun, generates a beam of electrons.
2. By using an EM field, the electron beam is accelerated and focused to the target specimen.
3. The incident electrons are scattered by the internal structure of the specimen as they transmit through the specimen.
4. The scattered electrons, after they transmit through the specimen, are focused by objective lens and amplified by magnifying lens to generate an enlarged image of the specimen's structure.
5. The electrons strike a phosphor screen, which generates light. This light in turn enables a user to see the image.

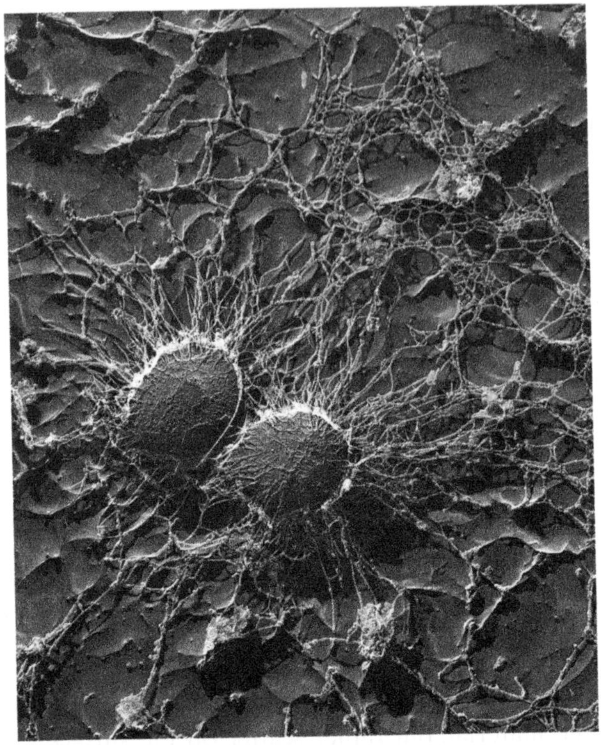

Figure 11.2. An image of bacterial cells of *Staphylococcus aureus* captured by a TEM. Image: coutesy of Agriculture Research Service, the research agency of the United States Department of Agriculture. This image was taken at 50,000X magnification.

As an example, Figure 11.2 presents an image of bacterial cells of *S. aureus*, a spherical bacterium, which is one of the causal agents of mastitis, inflammation of breast tissue in dairy cows, captured by a TEM. Its large capsule protects the organism from attack by the cow's immunological defenses.

The images generated by the microscope can be used to study the structure of the specimen. For example, the darker area of the image represents a thicker (denser) part of the specimen that did not allow many electrons to pass through it. In contrast, a lighter area represents a thinner (or less dense) part of the specimen that allowed many electrons to pass through it.

STUDY CHECKPOINT 11.2

By using an electron microscope called a scanning electron microscope (SEM), you have scanned the outer layer of the surface of a molecule. Can you reconstruct the three-dimensional (3D) image of the molecule by analyzing the scanning data?

Solution:

The answer is no, because scanning just one layer on the surface gives only two-dimensional (2D) information. However, inner layers can be scanned with SEM by using a technique called etching. This way, SEM is able to generate a 3D image.

As mentioned earlier, to understand the resolving power of an electron microscope, you need to recall that electrons can behave like both particles and waves, due to the wave–particle duality discussed in Chapter 3. Also from Chapter 3, you know that the wavelength, λ, of an electron moving with energy E is given by the following equation:

$$\lambda = \frac{h}{p} = \frac{h}{\sqrt{2mE}} \tag{11.3}$$

Note that the wavelength of an electron in the electron beam is inversely proportional to its energy, which means the higher the energy, the lower is the wavelength, and hence the higher is the resolving power of the microscope. To obtain the desired energy, E, of the incident electron beam in the TEM, the accelerating voltage can be adjusted in the range of 100 kV–1 MV, where V is voltage in units of volts (V). The relationship between V and E is given by the following equation:

$$E = \mathrm{e}V \tag{11.4}$$

where e is the electron charge.

It turns out that the resolution of the TEM is given by

$$R = AC_s^{1/4}\lambda^{3/4} \tag{11.5}$$

where A is a constant that depends on the operating conditions and C_s is the spherical aberration coefficient of the microscope objective lens. The constant A generally varies between 0.43 and 0.70 and C_s can be in the millions. For an example, see Problem 11.2.

PROBLEM 11.2

The incident electron beam in a TEM has been accelerated by applying a voltage of 200 kV. What is the resolution of the TEM? Assume that $A = 0.50$ and $C_s = 3.18 \times 10^6$.

Solution:

We can use the following equation to calculate wavelength:

$$\begin{aligned}\lambda &= (1.226/V^{1/2})\ \text{nm}\\ &= (1.226/200{,}000^{1/2})\ \text{nm} = (1.226/447.214)\ \text{nm}\\ &= 0.0027\ \text{nm}\end{aligned}$$

The TEM resolution is given by

$$\begin{aligned}R &= A\, C_s^{1/4} \lambda^{3/4}\\ &= 0.50(3{,}180{,}000)^{1/4}(0.0027)^{3/4}\ \text{nm}\\ &= 0.50 \times 42.23 \times 0.012\ \text{nm}\\ &= 0.25\ \text{nm}\end{aligned}$$

Because the electrons in the TEM penetrate the specimen, there is enough information to generate 3D images. A practical limit on the resolution of a TEM is imposed by the maximum energy of the electron beam that can be transmitted through a material without damaging the material. Taking this limit into account, the typical resolution of a TEM turns out to be 0.1–0.2 nm, a typical separation between two atoms in a solid material. This means a typical TEM can see an object that is 1000 times smaller than the smallest object an optical microscope can see and 500,000 smaller than the smallest object a naked human eye can see: a wonder of understanding and using quantum mechanics.

Electron microscopes can be used to determine the following properties of the target material:

- Properties, such as conductivity, strength, and melting point.
- Size and shape.
- Structure: arrangement of atoms and molecules.
- Texture and pattern.

Electron transmission spectroscopy uses the transmission of electrons through the material. There are techniques that use electrons to scan the surface of the material rather than transmitting the electrons through it. These techniques are called scanning microscopy. Electron microscopes use quantum mechanical effects, such as wave–particle duality and quantum tunneling. Another aspect of quantum mechanics at work in nanotechnology is quantum confinement.

11.3 NANOSCALE QUANTUM CONFINEMENT OF MATTER

You have learned about quantum confinement in this book. An electron in an atom is an example of a confined particle. The discrete piece of matter confined to the nanoscale level is called a nanoparticle or a nanostructure. Furthermore, these structures can also be used to confine other particles. A nanostructure is a structure that has at least one of the three spatial dimensions on the nanoscale (1–100 nm). There are already a wide spectrum of nanostructures under study and used in the field of nanotechnology. Obviously, the behavior of these structures is governed by quantum mechanics under most conditions. In this section, we give bird's eye view of a small subset of these structures, such as buckyballs, carbon nanotubes, nanocrystals, and quantum dots.

11.3.1 Buckyballs

A buckyball is a molecule composed of 60 carbon atoms (C_{60}) in the form of a hollow sphere ~1 nm in diameter. Buckyballs are an example of fullerenes, which are a type of carbon that consists of C_{60} and related molecules, a third form of solid carbon in addition to diamond and graphite. As illustrated in Figure 11.3, each carbon atom in a buckyball is bonded to three neighboring carbon atoms. The soccerball shaped buckyball molecule consists of 20 six-membered (hexagonal) rings and 12 five-membered carbon atom rings. These ball-like molecules bind with each other in the material (solid state) to form a crystal lattice with the centers of two adjacent molecules ~1 nm apart. These molecules are held together by a weak intermolecular force called the van der Walls force, a type of EM force.

The following characteristics of a buckyball make it attractive for nanotechnology applications:

1. Each carbon atom in a buckyball has an unused valence electron because it uses only three out of four available valence electrons to bond to the three neighboring carbon atoms.
2. Buckyballs are very strong and hard.
3. Their structure is large, they yet are small in size: ~1 nm in diameter.

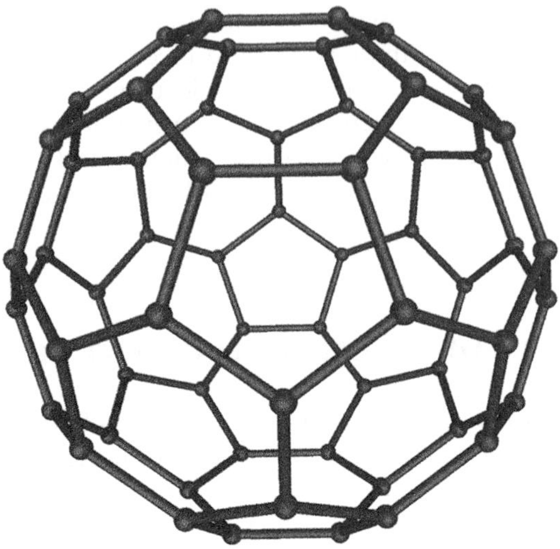

Figure 11.3. A 3D model of a C_{60} molecule called a buckyball. Courtesy of Michael Ströck.

4. Each buckyball has lots of empty space inside it that leaves room to trap other atoms to make useful molecules or simply to transport other molecules.

Because of these characteristics, several uses of buckyballs in nanotechnology are being explored including the following:

- A possible use in making armor due to their strength and hardness.
- Potential medical uses due to their small size and ability to trap other atoms include the following:
  - Use buckyballs as antioxidants to neutralize a free radical in the human body. A free electron from the buckyball makes a bond with the unpaired electron from the free radical, which can prevent certain diseases.
  - Bind certain antibiotics to buckyballs (making use of the space inside the ball), and target undesired cells in the body, such as cancer cells.
  - Use buckyballs to deliver drugs directly to the infected part of the body.
- If you dope metal atoms inside a buckyball, the buckyball can be used for medical imaging.
- Buckyballs can also be used to facilitate superconductivity, which is a state of matter in which the resistance to the electric current becomes zero. Due to the unused valence electrons inside the atoms of the molecule,

buckyballs offer some conductivity, which can be increased by adding metallic atoms to the buckyball.

Note: A free radical, discussed in the Chapter 7 opening section, is a molecule that has an unpaired electron, which makes this molecule very reactive. An antioxidant molecule can supply an electron to neutralize a free radical molecule. A certain level of free radicals are necessary for life because they play an important role in some necessary biological processes. However, a balance of antioxidants and free radicals is also necessary. This balance is disturbed by age, which results in a high level of free radicals in your body that may cause certain diseases.

A buckyball is certainly a quantum mechanical object and to take full advantage of it we should learn to handle it at the quantum level. For example, a group of experimental physicists, Arndt et al. at the University of Vienna has already confirmed the wave–particle duality of buckyballs, C_{60}. The wave nature of buckyball molecules was confirmed by observing the de Broglie wave interference of these molecules by diffraction at a material absorption grating. A material absorption grating is a platform that facilitates absorption and diffraction of material entitles, such as molecules and atoms. The observation of wave behavior in this relatively massive and complex object, with an atomic mass of 720 amu and a diameter of ~1 nm, is very significant. The de Broglie wavelength was found to be 2.5 pm. As Problem 11.10 shows, this corresponds to the buckyball speed of ~200 m/s, which is much less than the speed of light and hence nonrelativistic quantum mechanics considered in this book is justifiable for such structures.

Due to its many excited internal degrees of freedom and their possible couplings to their surrounding environment that would result in decoherence, a buckyball was considered by many to be an almost a classical object. This should ring the warning bell for molecular biologists, because it indicates that many biological molecules and their reactions, most likely also have quantum mechanical effects.

Buckyballs (C_{60}) are only one of several fullerenes that have been discovered, such as C_{70}, C_{80}, and C_{84}. These structures were named after Richard Buckminster Fuller, an American architect, author, designer, inventor, poet, and visionary, who designed the geodesic dome that resembles the structure of C_{60}, also called buckminsterfullerene, the first fullerene discovered by Robert Curl, Harold Kroto, and Richard Smalley. Generally speaking, a fullerene is any molecule composed entirely of carbon, in the form of a hollow sphere, ellipsoid, or tube. Also note that in both buckyballs and graphite, a carbon atom bonds with three neighboring carbon atoms. The same is true for what is called a carbon nanotube, or simply a nanotube. So, conceptually you wrap a graphite sheet into different shapes to make buckyballs and nanotubes.

Richard Smalley (1943–2005)
Courtesy of Rice University.

Richard Smalley, a physicist, chemist, and a nanotechnologist, was born on June 6, 1943 in Akron, Ohio. He attended Hope College in Holland Michigan before transferring to the University of Michigan, where he received his B.S. in chemistry in 1965. Between his studies, Smalley worked in industry as a research chemist with Shell, which helped develop his unique professionalism, managerial style, and practical intellectualism. He received his Ph.D. in Chemistry in 1973 from Princeton University. After postdoctoral work at the University of Chicago, Smalley began his teaching career at Rice University, Houston, Texas, in 1976. Here he became a professor of chemistry in 1982 and also a professor of physics in 1990. He was one of the founders of the Rice Quantum Institute in 1979, and served as the Chairman of this interdisciplinary Institute from 1986 to 1996.

In 1985, while investigating the constituents of astronomic dark matter (undetected matter that is believed to exist), Smalley, Curl, Kroto, and their students discovered the buckyball molecule (C_{60}), for which they were awarded the 1996 Nobel Prize in chemistry. After the discovery of the buckyball, Smalley's research focus turned to carbon nanotubes.

In the later years of his career, Smalley became very passionate about the importance of clean energy and education, believing that many of humanity's other pressing problems, such as poverty and shortage of food supply, would be much easier to solve by making affordable, clean energy available to all. In addition to researching paths to abundant, clean energy, Smalley also devoted time to educating politicians and other world leaders on the need for and a solution to the energy problem.

On October 28, 2005, Smalley lost a 6-year battle with chronic lymphocytic leukemia, and died at a rather young age of 62.

11.3.2 Carbon Nanotubes

A carbon nanotube is a one-atom thick sheet of graphite with a seamless cylindrical shape having a diameter on the order of 1 nm. Imagine you are a nanobeing living in the nanoworld. You pick up a graphite sheet in your hands and roll it up into a cylindrical shape. Congratulations! You have a nanotube. However, making nanotubes in the macroworld that we live in is not that trivial, as you will see later on in this section. To appreciate the uses of nanotubes, you need to understand their structure and strength.

Nanotubes are essentially one-dimensional (ID) structures with a length/diameter ratio exceeding 10,000. A nanotube that can be conceptually viewed as one graphite layer rolled into a cylindrical shape is called a single-walled carbon nanotube (SWCNT). A multiwall carbon nanotube (MWCNT) can conceptually be viewed as a coaxial assembly of cylinders of SWCNTs, one within the other. The separation between two neighboring tubes in a MWCNT is about the same as that between two neighboring layers in a natural graphite, that is, ~0.34 nm. As shown in Figure 11.4, the interior of a carbon nanotube is hollow, but can be loaded with a wide variety of molecules.

The strength of a material is its ability to withstand pressure or stress. Carbon nanotubes are the strongest and stiffest material known. They are about 20 times stronger than steel. Due to their amazing properties, such as their extraordinarily large strength and unique dimensionality and

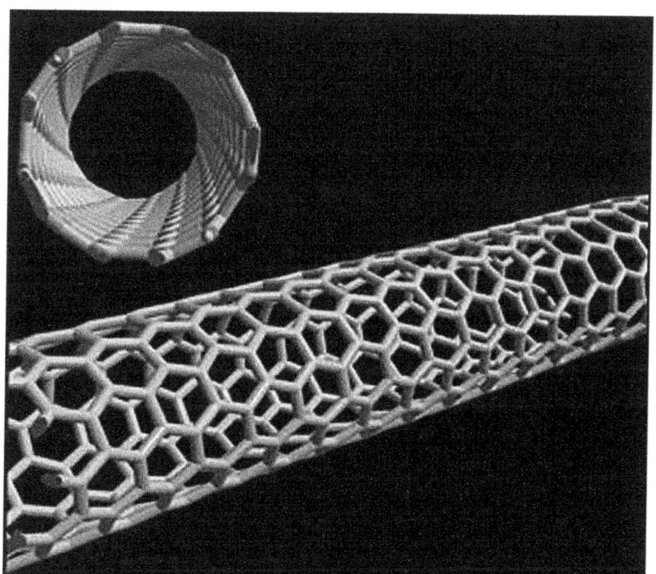

Figure 11.4. An example of SWCMT. Courtesy of R. Bruce Weisman, Ph.D., Rice University; National Cancer Institute.

conductivity, carbon nanotubes have enormous uses such as building bigger structures by using carbon nanotubes as building blocks. Some examples of these structures that can be built from nanotubes are computer chips, fibers, and machines, such as nanoelectromechanical systems and control tools.

The strength and flexibility offered by carbon nanotubes makes them potential tools for controlling other nanoscale structures. This means they will have a significant role to play in nanotechnology engineering.

So, nanotubes are, like buckyballs, carbon-based nanostructures. Another class of nanostructures that is of much interest is based on crystals, which are called nanocrystals.

11.3.3 Nanocrystals

A nanocrystal is a crystalline material with dimensions on the order of 10 nm. In other words, it is a nanoparticle with a crystalline structure. Nanocrystals are typically larger than molecules and smaller than bulk matter. Nanotechnology allows scientists to control the size and structure of nanocrystals, and therefore manipulate them in the following ways:

- Tune how a nanocrystal conducts charge.
- Change properties, such as their melting point.
- Tune how a nanocrystal displays (decipher) its crystalline structure.

> *Note*: A crystal is a solid in which the constituent atoms, molecules, or ions are packed in a regularly ordered and repeating pattern that extends in all three spatial dimensions. A crystal typically is a large structure on a macroscale level; An example of a crystal is a diamond. A not so romantic example is sodium chloride, NaCl (table salt), shown in Figure 11.5.

The uses of nanocrystals include the following:

- A nanocrystal can be studied to get information about the corresponding bulk (macro) material without having to deal with the grain boundaries and defects in the bulk material. A grain boundary is an interface between two grains in a crystalline material
- Nanocrystals made with minerals, such as zeolite, have many industrial application, for example, they can be used at oil refineries as filters to turn crude oil into diesel fuel.
- A layer of nanocrystals can be used to compose solar panels. Such solar panels have been found to be cheaper, more flexible, and more efficient than a large crystalline structure or a pure amorphous (noncrystalline) structure.

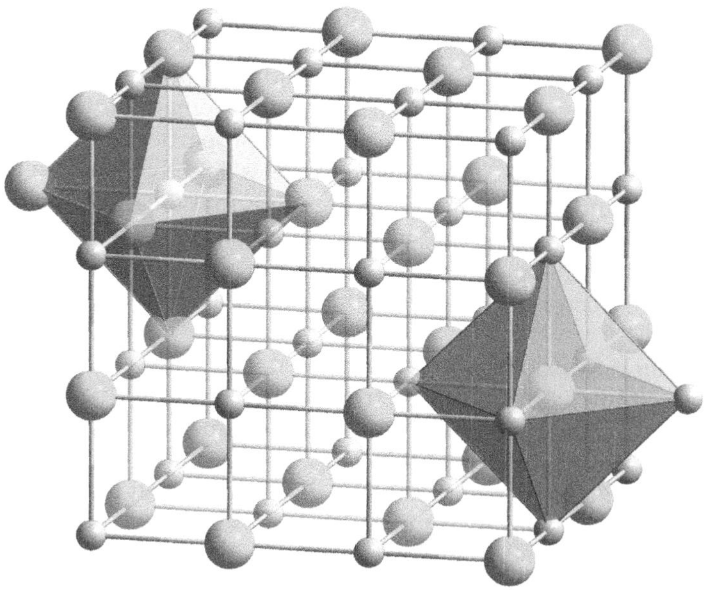

Figure 11.5. The crystal structure of sodium chloride. Each ion has six nearest neighbors, with an octahedral geometry. Courtesy of Bergamini Mills.

> *Note*: A grain, also called a crystallite, is a domain of the bulk material that has the same structure as a single crystal composing the material. A grain boundary is an interface where two grains of possibly different orientations meet.

Nanocrystals with sizes ≤10 nm are also called quantum dots. The term quantum is there to remind us that we are in the realm of quantum physics.

11.3.4 Quantum Dots

A quantum dot, also called a qdot or a nanodot, is a semiconductor nanocrystal that is ≤10 nm in size, and confines the motion of conduction band electrons in the three spatial directions. So, by definition, quantum dots are able to trap electrons. A given quantum dot contains a finite number of conduction band electrons typically in the range of 1–100. The energy levels of these electrons can be explained by using quantum physics, hence the name quantum dots.

> *Note*: A conduction band in a material is the range of energy in which an electron is free to conduct electricity through the material under the influence of an applied electric potential.

Just like an atom, a quantum dot binds some electrons in discrete energy levels, that is, it has a discrete energy spectrum. However, unlike an atom, the energy spectrum of a quantum dot can be engineered by controlling the geometrical size and shape of the nanodot, and therefore the strength of the electron confinement potential. This enables us to tune the properties of the quantum dot material. For example, the color of the material is directly related to the energy levels of the quantum dots. The rule of thumb is that the larger quantum dots have more energy levels that are more closely spaced. This allows a quantum dot to absorb photons containing less energy, that is photons closer to the red end of the EM spectrum. Conversely, the smaller a dot, the closer it will be to looking bluish.

The smaller dimensional size of quantum dots gives rise to a higher density of the states (how closely the electron energy levels are packed) than the structures with larger dimensions. This results in superior transport and optical properties. Owing to these properties, the quantum dots are being explored for uses in amplifiers, biological sensors, diode lasers, and quantum computation. Figure 11.6 presents a section of the mouse kidney visualized with the fluorescents quantum dot nanocrystals called quantum dot nanocrystal conjugates.

There are two other semiconductor nanostructures in the neighborhood of quantum dots, as described in Problem 11.3.

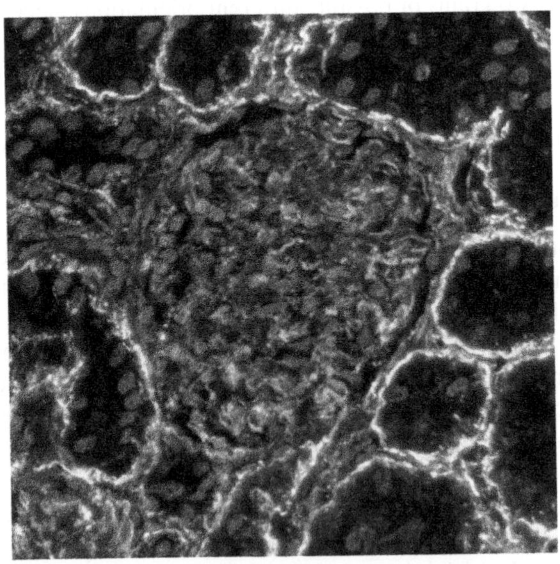

Figure 11.6. A mouse kidney section visualized with fluorescent Qdot® nanocrystal conjugates. Photo courtesy of Life Technologies, Carlsbad, CA.

PROBLEM 11.3

Calculate the number of quantum dots with 5-nm diameter that you can place back to back on the width of your thumb.

Solution:

Assume the width of your thumb, $w = 1\,\text{in.} = 2.5\,\text{cm} = 2.5 \times 10^{-2}\,\text{m}$
The diameter of a quantum dot, $d = 5\,\text{nm} = 5 \times 10^{-9}\,\text{m}$
Number of quantum dots you can place on this width $= w/d$

$$= 2.5 \times 10^{-2}\ \text{m} / 5 \times 10^{-9}\ \text{m}$$
$$= 5 \times 10^{6}$$

This equals 5 million.

Quantum Wire. This results from confinement in 2Ds. It is a 2D confinement, therefore the electrons in a quantum wire are only free to move in 1D, that is, along the length of the wire.

Quantum Dot. This results from confinement in 3Ds. Therefore an electron in a quantum dot is confined in a 3D box with the dimensions ranging from one to tens of nanometers. Recombination of carriers facilitates the emission of light with precisely defined wavelengths depending on the size of the dot. In this way, quantum dots can function as lasers in the junction layer of a diode.

The quantum mechanical confinement in spatial dimensions, called *dimensional confinement*, has a profound effect on the optical (and electrical) properties of materials. The optical properties of quantum confined nanostructures depend on the size (length) of confinement: The bandgap decreases with an increase in the dimensions of confinement. The quantum confined structures are remarkably efficient lasing media. For example, lasers supported by quantum wells hold a lion's share of the solid-state laser market. Furthermore, the optical (and electronic) properties of quantum confined structure exhibit strong dependence on the dielectric constant of the surrounding media. This offers us the opportunity to manipulate those properties for different purposes by tuning the dielectric constant.

Table 11.1 presents some examples of nanomaterials or nanostructures at the nanosize in 1D, 2D, and 3D.

Quantum mechanics governs the behavior of these nanostructures.

11.3.5 Quantum Mechanics for Nanostructures

The nanostructures discussed here are certainly in the realm of quantum mechanics. In other words, quantum mechanics is necessary to understand,

TABLE 11.1. Examples of Nanomaterials Corresponding to Their Size in Different Dimensions

Number of Dimensions <100 nm	Examples of Nanomaterials
One	Antistain coatings, applied films, and viruses
Two	Single-walled carbon nanotubes, and nanowires
Three	Microcapsules, nanoshells, and quantum dots

effectively fabricate, and use these structures. For example, electrons in nanotubes and buckyballs are under the influence of quantum confinement, as mentioned earlier. Such a quantum wire can conduct electricity in a truly metallic way, but due to its small diameter and sufficient crystalline nature, the electron in it moves as a coherent quantum wave down the wire. Due to quantum confinement, this is the only way it can move: Down the wire, no side motions are possible. In aluminum electric wires, there are so many electrons together, the quantum effects average out and the flow of electricity is pretty much a classical phenomena. However, in a quantum wire, the little quantum electron can only be there in its coherent wave packet. Now, here it gets even more interesting. Consider two nanotubes (or nanowires) next to each other. The electron in one wire can suddenly disappear and reappear in the other wire, and just keep on going, which is a type of quantum tunneling called resonate tunneling. We discussed quantum tunneling in several chapters in this book, including Section 6.4.5. To understand resonate quantum tunneling cited in this example, consider a classical analogy cited by Smalley by imagining that you are riding in one train and right next to you is an adjacent train running parallel to your train at the same speed. You blink and find yourself in the adjacent train, and you are still going in the same direction. So, electrons move along the quantum wire, extremely efficiently, and it is very hard to slow them down or stop, because the mechanisms to do that just do not exist. As mentioned earlier, they cannot be kicked to the side because there are no sides. Now, imagine a lot of wires in parallel and that the electrons in them can tunnel to the adjacent wire and keep on going without any loss. In the words of Smalley, quantum mechanics is going to save the world and rewire the grid. This quantum cable is just one of many possible applications of quantum mechanics understood and applied to nanostructures. Realize that the strength or stiffness of quantum wires enables this coherent transport of electrons at room temperature in the real world, because there are no bends in the quantum wires for an electron to localize.

You may have noticed that the quantum structures discussed so far are of some specific shapes; usually part of its name refers to the shape. An interesting science question to ask is Why are they balls, wires, and tubes as in buckyballs, quantum wires, and nanotubes, as opposed for example, to triangles and pyramids?

11.3.6 Favoring Balls and Tubes

In its spontaneous design and development, nature favors balls and tubes (shapes without edges) compared to shapes with edges. This is due to the two underlying physical principles that we have emphasized throughout this book:

1. A physical system tends to be in its equilibrium state, that is, the most stable state.
2. The most stable state is the state with minimum potential energy.

At the nanoscale level, energy at the edges matters. For example, even a tiny piece of graphite would have many atoms at its edge, which would lead to instability. Given the choice, a nanomaterial would naturally roll itself up into a ball or tube to get into a stable state by minimizing its potential energy, and hence its total energy. When you build nanostructures in the form of balls and tubes, you are taking advantage of the natural principles of stability and equilibrium. These principles are obvious to a physicist, and researchers in other fields can benefit by making them a natural component of their scientific thinking.

11.3.7 Fruits of Quantum Confinement

You know from previous chapters that confinement gives rise to discrete energy levels, that is, confined particles can only have certain discrete values of energies and the energy gaps in between these energies are not allowed. This quantum mechanical behavior has already been manipulated in many fields. For example, the fields of semiconductors, electronics, and computer industries are based on this. Before we can proceed along this line of thought, it is not out of place to review the following concepts:

- *Band.* The range of energy that an electron in a solid-state material is allowed to have or not have is a band. If this is the energy range in which an electron cannot exist, it is called a forbidden band.
- *Valence Band.* This is The highest energy band that the electrons bound to individual atoms in a solid can have is the valance band.
- *Conduction Band.* This band is a range of energy higher than the valence band. When an electron is in this range, it is free to move in the solid material under the influence of an external electric field, that is, the material can conduct electricity.
- *Band Gap.* The energy gap between the valence band and the conduction band of a solid-state material is the band gap. This is the gap in which an electron cannot exist. In other words, the electrons in the valence band must cross the bandgap to make the material electrically conductive.

- *Semiconductor.* A material that has electrical conductivity between that of a conductor and an insulator. Insulators have a larger bandgap than that of a semiconductor.

Electronics based on semiconductors are called microelectronics.

11.4 QUICK OVERVIEW OF MICROELECTRONICS

Most modern electronic devices are based on microelectronics, which is a branch of electronics that deals with very small electronic components. A multitude of components that used to be part of normal electronic design are now available in microelectronic form. Some examples are capacitors, conductors, diodes, inductors, resistors, and transistors. Microelectronic devices are generally made from semiconductors by using techniques such as lithography. With the improvement of techniques the size of these devices continues to decrease: a journey toward nanoelectronics.

11.4.1 Microelectronics: A Hindsight

Electronics, in general, is the branch of physics that deals with the flow of charge through various materials and devices, such as semiconductors and vacuum tubes. From the technical perspective, electronics is the workhorse in many fields including computing, communication, and consumer goods. The microchip, also called a chip, is the heart of the electronics age that we currently live in. It consists of an integrated circuit made of millions of tiny electronic components called transistors. There are many kinds of chips made for different purposes, such as central processing unit (CPU) chip, to execute the instructions given to the electronic machine, such as a computer and memory chip, to hold data. Modern chips and processors based on them are also called microchips and microprocessors, respectively. The advent of microprocessors opened the door to the world of small computers, such as desktop PCs and laptops. There are computers (or simply called chips) in almost any electronic device these days including handheld devices. From supporting the largest mainframe computers to hand-held computers, the microprocessor has gone through an evolution by following what is called Moore's law, which was originally based on an empirical observation made by Gordon Moore in 1965. Moore's law states that the number of transistors on a chip double about every 2 years, as depicted in Figure 11.7.

> *Note*: A semiconductor, as the name suggests, is a material that has the electrical conductivity halfway between that of a conductor, such as a metal, and an insulator, such as rubber. It can work as a conductor at high temperatures and as an insulator at low temperatures.

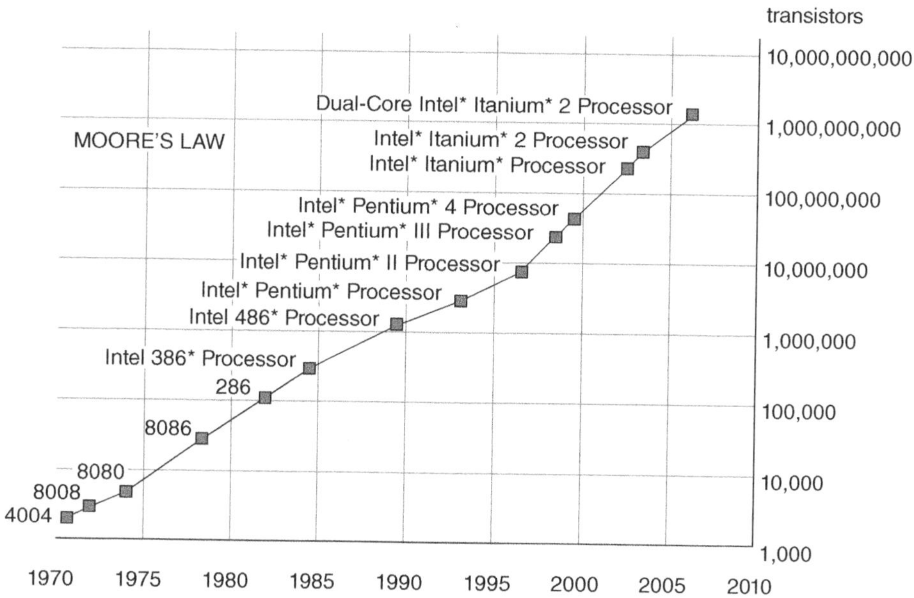

Figure 11.7. Illustration of Moore's law. Courtesy of Intel Corporation.

There are a number of implications to this law including the following:

- *Power.* As the number of transistors on a chip is increased, the power or performance of the chip (or the microprocessor) will also increase.
- *Size.* In order for the increased number of transistors to fit on the same chip, their size has to be decreased. In general, the sizes of individual electronic components offering different feature decrease as a result of an increase in the number of transistors.
- *Cost.* As the number of transistors on a chip increases, the cost per transistor reduces. That means more power (performance) with less money.

A chip consists of an integrated circuit that contains millions of transistors. What is a transistor, anyway? A transistor is a device that acts as a current amplifier or an electrically controlled switch. On a chip, a transistor acts like a switch that can store one bit of information. For example, the switch state "on" corresponds to binary bit 1, and the switch state "off" corresponds to binary bit 0. As the size of a transistor decreases the number of electrons available in the transistor also decreases. As you keep decreasing the size, you run into the hard limit of one electron needed to store one bit. The other problem where Moore's law runs into trouble is as follows: as the size passes the limit of 0.1 μm (i.e., 100 nm), you enter the quantum regime where different laws of physics are at work. This is where nanotechnology comes to the rescue in the form of nanoelectronics.

As already mentioned, there are different kinds of chips, such as processor and memory chips. Let us understand a very basic point of how the information is stored on a chip. The information on computers is stored in binary language, which has only two alphabets called binary bits: 1 and 0. Think of a transistor on a chip as a switch, which at a given time can be in one of the two possible states: on or off. The switch-on state can represent 1 and the switch-off state can represent 0. In the American Standard Code for Information Interchange (ASCII) standard, a character on your computer keyboard, such as A or 3 takes 8 bits, called a byte, to store. Each of the eight transistors will represent a bit by being in an on or off state. For example, the character K is represented by the following binary number: 01001011. It takes eight transistors to store this number. The first transistor is off and represents 0, the second transistor is on and represents 1, and so on. In ASCII, this number is a byte. So, when somebody says that a computer has 1 GB (gigabyte) of memory, it means it can store 1 billion bytes, that is, equivalent to 1 billion characters of information.

> *Note*: A transistor can be put into on and off states by changing the applied voltage across it. For example, the current flowing through the transistor may represent the on state, and no current may represent an off state.

The transistors form the integrated circuit on which a microchip is based.

11.4.2 Basics of Microchips

To understand how a microchip works, you need to understand some related concepts discussed in this section.

Insulator. An insulator is a material that does not allow the flow of charge (electric current) because it has no free electric charge, such as free electrons. In a good insulator, such as glass or quartz, almost all electrons are bound to the nuclei. When a voltage is applied across an insulator, no electric current flows through it. So, insulators are very poor conductors.

Conductor. A conductor is a material that allows the flow of charge (electric current) through it because it has free charge, such as free electrons. For example, each atom of iron (Fe) has 26 electrons and each nucleus of an Fe atom has 26 protons. This makes Fe electrically neutral as a whole. However, 1–3 electrons in an iron atom are free to move. That means when a voltage is applied across an Fe material, such as a wire, the electric current will flow through it. All metals are good conductors, but copper (Cu) and silver (Ag) are the best.

Semiconductor. A semiconductor is a material with conductivity in between that of an insulator and a conductor; that is, the conductivity of a semiconductor is better than that of an insulator and worse than that of a conductor. Silicon (Si) is commonly used to commercially create semiconductors. Modern electronic components are made of semiconductor materials. The conductivity can be switched on and off.

Integrated Circuit. An integrated circuit is an electronic circuit made of tiny components called transistors. A chip is based on an integrated circuit.

Transistor. A transistor is a semiconductor device that can act as an amplifier or a switch in an electric circuit. It is a basic building block of the circuitry used in modern electronic machines, such as computers and cellular phones. Transistors can be used (packaged) individually or as parts of an integrated circuit, which can consist of more than a billion of them.

Binary Algebra. Digital electronics and computer chips use the logic of binary (also called boolean) algebra, which is based on two digits: 1 and 0. Here are some basic terms used in binary algebra that you need to know:

- *Bit.* A bit is a binary digit that can take one of two possible values 1 or 0.
- *Binary Number.* A binary number is a number written in the binary numeral system.
- *Binary Numeral System.* A binary numeral system is a base-2 number system in which numbers are written in terms of bits, that is, 1 and 0.

We are used to the decimal numeral system, which is a base-10 number system. For example, two hundred and fifty six will be written as: 256, which is equivalent to:

$$2 \times 10^2 + 5 \times 10^1 + 6 \times 10^0 = 200 + 50 + 6 = 256$$

The general process to obtain the value from a number written in a combination of digits is

1. Find the value of each digit in the number, which is given by $C_i \times 10^i$ where i is the position of the digit in the number counted from the right with the position of the first digit being 0. For example, the value if i is 0 for 6, 1 for 5, and 2 for 2 in the number 256. C_i is the digit itself.
2. Add the values of all the digits in the number.

For example the number 1026 can be written as:

$$1 \times 10^3 + 0 \times 10^2 + 2 \times 10^1 + 6 \times 10^0 = 1026$$

Because we know that 10^0 is equal to 1 and anything multiplied by 0 is 0, we could have simply written it as:

$$1 \times 10^3 + 2 \times 10^1 + 6 = 1026$$

OR

$$10^3 + 2 \times 10^1 + 6 = 1026$$

because multiplying by 1 does not change anything.

Now, consider an example of a number written in binary numeral system:

1111111

Because it has base 2 instead of base 10, the value of each digit (bit in this case) will be determined by the following formula:

$$C_i \times 2^i$$

The value of C_i is either 1 or 0. If it is 0, we can simply ignore that digit in the sum because its contribution to the sum will be zero.

Following this the value of 1111111 is

$$1\times 2^7 + 1\times 2^6 + 1\times 2^5 + 1\times 2^4 + 1\times 10^3 + 1\times 2^2 + 1\times 2^1 + 1\times 2^0$$
$$= 128 + 64 + 32 + 16 + 8 + 4 + 2 + 1$$
$$= 255$$

How many numbers can we represent with a given number of bits. The answer is as many unique combinations as you can make from the given number of bits. Each unique combination will represent a unique number. If you have only one bit, you can only represent two numbers because a bit can be 1 or 0. So, two unique combinations, that is, 2^1. If you have two bits you can have four unique combinations:

11
00
10
01

Therefore, you can represent four, that is 2^2, numbers. Similarly, if you have three bits, you can make eight unique combinations out of them:

111
000
001
010
100
101
110
011

Therefore, the total number of unique combinations that can be made with 3 bits = 2^3. You can generalize this approach to make the statement that n bits can represent 2^n unique numbers.

PROBLEM 11.4

A classical chip has four bits to represent numbers.

(a) How many numbers can be represented by this chip?
(b) What is the greatest number it can represent?
(c) What is the number represented by 1011?

Solution

(a) Each of the four bits can acquire two possible values: 1 or 0.
Therefore the total number of arrangement of four bits: $2^4 = 16$
Therefore, four bits can represent 16 numbers.
(b) The largest number represented by four bits is

$$1111 = 2^3 + 2^2 + 2^1 + 2^0 = 8 + 4 + 2 + 1 = 15$$

(c)

$$1011 = 1\times 2^3 + 0\times 2^2 + 1\times 2^1 + 1\times 2^0 = 8 + 0 + 2 + 1 = 11$$

Note: Just like a micro/macrocircuit, a nanocircuit also has three elements: transistors, interconnections, and architecture. All of these elements are dealt within the nanoscale.

To summarize, a microchip is based on an integrated circuit (IC) that consists of millions of semiconductor components, such as transistors. Instead of making one semiconductor component at a time, the fabrication techniques are used to manufacture these millions of components and interconnect them on a semiconductor substrate. One such technique is called lithography.

The electrical properties of semiconductors can be controlled by controlling the bandgaps. Various quantum confined structures developed from semiconductors by manipulating the bandgaps are already in the market place. Different types of confinement used give rise to different structures, such as quantum wires, quantum dots, and quantum wells, already discussed in this chapter. How about putting the quantum properties of these particles to work in nanoelctronic machines? Now, we are talking about quantum computing.

11.5 QUANTUM COMPUTING

A quantum computer is a computer that uses quantum properties of particles to represent, store, and process data. In quantum computing, data is represented by quantum bits, called qubits, as opposed to bits in conventional computing. The famous physicist and a noble laureate Richard Feynman proposed the idea of quantum computing in 1982. Quantum computing involves controlling and manipulating atoms at individual levels. The researchers believe that a quantum computer, once built, will be exponentially faster than a classical (that is conventional) computer. This is made possible by a quantum process called quantum parallelism, which is a quantum mechanical process based on the superposition of base states of a particle that can be exploited to perform calculations in parallel.

Recall the double slit experiment discussed earlier with two slits A and B. Consider a particle (e.g., an electron or a photon) emitted by a source that arrives at a pair of slits A and B. There is a probability that the particle will pass through slit A and some probability that it will pass through slit B. The two behaviors of the moving particle reflecting these two probabilities exist in parallel. In one interpretation, it translates into saying that the particle will pass through both slits, which makes no sense in the classical world (that is why classical physics fails in the micro- or nanoworld). In other words, the particle exists in a state that is a superposition of two "classical" states. Such a state can exhibit the effects of interference between the two component states or classical realities. We already discussed an example of this in Section 5.9: How chemists have found that a benzene molecule can be represented as a resonance hybrid of two structures. Along the same lines, a hydrogen bond can be considered as a superposition of two covalent bonds. The key point here is that two or more states are active simultaneously and are in parallel to one another. This parallelism is believed to be exploited in quantum computing to perform calculations in parallel, and therefore improve performance. The key point in superposition is that if a quantum system is capable of existing in any of a number of individual states, it can also exist in any linear combination of those individual states.

As you learned earlier in this chapter, data (or memory) in classical (conventional) computers is made of bits, and each bit can hold one of two values: 1 or 0. The data in a quantum computer is made up of qubits, where a qubit can hold one of many (possibly infinite) values at the same time. The qubit may represent any observable quantity that may have at least two final discrete values and is conserved over time. For example, an electron may have a spin of +1/2 or −1/2, called up or down, respectively. The state of an electron at a given time is a combination (superposition) of up or down with some probabilities attached to each of these two discrete values, and can be represented by

$$\Psi = a\Psi_{\text{up}} + b\Psi_{\text{down}} \tag{11.6}$$

where the square of a is the probability that the electron has spin +1/2 and the square of b is the probability that it has spin –1/2. The values of a and b in Eq. 11.6 can be varied in an infinite number of ways and yet satisfy the equation. Therefore, there are an infinite number of possible states (superposition states) for the electron, and therefore an infinite number of possible values for the qubit that represents the spin of the electron in terms of the superposition of two spin states. Recall that there are only two "classical" spin states. However the quantum mechanical superposition of these two "classical" states gives rise to infinite quantum mechanical states.

There is another way of looking at the qubit in terms of bit in classical computing, where a bit represents 1 or 0. Instead, a qubit can represent 1, 0, or both simultaneously. What do we mean by "both simultaneously" ? It means a qubit can represent 1 and 0 with some probabilities attached to each representation. In other words, it is a superposition of two representations (or states). Assigning different values to these probabilities gives rise to different states and representations. As you can realize, multiple qubits in this process work simultaneously on the same computation (or problem), and this parallelism improves the performance. Remember, however, that the final result (or solution) of a computation (or measurement) will be a definite result and not a combination of probabilities. This is consistent with what you learned about quantum mechanics in Chapter 5 (e.g., Sections 5.3 and 5.8).

Tables 11.2 and 11.3 should give you a feel for the difference between a classical and a quantum computer. Table 11.2 presents examples of how three bits can be put together to represent numbers, and Table 11.3 shows how three qubits can be put together to represent numbers.

Some of the challenges facing the advancement in quantum computing include the following:

- *New Techniques.* If quantum computers were to be used to facilitate measurements, the measurement techniques must lend themselves to quantum computing. So, we will need to discover new techniques compatible with quantum computing.

TABLE 11.2. Three Bits in a Classical Computer Representing Numbers between 0 and 7

Bit 1	Bit 2	Bit 3	Number Represented	Calculation
0	0	0	0	
0	0	1	1	$2^0 = 1$
0	1	0	2	$2^1 = 2$
0	1	1	3	$2^1 + 2^0 = 3$
1	0	0	4	$2^2 = 4$
1	0	1	5	$2^2 + 2^0 = 5$
1	1	0	6	$2^2 + 2^1 = 6$
1	1	1	7	$2^2 + 2^1 + 2^0 = 7$

TABLE 11.3. Examples of Three Qubits Representing Numbers between 0 and 7

Qubit 1	Qubit 2	Qubit 3	Number Represented
Ψ_0	Ψ_0	Ψ_0	0
Ψ_0	Ψ_0	$a\Psi_0 + b\Psi_1$	0 and 1 simultaneously
Ψ_0	$a\Psi_0 + b\Psi_1$	Ψ_0	0 and 2 simultaneously
Ψ_0	Ψ_1	$a\Psi_0 + b\Psi_1$	2 and 3 simultaneously
Ψ_0	$a\Psi_0 + b\Psi_1$	$a\Psi_0 + b\Psi_1$	0, 1, 2, and 3 simultaneously
$a\Psi_0 + b\Psi_1$	Ψ_0	Ψ_0	0 and 4 simultaneously
Ψ_1	Ψ_0	$a\Psi_0 + b\Psi_1$	4 and 5 simultaneously
$a\Psi_0 + b\Psi_1$	Ψ_0	$a\Psi_0 + b\Psi_1$	0, 1, 4, and 5 simultaneously
$a\Psi_0 + b\Psi_1$	$a\Psi_0 + b\Psi_1$	Ψ_0	0, 2, 4, and 6 simultaneously
Ψ_1	Ψ_1	$a\Psi_0 + b\Psi_1$	6 and 7 simultaneously
Ψ_1	$a\Psi_0 + b\Psi_1$	$a\Psi_0 + b\Psi_1$	4, 5, 6, and 7 simultaneously
$a\Psi_0 + b\Psi_1$	$a\Psi_0 + b\Psi_1$	$a\Psi_0 + b\Psi_1$	0, 1, 2, 3, 4, 5, 6, and 7 simultaneously
Ψ_1	Ψ_1	Ψ_1	7

- *New Algorithms.* We will need new algorithms for solving problems that will take advantage of quantum computing.
- *Decoherence.* Because qubits are superpositions of all the possible states of a system, they are correlated with each other. So, if one qubit is corrupted by some external phenomenon, such as passing cosmic rays, the whole computation will become useless. This situation is called decoherence because all the qubits are not coherent with each other after something specific happens to one qubit.

> *Note:* In general, quantum decoherence is defined as the mechanism is which different states of a system seem to collapse as a result of their interaction with the external environment.

Not only nonliving material, but also living things, the organisms, are at the fundamental level composed of atoms and molecules, which also play an important role in maintaining life. Therefore quantum mechanics governs the living world as much as the nonliving world. Recognition of the importance of quantum mechanics in understanding biological systems has given rise to a new field called quantum biology.

11.6 QUANTUM BIOLOGY

Although biologists have in some way dealt with biological molecules since the time of Gregor Mendel (1822–1884), molecular biology as a field was really launched with the discovery of the DNA structure in the 1950s. Beginning in the late 1950s and early 1960s, molecular biologists learned and devised techniques to characterize, isolate, and manipulate to some extent the biological molecules including DNA and proteins. Nevertheless, molecular biology is a relatively new field and like most of biology is still largely untouched by physics, that is, until now. Obviously, the physics useful for molecular biology will be quantum physics. Not only molecular biology, but most of biology so far has managed successfully to make progress solely on the basis of empirical studies and research. In other words, biology (biotechnology included) has a large cookbook component: mix up different things and see what happens. Process certain foods or drugs through a set of bodies and not through another set of bodies, and compare the results. This kind of research is of course important, but can only go so far. Here is the good news: Recently, a wealth of evidence has emerged to underline the importance of quantum mechanics in understanding biological systems. Consequently, a new field called quantum biology is emerging.

To start with, making and breaking of chemical bonds is constantly happening inside living organisms, which are undoubtedly quantum phenomena. Here, for example, are a few other quantum mechanical phenomena occuring in living organisms: absorbance of frequency-specific radiation during photosynthesis and vision, conversion of chemical energy into mechanical motion (e.g., during adenosine triphasphate, ATP cleavage), *magnetoreception* in animals, and single electron transfers through biological polymers (e.g., DNA or proteins). Capabilities provided by nanotechnology will certainly help remove the barrier between quantum physics and biology.

> *Note*: Magnetoception refers to the ability to detect a magnetic field in order to perceive direction, altitude, or location. This ability (or sense) plays a role in the navigational abilities of several animal species and has been postulated as a method for animals to develop regional maps. For example, sensing of the Earth's magnetic field is important to the navigational abilities of birds during migration. Megnetoreception ability has also been observed in many other animals including bacteria, fruit flies, fungi, honeybees, lobsters, sharks, and turtles.

There are trillions of quantum machines in your body called cells dealing with four fundamental nanostructures of life.

11.6.1 Four Fundamental Nanostructures of Life

One of the fundamental secrets of nature is the unity behind diversity. What we see at first sight is the diversity of nonliving and living things from a unicell bacterium to humans. For example, there are at least 10^{31} microorganisms, ~3 billion species of microorganisms, and ~6 billion humans on our planet. Behind this diversity, there is an underlying unity: All living organisms, animals and plants, are made of cells. Here is another case of diversity: A bacterium is made of one cell, whereas an average human being is made of ~10 trillion cells. Now, what is the unity behind this diversity? Well, all cells are largely made up of and run by four fundamental nanostructures of life, the four types of molecules as listed in Table 11.4 and described below:

TABLE 11.4. Summary of the Molecules of Life

Category of Molecules	Components	Examples and Functions
Carbohydrates	Monosaccharide (simple sugar) monomers	**Monosaccharide** Glucose: Source of energy. **Disaccharide** Sucrose: Most common form of sugar **Polysaccharides** Cellulose: Strengthens cell walls in plants Glycogen: Stores energy in form of glucose in animals Starch: Stores energy in form of glucose in plants
Lipids	Glycerol: Composed of three fatty acids Phospholipids: Phosphate group with two fatty acids Steroids: Four fused carbon rings with chemical groups attached to them.	Triacylglycerols (fats and oils): Storage of energy Phospholipid bilayers: Key components of cell membranes Cholesterol (steroid): Components of cell membranes Hormones (steroids): Signals that travel through the body.
Nucleic Acids	Chains of monomers called nucleotides with each nucleotide compoased of a five-carbon sugar, phosphate group, and nitrogen-containing base.	DNA: Stores all hereditary information. RNA: Carries instructions to make protein from DNA to protein-making machinery
Proteins	Composed of one or more polypeptide chains; each chain consists of covalently linked amino acids	Collagen: Structural component of bones. Defensive proteins: protect against disease Enzymes: Catalyze chemical reactions. Keratin: Structural component of hair and nails. Motor proteins: Facilitate cell movement

Carbohydrates. Carbohydrates are largely the energy molecules. Carbohydrates serve mostly as a fuel, but some of them are also a building material, as they include sugars and polymers of sugar. These are the organic molecules that contain carbon, hydrogen, and oxygen in a 1:2:1 ratio. In cells, different kinds of carbohydrates are used as sources of internal energy. Table sugar and glucose are some examples.

Proteins. Proteins are the workhorses of life. Proteins support and facilitate almost all the dynamic functions of a living organism. These are large, complex molecules made up of smaller subunits called amino acids, but are not large enough to escape quantum mechanics. Chemical properties that distinguish the 20 different amino acids cause the protein chains to fold up into specific 3D structures that define their particular functions in the cell. Proteins play a variety of roles in the cell, including biochemical (enzymes), cell signaling (hormones), mechanical (muscle), structural (cytoskeleton), and transport. Enzymes, the production machines, are a special kind of proteins that catalyze biological reactions.

Protein molecules are highly structured molecules and their structure defines their functionality. For example, Figure 11.8a represents a 3D structure of a protein molecule called a myoglobin showing helices called alpha helices. This was the first type of protein molecule to have its structure revealed by using X-ray crystallography. Hemoglobin, a transport protein shown in Figure 11.8b, transports oxygen from the lungs to other parts of the body.

(a) (b)

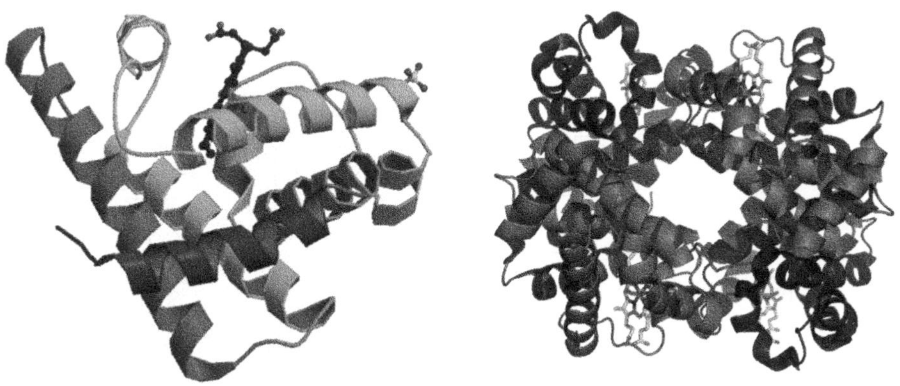

Figure 11.8. Representation of the 3D structure of protein molecules called myoglobin (a) and hemoglobin (b) at 1.6-Å resolution showing helices called alpha helices. Courtesy of the Protein data bank http://www.rcsb.org/pdb.

DNA/RNA. These information molecules carry the blueprint (genetic code) to develop the workforce: the proteins. All the instructions needed to direct the activities of a cell are contained within the molecule called DNA. The structure of the DNA molecule is called a double helix structure because it consists of two strands that wind around one another to form a shape known as a double helix. As illustrated in Figure 11.9, each of the two strands has a backbone made of alternating sugar (deoxyribose) and phosphate groups. Attached to each sugar is one of four bases: adenine (A), cytosine (C), guanine (G), and thymine (T). The two strands are held together by bonds between the bases called hydrogen bonds; adenine only bonds with thymine and vice versa, and cytosine only bonds with guanine and vice versa. The sequence of the bases along the backbones serves as instructions for assembling protein and RNA

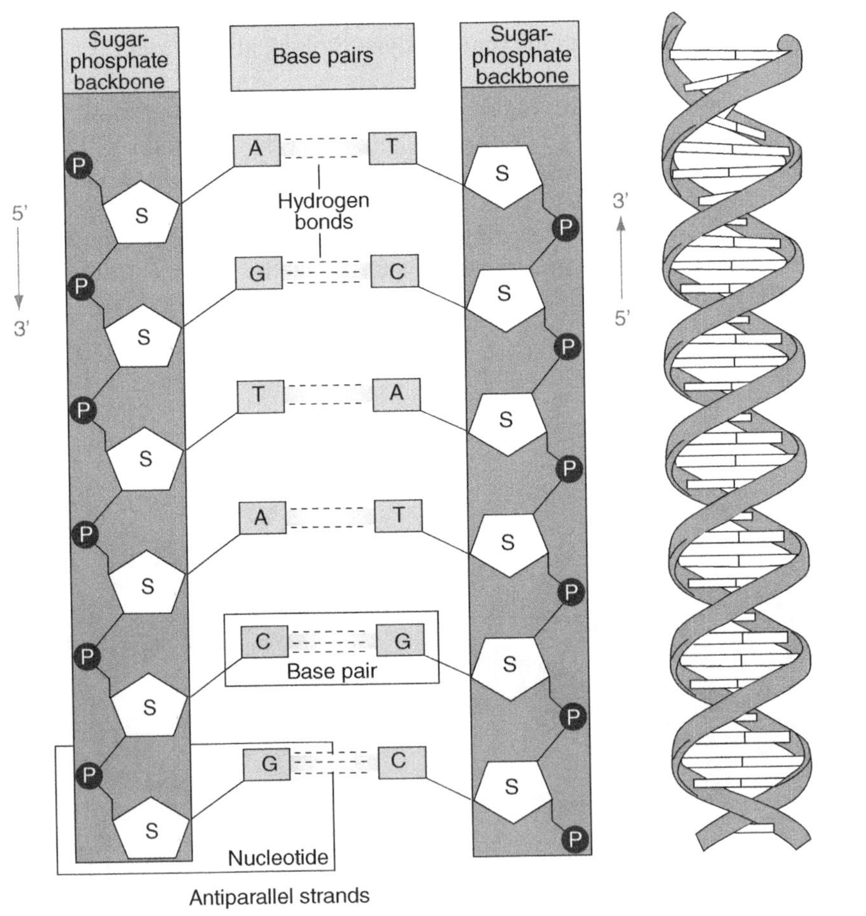

Figure 11.9. Illustration of the double helix structure of the DNA molecule that carries genetic instructions in all living things. Courtesy of the National Institute of Health.

(ribonucleic acid) molecules. The DNA sequence is the particular side-by-side arrangement of bases along the DNA strand (e.g., ATTCCGGA) built from a total of four bases: A, T, C, and G, already mentioned. This order spells out the exact instructions required to create a particular organism with its own unique traits.

Lipids. Lipids are the infrastructure molecules. Lipids form a diverse group of hydrophobic molecules, that is, not soluble in water. These molecules make fatty, oily, or waxy organic compounds. Fats and cholesterol are examples. Phospholipids, an important components of the cell membrane, is another example. These are used as basic building blocks to develop sex hormones, gender traits, cell membranes, and also to store energy.

> *Note*: In nature, only living cells make complex carbohydrates, complex lipids, proteins, and nucleic acids. Through eating, we send in the raw material to make these molecules.

These molecules make and run the cell. But, how are these molecules themselves built? Generally speaking, these molecules are polymers made of smaller building blocks called monomers through chemical reactions that link two monomers together:

$$\text{monomer} + \text{short polymer} \rightarrow \text{longer polymer} + \text{water molecule}$$

Figure 11.10 presents the unified picture of life according to molecular biology developed so far. Cells, the smallest living systems with the size on the microscale level, are the fundamental working units of every living system. What directs the cell to function the way it does? All the instructions needed to direct the activities of a cell are contained within the DNA molecule called: deoxyribonucleic acid, the building block of all life. The DNA from all organisms is made up of the same chemical and physical components. The genome is an organism's complete set of DNA. Genomes vary widely in size: The smallest known genome for a free-living organism (a bacterium) contains ~600,000 DNA base pairs (bp), while human and mouse genomes have some 3 billion. Except for mature red blood cells, all human cells contain a complete genome: A complete set of genes for the individual, that is, the entirety of the hereditary information for the individual. The DNA (that contain genes) in the human genome is arranged into 23 distinct chromosomes: physically separate molecules that range in length from ~50–250 million bp. Each chromosome contains many genes, the basic physical and functional units of heredity. Genes are specific sequences of bases that encode instructions on how to make proteins. Genes comprise only ~2% of the human genome; the remainder consists of noncoding regions, whose functions may include providing chromosomal structural integrity and regulating where, when, and in what quantity proteins

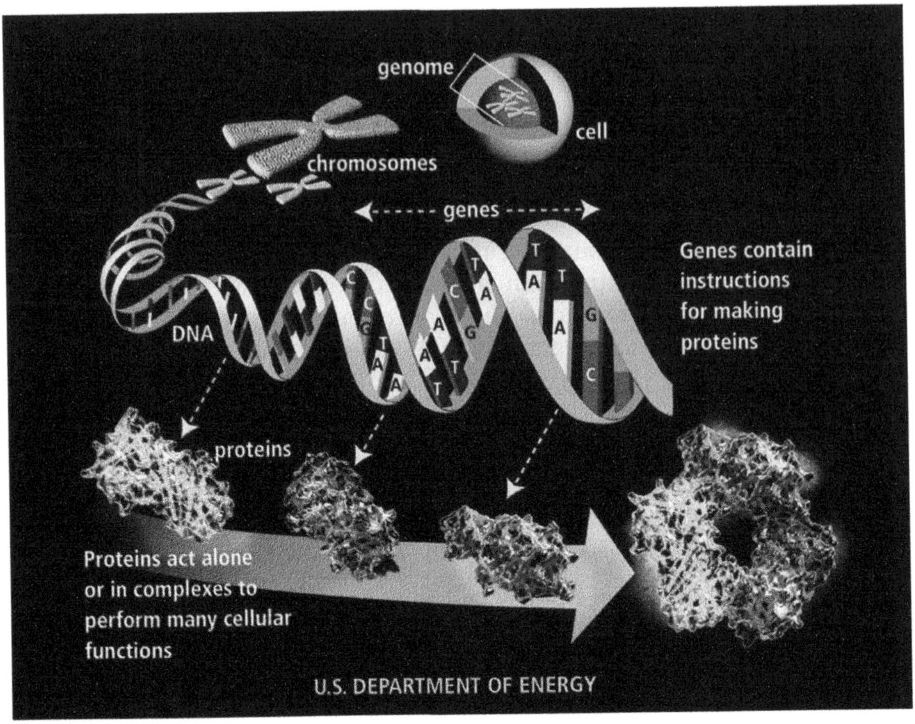

Figure 11.10. A living cell functioning as a quantum machine. Courtesy of Genome Management Information System, Oak Ridge National Laboratory.

are made. The function of a large part of the human genome is not yet well understood.

> *Note*: A chromosome is a packaging structure that carries DNA in the nucleus of a cell. Each chromosome consists of a large molecule of DNA and the proteins associated to this molecule that help in the functioning of the DNA.

The human genome is estimated to contain 25,000 genes. Although genes receive a lot of attention, it is the proteins that perform most life functions and even make up the majority of cellular structures.

> *Note*: A few types of major chromosomal abnormalities, including missing or extra copies or gross breaks and rejoinings (translocations), can be detected by microscopic examination. Most changes in DNA, however, are more subtle and require a closer analysis of the DNA molecule to find perhaps single-base differences.

The basic biology courses these days are full of stories based on empirical study and research. One of the most popular stories is called the central dogma of molecular biology.

STUDY CHECKPOINT 11.3

What is the smallest unit of life?

Solution:

Cell

11.6.2 Central Dogma of Molecular Biology

The central dogma of molecular biology explains how protein is made in a cell and is illustrated in Figure 11.11.

In the cell nucleus, RNA is produced by a process called transcription, in much the same way that DNA replicates itself. The RNA, however, substitutes the sugar ribose for deoxyribose (in DNA) and the base uracil for thymine (in DNA), and is usually single-stranded, as opposed to double-stranded DNA. One form of RNA, the messenger RNA (mRNA), takes the DNA recipe for protein synthesis to the cell cytoplasm. There it binds temporarily to a cytoplasmic particle known as a ribosome. Each of the three-base sequence units, called codons, of the mRNA link to a specific form of transfer RNA (tRNA)

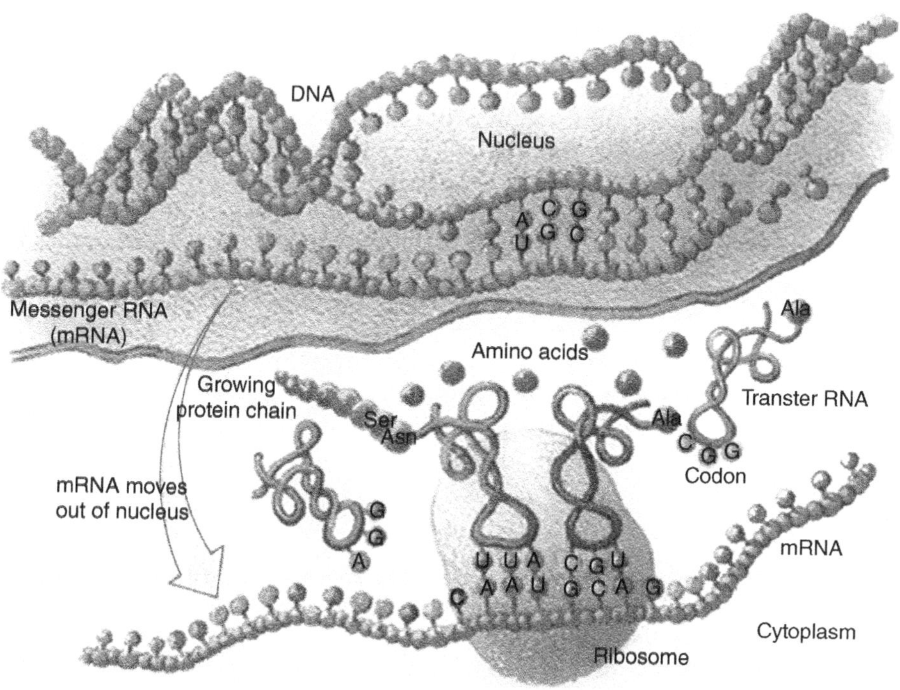

Figure 11.11. An illustration of the central dogma of molecular biology. Courtesy of the Genomone Management System, Oak Ridge National Laboratory.

containing the complementary three-base sequence. This tRNA, in turn, transfers a single amino acid to a growing protein chain. This way each codon unambiguously directs the addition of one amino acid to the growing protein chain. However, the same amino acid can be added by different codons because multiple codons code for a given amino acid. For example, the mRNA sequences GCA and GCC are both specifying the addition of the amino acid alanine, (Ala).

Currently, molecular biology is based on this central dogma, which itself is based on empirical studies and research. Understanding molecular biology in terms of the central dogma itself has already helped biotechnology. The next big step in molecular biology will be to uncover the hand of quantum mechanics behind the central dogma. This step will enable us to manipulate and control the basic processes of life in a more useful and powerful way.

As mentioned earlier, classical mechanics is a good approximation of quantum mechanics in the macroworld, that is, for objects of large size. On a smaller scale, the classical approximation breaks down and we must use quantum mechanics to understand and control objects at this scale. Therefore, to appreciate the need for quantum mechanics to study the basic building blocks of life, it is important to realize their sizes.

11.6.3 Sizes of Biological Particles

In Section 11.6.1, we reviewed the role of the basic building blocks of life, such as amino acids, proteins made of amino acids, nucleotides, DNA made of nucleotides, and cells, largely made of proteins and DNA. Most of these, such as proteins and nucleotides, are the nanoparticles and their sizes along with their molecular masses are listed in Table 11.5.

Amino acids, which are nanoparticles, compose proteins, which will obviously be larger in size than amino acids. Similarly, DNA that is composed of nucleotides, the nanoparticles, is also expected to be larger in size than the nucleotides. Both cells and some structures in the cells fall into the microscale size. The sizes of these biological microparticles are listed in Table 11.6.

STUDY CHECKPOINT 11.4

Can you see human cells, such as blood cells, through an optical microscope? How about a DNA molecule?

Solution:

Yes, you can see cells through a light microscope because their size is a few micrometers. Actually, biologist have been studying cells and bacteria, which are unicellular organisms, by using optical microscopes.

The DNA molecule has a diameter of ~2 nm. So, to inspect a DNA molecule, you will need an electron microscope.

TABLE 11.5. Typical Sizes of Various Biological Particles–Structures at the Nanoscale Level

Particle	Molecular Mass (Da) Dalton = g/mol	Size: d (nm)	Type	Comment
DNA	6,569	2.5	Nucleic acids	The molecular mass is for a DNA molecule of 20 nucleotides and the size refers to the diameter of the double helix.
Glycine	75	0.42	Amino acids	Simplest of the 20 standard amino acids.
Tryptophan	204	0.67	Amino acids	One of the 20 standard amino acids. Essential in human neutrition.
Cytosine monophosphate	309	0.81	Nucleotides	Smallest DNA nucleotide
Guanine monophosphate	361	0.86	Nucleotides	A DNA nucleotide on the larger end of size spectrum
Insulin	58,00	2.2	Proteins	A polypeptide hormone that regulates carbohydrate metabolism
Hemoglobin	68,000	7.0	Proteins	Carries oxygen in human blood
Lipoprotein	1,300,000	20	Lipids and proteins	Carry fats around the body
Fibrin	400,000	50	Proteins	Involved in clotting of blood

TABLE 11.6. Typical Sizes of Various Biological Particles–Structures on a Microscopic Scale

Particle	Size: d (μm)	Type	Comments
Chloroplasts	2–10	Organelle	Structures in the cells
Human blood platelet	1.5–3	Cells	Circulating in the blood and involved in formation of blood clots
Leukocytes	8–15	Cells	White blood cells
Human chromosome	9	DNA molecules	Constitutes an organized form of DNA in a cell

A cell is the basic unit of life in which most biochemical reactions happen, for example, those that produce nanostructures like DNA (replication) and proteins, in order to develop and maintain a living organism. So, it will not be a surprise, if scientists ultimately find that at its very core or bottom, cell is a quantum machine.

11.6.4 Diving Deeper into the Cell with Quantum Mechanics

By using quantum mechanics, we can probe deeper into the cell, in addition to studying the cell phenomenon involving the basic molecules of life. As Aristotle said: All life involves movement. This is true even when the living organism is at total rest. In that case, there is movement at the microscopic level, the movement in the cells. Where there is a movement, there is physics to explain this movement; and the physics of movement on the microscal (or nanoscale) level is quantum physics. You have already heard a micromovement story from your biology class: the central dogma of life. Here is another story of micromovement in plant cells, which you might have heard of in a biology class as well.

This is the story of photosynthesis, the mother of all processes of life to the extent that animals feed on plants to get their energy in the form of carbohydrates, and plants make carbohydrates by using photosynthesis. Photosynthesis takes place in the chloroplast in a plant cell, which is much like the mitochondria in an animal cell. The story begins with photons, particles of light discussed in this book. Photons are captured in this process to make ATP. Chloroplasts use a proton pump (a protein in a cell membrane that transports protons out of the cell), which is driven by electrons to generate a proton gradient that drives an ATPase turbine engine. These electrons, originally captured from water by the magnesium ion, need more energy to come into action. This is where photons come into the picture to help the electrons. The plant-pigment molecules capture the photons and funnel them toward a magnesium atom in the center of a chlorophyll molecule. The outermost electron in the magnesium atom absorbs a photon of sufficient energy to knock it out of the atom. The ionized magnesium atom plucks an electron from a water molecule to generate a proton (H^+) and oxygen. The electrons, that gobbled up photons, are now energized enough to drive the proton pump. Plant's enzymes direct this motion to harness this energy into an ATP molecule. Because we are talking here about the motion of quantum mechanical entities, such as protons, electrons, and pigment molecules; it is a safe bet to claim that quantum mechanics is at work in turning sunlight into food through photosynthesis. As a matter of fact, scientists have already gathered a good evidence that natural photosynthesis (in plants and also in some bacteria) gets its remarkable efficiency, unexpected from classical physics predictions, from quantum mechanical phenomena. An example is discussed in Living in the Quantum World, Section 11.7.

Note: ATP is an acronym for adenosine triphosphate, a nucleotide molecule composed of an adenine base, the five carbon sugar ribose, and three phosphate groups. This is the main energy carrier between reaction sites in cells. The ATPases are a class of enzymes that catalyze the decomposition of ATP in order to release the energy needed to drive other necessary chemical reactions in the cell and to create motion, such as driving our muscles.

In animal cells, the energy is harnessed from the chemical energy of food molecules, such as glucose (a carbohydrate), but photons are not needed to make ATP. The part of the cell where ATP formation occurs in animal cells is called the mitochondria, which takes the electrons required to drive the proton pump from a molecule called reduced nicotinamide adenine dinucleotide (NADH).

As we mentioned in Chapters 2 and 3, photons and electrons are the particles whose behavior was responsible for the discovery of quantum mechanics. If there is a physicist in the class and you tell a story, such as photosynthesis, without mentioning quantum mechanics, you will have a lot of explaining to do. If you are a biology student, you know that biology is full of stories casting (if not starring) protons, electrons, and ions, such as Na^+ and Mg^{2+}. In fact, the directed movement of fundamental particles is involved in all biological activity at the level of the living cell including digestion, metabolism, growth, photosynthesis, replication of DNA, and division, which is actually multiplication, of cells.

Therefore, the movements of these quantum mechanical (fundamental) particles in the cell, are essential for all living cells, and therefore for life. This underlines the importance of quantum mechanics in understanding the cell at its very fundamental level. Here are some more examples of how scientists are working at finding the hand quantum mechanics plays in biological systems and processes:

11.6.4.1 It Smells Quanta. Why do the things smell the way they do? The prevailing notion to explain smell has been based on classical thinking: The odorant molecules fit into specific receptors in our nostrils compatible with their size and shape; which triggers the sensation of different smell, called olfaction. This sense of smell stimulates the action potential that olfactory nerves transmit to the brain. In this approach, different smells are attributed to different structures and shapes of the odorant molecules. Of course, the structure and shape of a molecule may play an important role, but there is evidence against this being the only factor or even the decisive factor in determining smell. For example, odorant molecules with a similar structure have been found to create different smells. This suggests that it is the interaction between the odorant molecules and the receptor neurons, rather than the structure of the molecules, that is responsible for creating the sensation of a specific smell.

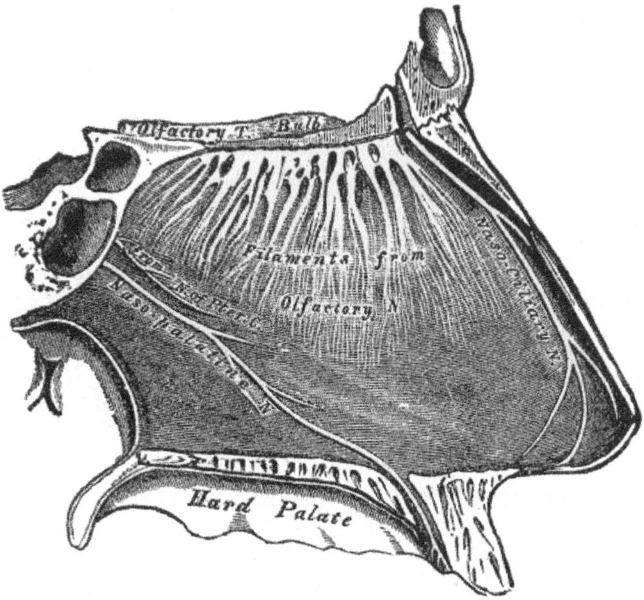

Figure 11.12. The olfactory nerve.

> *Note*: The olfactory nerve, also called cranial nerve I, illustrated in Figure 11.12, is the first of 12 cranial nerves. It is instrumental in the sense of smell. The specialized cells of the olfactory nerve called olfactory receptor neurons are located in the olfactory mucosa of the upper parts of the nasal cavity. Sense of smell is created by the interaction of odorant molecules with the receptor neurons.

For example, in 1996, a biophysicist Luca Turin, then teaching at University College London, presented a hypothesis based on quantum mechanics to explain the creation of smell. Turin's hypothesis proposes that binding of an odorant to receptors mediates electron transfer through the odorant by using quantum tunneling, which in turn causes the odorant to vibrate. This vibration pattern will be specific to a given odorant molecule and even a small change in shape or structure of the molecule will cause a significant change in the vibrational spectrum. These different patterns of vibrations cause different smell from different types of odorant molecules. Turin's hypothesis has correctly predicted the smell of some chemicals before they were synthesized.

11.6.4.2 Quantum Elimination of Radicals with Catechin. This classical example illustrates how the empirical research on which most of the today's biology is based is only a starting point in understanding processes. It has been proven, through empirical research based on injecting mice and population-based clinical studies, that green tea is effective in preventing many diseases related to free radicals. It is believed that green tea accomplishes this by playing the role of an antioxidant. A fundamental research question will be Exactly what is the mechanism through which green tea accomplishes this task? We have already discussed in this chapter that free radicals are molecules in our body, for example, byproducts of the breakdown of food by metabolism, that have an unpaired electron that makes the molecule very reactive. These molecules travel through the blood stream and cause damage due to their reactive nature. It has been proposed by biochemists that an electron from the catechin molecule can quantum mechanically tunnel through the gap to get to the free radical and unarm it by pairing with the unpaired electron. Green tea along with some other products, such as red wine and some fruits and vegetables, contain the catechin molecules or compounds.

11.6.4.3 Quantum Hope for the Hopeless Enzymes. Enzymes are the types of proteins that catalyze the biochemical reactions in the cells of our body. Enzymes need to have certain energy in order to support a given reaction. It has been found that some enzymes that have less energy than expected from classical physics to complete the reactions, were able to do their job anyway. This can only be explained through quantum phenomena, such as wave–particle duality and quantum tunneling.

11.6.4.4 Brain, a Quantum Machine? It is often said that there are as many neurons (specialized cells) in our brain as there are stars in our galaxy, called the Milky Way. We are still at the very beginning of understanding how this 3 lb of matter that defines who we are actually works. For example, how does consciousness work? Classical mechanics has been unable to explain consciousness. Various hypotheses based on quantum mechanics have been proposed to explain consciousness. For instance, a convincing hypothesis was proposed by Stuart Hameroff, an anesthesiologist at the University of Arizona at Tucson based on the dynamics of microtubules. Microtubules, illustrated in Figure 11.13, are long hollow cylinders with a diameter of ~25 nm and a length varying from 200 nm to 25 μm. Microtubules, components of the cytoskeletons of eukaryotic cells, form a dynamic framework for many cell activities by rapidly assembling when required and disassembling when not. Microtubules are involved in many cellular processes including cytokinesis, mitosis, and vesicular transport.

Each neuron contains hundreds of microtubules. Free electrons in a microtubule play an important role in the dynamics of the microtubule. According to classical mechanics when the electrons go to one side of the microtubule, they squash that side to cause the other side to elongate, and when they get to the other side, the process is reversed. However, according to quantum mechanics, as Hamroff proposed, electrons can be on both sides of the tube simultaneously causing it squashed and elongated at the same time giving rise

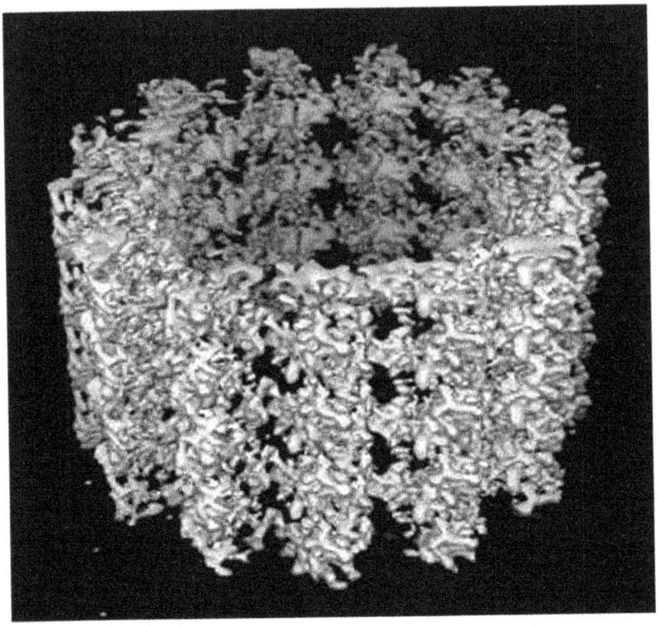

Figure 11.13. Space-filling model of a microtubule segment derived from cryoelectron microscopy. Courtesy of the U.S. Department of Energy.

to sophisticated quantum mechanical dance out of which consciousness is born. Anesthetics, a drug that turns consciousness off, interacts with the quantum dance of the electrons as an environment (observer) and causes decoherence or the collapse of the wave function, as discussed in Chapter 5; it is also called quantum entanglement. Hence, the quantum process is stopped and as a result consciousness is turned off while the brain still remains active. Such hypotheses, although making sense, still need to be tested; although it is a good beginning.

I can give more examples, but you get the point: A new field, call it quantum biology, is in the making.

In a nutshell, from the tunneling effect to entanglement, the unique features of quantum mechanics allow events to unfold at speeds and efficiencies that would be unachievable by applying laws of classical physics alone. For years, experts in biology have doubted that quantum mechanics could be driving some of the most elegant and inexplicable processes of life. Yet experiments keep finding quantum effects at play in biological systems and processes. Here is the key to generating new ideas in biology: There are many things that atoms and molecules can do, which cannot be understood in terms of what living organisms do; whereas there is nothing that living organisms do that cannot be understood in terms of what atoms and molecules can do. Quantum physics is the science to understand what atoms and molecules do.

11.7 EXPLORING THE INTERFACE OF CLASSICAL MECHANICS AND QUANTUM MECHANICS

There is a widespread belief in the biological community that the quantum mechanical effects in biological particles and reactions are negligible. One of the arguments given in support of this belief is that the biological molecules are macromolecules, that is, molecules with high molar mass. It is true that when the de Broglie wavelength of a particle becomes too small, the quantum mechanical effects of the particle becomes negligible and the particle can be treated as a classical particle: classical approximations are valid.

As mentioned earlier, buckyballs (Section 11.3.1) with molar mass of 720 Da have demonstrated an observable wavelength. As Table 11.5 shows, biology offers a wealth of particles with masses over a very wide range with molar mass ranging from a few to millions of daltons. These two facts point to the following:

1. Some biological particles do function under the influence of quantum mechanics.
2. The wealth of biological particles can be used to discover and explore the interface between quantum and classical mechanics.

11.8 LIVING IN THE QUANTUM WORLD

Biology, like physics, demonstrates that only a few basic building blocks can assemble themselves into complex structures. With the help of quantum physics, it is important to understand these basic building blocks and how they self-assemble to compose more complex structures. This understanding helps the scientists and technologists to use these building blocks and structures to build new structures and applications.

One of the applications of nanotechnology is clean technology, which has turned into a revolution by its own virtue. Clean technology refers to the products, services, or processes and tools that deliver value, such as energy, by using mostly renewable resources and creates less waste than conventional means. Cleantech usually refers to the energy derived from renewable resources. The energy derived this way is called renewable or alternative energy. Renewable resources are the resources that are naturally replenished and you do not run out of them even after you use them to create energy. Examples of renewable energy resources are sunlight, wind, tides, and geothermal heat. Examples of nanotechnologies are solar power, wind power, hydroelectricity, biomass, and biofuels. Currently, a considerable attention is being paid to nanotech's growing impact on clean technology in terms of energy and environmental innovation both in the product and the process area.

Figure 11.14. The largest photovoltaic solar power plant in the United States is at Nellis Air Force Base with the size of 70,000 solar panels worth 15 mega watt of power. Using about 140 acres of unused land. Courtesy of the U.S. Air Force, Neils Air Base.

One of the examples of a clean energy device is the photovoltaic cell, which converts light energy into electrical energy by using the process called the photoelectric effect. An example of photovoltaic solar power is presented in Figure 11.14.

In the process of the photoelectric effect, photons of light knock electrons out of the atoms of a semiconductor material, such as Si. These free electrons are collected as electric current, that is, electricity. Poor efficiency of a solar cell based on this process is a major concern. However, to understand and manipulate processes like these at the quantum level can help us improve our devices and applications. As an example, photosynthesis, the process by which plants and some microbes convert water, carbon dioxide, and sunlight into oxygen and energy in the form of carbohydrates, was studied at the quantum level by a team of scientists led by Graham Fleming at the University of California Berkeley in 2007. To study the photosynthesis in green sulfur bacteria, the team examined the protein scaffold connecting the bacteria's external solar collectors, called the chlorosome, to the reaction centers deep inside the cells. Unlike electric power lines, which lose as much as 20% of their energy in transmission, these bacteria transmit energy at a staggering efficiency rate of 95% or better. The team concluded that the reason for this excellent efficiency is quantum reality, which is based on wave–particle duality, superposition of quantum states, uncertainty principle, and collapse of the wave function. Inspired by these findings, researchers are looking to mimic nature's quantum ability to build solar energy collectors that work with much better efficiency than achieved by traditional photovoltaic cells. For example, Alán Aspuru-

Guzik's team at Harvard University is performing research to find ways to apply the results of the quantum study of photosynthesis into organic photovoltaic solar cells.

11.9 SUMMARY

Quantum mechanics is entering its golden age. Multiple disciplines in science have arrived at the stage of their development where they are studying materials and processes at the nanoscale level. This has given rise to a broadly unified field called nanoscience and nanotechnology: a juncture for all these fields. Without a doubt, quantum mechanics governs the structure and behavior of matter at the nanoscale level. Empirical studies and research on which some disciplines, such as biology, have been based can only go so far. The time has come that these fields need to use quantum mechanics to make further progress in areas, such as molecular biology.

Electrons, photons, atoms, and molecules are not some exotic state of matter conjectured to offer theoretical explanations for complex phenomena or explain results from colossal high-energy physics experiments, such as those at Large Hadron Collider (LHC) at CERN. These particles are in action in nanostructures, and these particles are in action right in our bodies: in our eyes through which we see, in our brains through which we are conscious of what we see, and so on.

In a nutshell, quantum mechanics is the science that governs the fundamental reality of matter including the matter of life. Nanoscience and nanotechnology marks the beginning of the molecular age that will be governed by the laws of quantum physics.

11.10 ADDITIONAL PROBLEMS

Problem 11.5 Show that the wavelength of an electron moving through the electric potential V is given by

$$\lambda = \frac{1.23}{\sqrt{V}}$$

Problem 11.6 Calculate the speed of the electron moving in 60 kV voltage difference.

Problem 11.7 The incident electron beam in a TEM has been accelerated by applying a voltage of 200 kV. What is the electron wavelength of this beam?

Problem 11.8 Compare the diameter of a buckyball to the thickness of a human hair.

Problem 11.9 A quantum chip has four qubits to represent numbers. Which numbers are represented by the following arrangement of qubits?

First bit: Ψ_1

Second bit: $a\Psi_0 + b\Psi_1$

Third bit: Ψ_0

Fourth bit: $a\Psi_0 + b\Psi_1$

Problem 11.10 Calculate the speed of a buckyball corresponding to its de Broglie wavelength of 2.5 pm.

BIBLIOGRAPHY

A short list of books that cover the topics, which are the prerequisite for the material covered in this book. These books are available in most college libraries.

CLASSICAL PHYSICS

H. D. Young and R. A. Freedman. 2007. *University Physics*, 12th ed.: Addison-Wesley, Reading, MA.

P. A. Tipler and G. Mosca. 2007. *Physics for Scientists and Engineers*, 6th ed.: W. H. Freeman, New York.

CALCULUS

H. E. Campbell and P. F. Dierker. 2007. *Calculus with Analytic Geometry*, 3rd ed.: Prindle Weber & Schmidt, Boston, MA.

J. Stewart. 2007. *Calculus*, 6th ed.: Brooks Cole, Belmont, CA.

INDEX

Printed and bound by CPI Group (UK) Ltd, Croydon, CR0 4YY
16/09/2026
14963409-0001